Membrane Technology for CO_2 Sequestration and Separation

Editor

Zeinab Abbas Jawad

Faculty of Engineering and Science
Department of Chemical Engineering
Curtin University
Miri, Sarawak
Malaysia

CRC Press
Taylor & Francis Group
Boca Raton London New York

CRC Press is an imprint of the
Taylor & Francis Group, an **informa** business
A SCIENCE PUBLISHERS BOOK

Cover illustration reproduced by kind courtesy of the editor, Dr. Zeinab Abbas Jawad

CRC Press
Taylor & Francis Group
6000 Broken Sound Parkway NW, Suite 300
Boca Raton, FL 33487-2742

Printed on acid-free paper
Version Date: 20190111

International Standard Book Number-13: 978-1-138-50450-9 (Hardback)

Library of Congress Cataloging-in-Publication Data

Names: Jawad, Zeinab Abbas, editor.
Title: Membrane technology for COÒ sequestration and separation / editor, Zeinab Abbas Jawad (Faculty of Engineering and Science, Department of Chemical Engineering, Curtin University, Miri, Sarawak, Malaysia).
Other titles: Membrane technology for carbon dioxide sequestration and separation
Description: Boca Raton, FL : CRC Press, Taylor & Francis Group, [2018] | "A science publishers book." | Includes bibliographical references and index.
Identifiers: LCCN 2018059416 | ISBN 9781138504509 (hardback)
Subjects: LCSH: Carbon dioxide mitigation. | Carbon sequestration. | Membranes (Technology) | Gas separation membranes.
Classification: LCC TD885.5.C3 M46 2018 | DDC 628.5/32--dc23
LC record available at https://lccn.loc.gov/2018059416

Visit the Taylor & Francis Web site at
http://www.taylorandfrancis.com

and the CRC Press Web site at
http://www.crcpress.com

Preface

According to the Intergovernmental Panel on Climate Change, the global average temperature is increasing dramatically in the range of 4–7°C annually with major impacts on the environment and human life. The acceleration of economic growth is also partly responsible for today's expanding demand for energy. The increasing use of fuels, especially conventional fossil fuels that have developed into an indispensable energy source is an undeniable consequence of economic activity since the industrial revolution. Therefore, the extensive use of fossil fuels has become a matter to be concerned due to their unfavorable effects on the environment, especially the emission of carbon dioxide (CO_2), which is a primary anthropogenic greenhouse gas. Hence, some initiatives have been taken to reduce the release of CO_2 to the atmosphere.

The membrane separation process is one of the approaches that has received much attention ways to aid in the removal of CO_2 from the environment. It is a composite organic and/or inorganic membrane, which is a thin selective layer that bonds to a thicker, non-selective and low-cost layer that gives mechanical support to the membrane. CO_2 is only allowed to pass through the membranes, and at the same time eliminates the other components of the flue gas from passing through. Although there is significant advancement in the gas separation membrane systems, there is still a great deal to be done to realize the potential in this technology.

This book aims to provide a clear idea on the practical aspects of CO_2 sequestration in deep geological formations which is often referred to as carbon capture and storage. In addition, it discusses the main challenge that take place in membrane-based gas separation representing by the trade-off relationship between the membrane's permeability and selectivity that wards off this technology from achieving the desired outcome.

Content

1

An Introduction to Carbon Capture and Storage Technology

Arshad Raza,[1] *Raoof Gholami,*[1,*] *Vamegh Rasouli,*[2] *Reza Rezaee,*[3] *Chua Han Bing*[4] and *Ramasamy Nagarajan*[5]

INTRODUCTION

Annually, a huge amount of greenhouse gases, ranging from carbon dioxide (CO_2) and methane (CH_4) to nitrogen Oxide (N_2O), are released in the atmosphere by energy supply and consumption sites. According to the latest studies conducted by the US government, this significant emission of greenhouse gases, particularly CO_2, is the main reason behind what is called "global warming". There have been many studies referring to the rise of average sea level and severe melting of ice in Antarctic as the signs of global warming and climate change over the past decades. Figure 1.1 shows the increase of atmospheric CO_2 recorded by the meteorological station. The atmospheric abundance of CO_2 has increased by an average of 1.80 ppm per year over the past 38 years (1979–2016).

[1] Curtin University, Department of Petroleum Engineering, CDT 250, Miri 98009, Sarawak, Malaysia. Email: arshadraza212@gmail.com
[2] University of North Dakota, Department of Petroleum Engineering, USA. Email: vamegh.rasouli@UND.edu
[3] Curtin University, Department of Petroleum Engineering, Bentley WA 6102, Australia. Email: R.Rezaee@curtin.edu.au
[4] Curtin University, Department of Chemical Engineering, CDT 250, 98009 Miri, Sarawak, Malaysia. Email: chua.han.bing@curtin.edu.my
[5] Curtin University, Department of Applied Geology, CDT 250, 98009 Miri, Sarawak, Malaysia. Email: nagarajan@curtin.edu.my
* Corresponding authot: raoof.gholami@curtin.edu.my

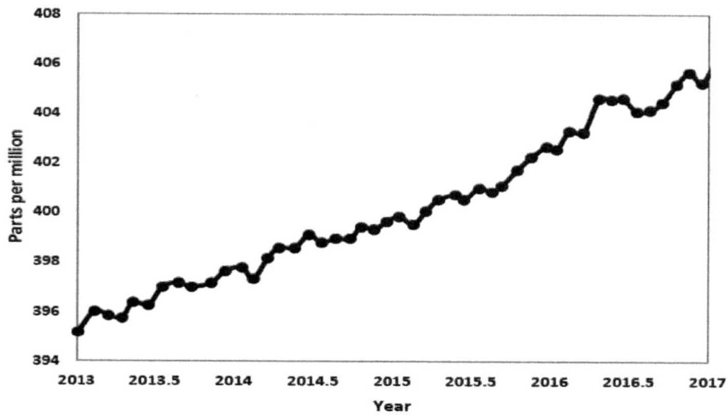

Figure 1.1 Level of CO_2 concentrations in the atmospheric since 2013 (modified after NOAA 2017).

There are several action plans proposed to avoid or at least mitigate climate changes. According to these plans, the global temperature might be stabilized if the emission of greenhouse gases could be reduced to as much as 80% by 2100 (Bennaceur et al. 2008). In order to achieve this objective, many studies were carried out and several approaches were proposed including Carbon Capture and Storage (CCS), energy efficiency improvement, consuming less carbon-intensive fuels, switching to renewable energy sources, enhancing biological sinks, and reduction of non-CO_2 greenhouse gas emissions, among which CCS appears to be the most successful technique due to its cost effective and flexibility characteristics. However, successful implementation of CCS technology depends, to a great extent, on technical maturity, costs, overall potential of chosen sites, development of the technology in potential countries, regulatory aspects, environmental issues and public awareness.

There are more than 800 worldwide sedimentary basins which have been recognized as suitable sites for implanting the CCS technology and a number of CO_2 sequestration projects have already been initiated in advanced countries. For instance, the Sleipner CO_2 storage project commenced in 1996 in Norway which successfully sequestered more than 16 million tonnes (Mt) of CO_2 in depleted storage reservoirs. In a different case, the Quest project in Alberta, had effectively captured and stored more than one million tons of CO_2 into a saline aquifer. Figure 1.2 indicates the on-going and planned CCS in worldwide projects.

However, there are many concerns related to the success of CCS technology such as selection of the best capturing and separation methods, proper characterization of storage sites, structural integrity of seals, etc., which may need a deeper knowledge as to how CO_2 can be captured and stored in subsurface layers for thousands of years. This chapter presents some basic concepts related to the CCS technology.

Figure 1.2 An overview of the key CCS projects and milestones around the globe (Institute 2016).

Principle of Carbon Capture and Storage (CCS)

As it was mentioned earlier, carbon capture from large emission sites and its storage in subsurface geological basins has been recognized as the best large-scale opportunity to reduce the amount of greenhouse gases, especially CO_2, into the atmosphere. This is mainly because CCS uses the in-practice technology and does not pose extra cost to the projects. A CCS practice involves capturing of carbon dioxide from emission sites, transporting it to a suitable site and injecting it into a chosen geological storage formation. Here, different parts of a CCS technology ranging from capturing and separation to storage are discussed in detail.

Carbon dioxide

There are many natural or manmade chemical compounds in the atmosphere which are often referred to greenhouse gases. These gases can weaken the atmosphere and provide a direct path for sunlight to reach the surface unhindered. They also absorb the heat energy of the sun and store it at the lower atmosphere causing the temperature to rise up.

Carbon Dioxide (CO_2) is one of the major components of the greenhouse gas family contributing more than 60% in global warming due to its increasing concentration in the atmosphere as a result of burning fossil fuels in the industrial sites to generate energy. CO_2 has a very complicated phase diagram and appears under different phases when it is subjected to various pressure and temperature regimes as shown in Fig. 1.3. These changes in phase may cause difficulty during capturing, separation and storage where certain pressure and temperature must be maintained to achieve a favourable result.

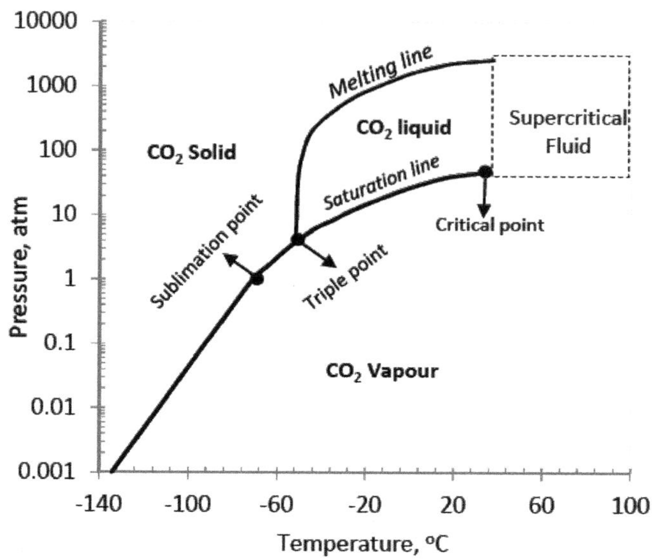

Figure 1.3. Phase diagram of CO_2 (modified from IPCC 2005).

For instance, CO_2 must appear under a supercritical phase to have a low density for large volume storage. Supercritical CO_2 is a non-polar solvent for organic compounds and its density and viscosity vary as a function of temperature and pressure. The density of CO_2 can even be further reduced due to contamination induced when CO_2 is mixed with methane (CH_4) during capturing or separation (Oldenburg and Doughty 2011).

CO_2 Capturing Techniques

The first phase of CCS technology includes capturing CO_2 from emissions sources such as fossil fuel and processing plants, particularly those manufacturing iron, steel, cement and bulk chemical materials. Generally, four main technological options are available for CO_2 capturing which include post-combustion, pre-combustion, oxy-fuelling, and chemical looping (Songolzadeh et al. 2014). Here these capturing techniques are presented and discussed.

Post-combustion

In the post-combustion technology, CO_2 is captured from the combustion process by a chemical solvent which is typically an amine solution. New technologies in which membranes and solid materials can be used rather than liquid solvents have been under investigation in the past few years but have not been properly tested in the industrial sites. Post-combustion is an ideal choice for retrofitting existing power plants and has been found as a very feasible option at small-scales where CO_2 is recovered at the rate of 800 t/day, regardless of associated large parasitic loads (de Visser et al. 2008). However, the post-combustion capturing method may raise the price of electricity production. Currently, two projects out of 16 large scale integrated CCS projects worldwide are employing post-combustion technology for capturing CO_2 (Institute 2012).

Pre-combustion

Pre-combustion technology is used to remove CO_2 from fossil fuels under high pressure high temperature conditions before combustion. In this process, H_2-rich fluid remained is safely combusted and CO_2 is captured, transported, and ultimately sequestered (Olajire 2010). Compared to post-combustion technology, where CO_2 is removed at low pressure, pre-combustion generates a CO_2 rich synthesis gas stream under high pressure conditions. Due to its efficiency in capturing CO_2, pre-combustion is typically a better approach than post-combustion but its related processes are more complex. Thus, the pre-combustion capturing method can be used for the situation where coal is used as a fuel in order to bring the efficiency loss to as low as 7–8%. Physical or chemical adsorption processes are two techniques used as part of this capturing method which may cost around US$60/tonne to capture CO_2.

Oxyfuel combustion

The oxy-fuel combustion gives a flue gas by burning oxygen instead of air to provide a better separation process (Buhre et al. 2005). The major composition of the flue gas would be CO_2, water, particulates and sulphur dioxide (SO_2), if pure oxygen is used for the combustion. In this process, particulates and SO_2 are separated from the rest of the flue gas by general electrostatic precipitators and desulphurization methods, respectively. If regulations and geochemistry permit, the remaining dehydrated gas with a large quantity of CO_2 (more than 90%) is compressed, transported and stored. Otherwise, the impurities such as O_2, N_2, and Ar are removed by reducing the temperature at which CO_2 can condense. Technically, oxy-fuel combustion is feasible, but a large quantity of oxygen is required for separation which may pose a huge cost on the CCS projects (Burdyny and Struchtrup 2010).

Chemical looping combustion

Chemical Lopping Combustion (CLC) is referred to the separation of CO_2 from raw natural gas using the same methodology as that of the post-combustion method. In this method, oxygen is used as a carrier and the combustion is divided into reduction and oxidation reactions in multiple reactors at an elevated temperature. There would be CO_2 and H_2O stream which can be further purified before getting compressed, and sent to storage sites (Songolzadeh et al. 2014). The CLC method is still under development and have not been largely implemented at any CCS projects. Figure 1.4 simplifies the mechanism involved in different CO_2 capturing methods presented here.

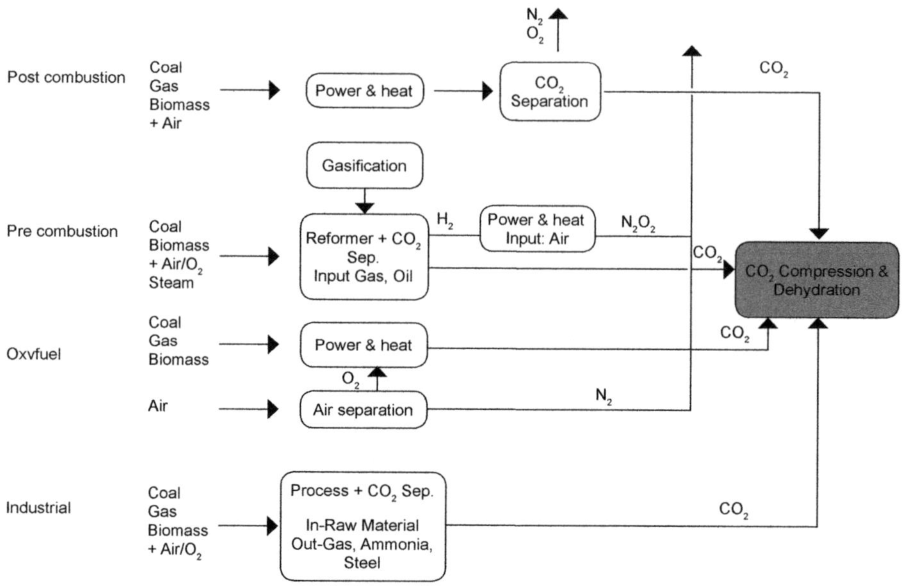

Figure 1.4. Different CO_2 capturing method (after IPCC 2005).

CO_2 Separation

Separation is the second stage of the CCS technology in which greenhouse gases such as CO_2 are separated from different emission sources. To date, there have been several gas separation technologies developed for the post-combustion capturing which are mainly based on different physical and chemical processes including: (a) absorption, (b) adsorption, (c) cryogenic distillation, and (d) membrane separation. Selection of these techniques, however, depends on the characteristics of the flue gas stream being produced from the power plants. Here, these separation methods are briefly discussed, and their advantages and disadvantages are highlighted.

Absorption

There are chemical and physical absorption methods which can be used for CO_2 separation. In chemical absorption, which is known as the most established method of separation, a liquid sorbent is used to weaken the bound between CO_2 and the flue gas. This weak bound is then broken down by heat to provide a pure CO_2 stream. However, implementation of the absorption separation method in large scale is challenging due to the significant amount of amine degradation which may cause solvent loss, equipment corrosion and generation of unstable degradation compounds (Rochelle 2012). The amine produced during separation may also degrade into nitrosamines and nitramines which are toxic.

In the physical absorption, an organic solvent is used to physically absorb CO_2 based on Henry's Law. The absorbed CO_2 is then removed according to its solubility in the solvents which depends, to a great extent, on the pressure and the temperature of the feed gas (Olajire 2010).

Adsorption

Unlike absorption in which separation is done using a liquid absorbent, a solid sorbent is employed to separate CO_2 during the adsorption process. In this method, the main criteria for the selection of the sorbent is its specific surface area, selectivity and regeneration ability. The common solid sorbent often used in the adsorption process are molecular sieves, activated carbon, zeolites, calcium oxides, hydrotalcites and lithium zirconate. The adsorbed CO_2, under these circumstances, is then separated from the sorbent by fluctuating the pressure or temperature (Takamura et al. 2001). The residues of industrial and agricultural units have recently been used to make sorbents for CO_2 capturing process which may decrease the total cost of the capturing posed by the adsorption technique (Maroto et al. 2008).

Chemical looping combustion

Chemical Looping Combustion (CLC) is an important and a new class of technology developed in recent years with applications for direct combustion and gasification. In this separation method, a solid is used to bring oxygen to the fuel gas for the combustion. During the combustion, the metal oxide is transformed into metal whereas the fuel is oxidized to CO_2 which is suitable for sequestration. Subsequently, the metal is oxidized in another stage and recycled in the process so that the technology effectively achieves oxygen separation from air without any need for cryogenic process or membrane technology (Ishida 2002).

Membrane separation

Membrane technology has been recognized as an established process for the separation of carbon dioxide (CO_2) since 1981. It has been mainly utilized for natural gas sweetening and Enhanced Oil Recovery (EOR), where CO_2 is separated from natural gas stream and reinjected into hydrocarbon reservoirs to improve the recovery (Dortmundt and Doshi 1999). This technology has also been employed lately to remove O_2 from N_2, and CO_2 from natural gas systems (Leung et al. 2014).

Membranes are especially designed materials to allow the flow of selected gas based on the principal of the pressure differential. Thus, high-pressure streams are generally favoured for the membrane separation (Roy 2016). In fact, membranes can be designed such that only CO_2 can flow through them by eluding other components associated with the flue gas. The membrane is manufactured by composite polymers which consist of selective layers for providing mechanical support to the membrane. It is now known that the manufacturing of ceramic, metallic and polymeric membranes would provide more effective means for CO_2 separation compared to the liquid absorption processes (Leung et al. 2014). Moreover, these membranes may improve the process of separating H_2 from a fuel gas stream, CO_2 from a range of processes or O_2 from air (IPCC 2005).

Hydrate-based separation

Hydrate-based separation is a newly established method to remove CO_2 from the exhaust gas by generating hydrate at high pressure conditions. In this process, CO_2 is trapped inside the hydrate cages and subsequently separated from other gases according to the differences of phase equilibrium. This technology does not pose a huge cost on the CCS projects due to low energy consumption but its efficiency is increased by enhancing the rate of hydrate formation and reduction of the hydrate pressure (Fan et al. 2011).

Cryogenic distillation

Cryogenic distillation works based on distillation at very low temperature and high-pressure conditions. It is often only applied to remove the components of gaseous mixture based on their boiling points. In this process, the temperature of the flue gas is brought to –100–135°C and it is solidified. This solidified CO_2 is then removed from other gases and compressed under atmospheric pressure before sending for storage. It is a cost effective practice with high efficiency in separation of CO_2 (Gottlicher and Pruschek 1997).

Table 1.1 compares the advantages and disadvantages of different separation technologies developed so far. Looking at this table, it may be concluded that fuel

Table 1.1. Comparison of different CO_2 separation technologies (Leung et al. 2014).

Technology	Advantage	Disadvantage
Absorption	- High absorption efficiency (> 90%) - Sorbents can be regenerated by heating and/or depressurization - Most mature process for CO_2 separation	- Absorption efficiency depends on CO_2 concentration - Significant amounts of heat for absorbent regeneration are required - Environmental impacts related to sorbent degradation have to be understood
Adsorption	- Process is reversible and the absorbent can be recycled - High adsorption efficiency achievable (> 85%)	- Require high temperature adsorbent. - High energy required for CO_2 desorption
Chemical looping combustion	- CO_2 is the main combustion product, which remains unmixed with N_2, thus avoiding energy intensive air separation	Process is still under development and there is no large-scale operation experience
Membrane separation	- Process has been adopted for separation of other gases - High separation efficiency achievable (> 80%)	Operational problems include low fluxes and fouling
Hydrate-based separation	- Small energy penalty	- New technology and more research and development is required
Cryogenic distillation	- Mature technology - Adopted for many years in industry for CO_2 recovery	- Only viable for very high CO_2 concentration > 90% v/v - Should be conducted at very low temperature - Process is very energy intensive

gas properties can be used for the selection of a suitable process by which CO_2 can be separated. Technically speaking, cryogenic distillation and membrane processes appear to be efficient methods for gas streams with a high CO_2 concentration. Comparatively, the chemical absorption provides the lowest capturing cost (i.e., US$20 to US$42) compared to adsorption (i.e., US$40 to US$63) and membrane separation (i.e., US$25 to US$217) Hongjun et al. (2011).

CO_2 Storage

There are generally three methods to dispose CO_2 including geological storage, ocean storage and surface mineral carbonation (Bachu 2008). The ocean storage is done at deep intervals where CO_2 dissolves or generates hydrates at the bottom of the ocean. The surface mineral carbonation is the process in which CO_2 is converted into solid inorganic carbonates using chemical reactions. Compared to surface mineral carbonation and ocean storage, the geological storage of CO_2 currently represents the most effective means of storing CO_2 (Bennaceur et al. 2008). However, the scale of geological storage worldwide depends on affordability, the emission cap and climate change policies of each country.

The basin-scale suitability assessment is the first step of geological storage evaluation which is done by considering geological, geothermal and hydrodynamic aspects of the chosen reservoirs together with basin maturity and few other criteria such as feasibility and societal concerns (Bachu 2001). Having said that, the potential media for storage in subsurface geologic basins are depleted oil and gas reservoirs, deep coal beds, and deep saline aquifers, among which the oil and gas reservoirs and deep aquifers are often the best choice considering the fact that they mainly composed of sandstone or carbonates (Carroll et al. 2011, Raza et al. 2016a, Raza et al. 2016b). Technically speaking, to choose a safe, feasible and environmentally secure storage medium, three basic criteria must be fulfilled: (i) the selected medium must have a sufficient pore volume to effectively store CO_2, (ii) chosen sites must have favourable petrophysical characteristics, and (iii) seals and reservoirs must be completely secure without having any integrity issues (Cooper 2009).

Prior to storage site selection, however, a preliminary assessment must be done to evaluate the key CO_2 storage aspects including reservoir and well types, types of minerals, residual gas and water saturations, reservoir conditions, rock types, wettability, CO_2 properties, and sealing potentials (Raza et al. 2015). After the preliminary stage evaluation, a comprehensive characterization of the reservoir before CO_2 injection is carried out using experimental or numerical methods to ensure the suitability of the chosen reservoir for a safe storage, in which capacity, injectivity, trapping mechanisms (structural, capillary, dissolution, and mineral) and containment are evaluated in details.

Some other aspects such as the condition of the injection well and layers (Raza et al. 2017b, Raza et al. 2017c) and wellbore cement stability (Abid et al. 2015) are also essential parameters to evaluate at any CCS practices. The operating cost is perhaps another phase which should not be assessed based on locations (e.g., onshore or offshore), the size and composition of pipelines and operating conditions. Last but not least is the monitoring and risk assessment of a storage site is done to ensure that injection and storage of CO_2 is safe, effective, and permanent in the geologic media chosen (Leung et al. 2014).

Trapping mechanisms

There are generally four trapping mechanisms taking place upon CO_2 injection. The first trapping mechanism, which is called structural/stratigraphic trapping, occurs when the injected CO_2 migrates literally and vertically towards the caprock and stops there. During this migration, the capillary (residual) tapping of CO_2 as a residual gas also takes place when brine invades the CO_2 plume. Subsequently, CO_2 dissolves into the aqueous phase, creating solubility trapping and the dissolved CO_2 reacts with rocks in a long term, leading to mineral trapping, or adsorbed on the rock surface because of adsorption trapping (Bachu et al. 2007). Comparatively, the capillary trapping is recognized as a rapid, effective and safe mechanism to immobilize CO_2 in subsurface formations (Herring et al. 2015). It may also offer more CO_2 entrapment compared to other trappings such as the dissolution and mineral trappings. Although capillary, solubility and mineral trapping mechanisms are very slow rate reactions, they play important roles in increasing the security and safety of geological storage after stoppage of injection. However, these trapping mechanisms are mainly controlled by temperature, pressure and salinity (Tambach et al. 2015). Raza et al. (2017a) recently reported on the behaviour of the trapping mechanisms against pressure, temperature and salinity. Their results indicated that the amount of CO_2 saturation in a mobile phase (structural trapping) and residual CO_2 saturation (capillary trapping) is directly linked to the increase of salinity and temperature. It was also revealed that free gas saturation increases up to a certain pressure, depending on the temperature and salinity level, and then begins to decline till the effective pore volume is filled by CO_2. The increase of solubility trapping, at a particular temperature and salinity, with the increase of pressure was another finding reported by Raza et al. (2017a).

CO_2 Monitoring for Safety

There is a risk of leakage from CO_2 storage sites which may cause the seepage and hazards to human, ecosystem and ground water resources. As mentioned earlier, CO_2 injection into a storage site may result in changes of the formation mechanical properties, i.e., changes of stress in the reservoir and adjacent formations, uplift of ground, opening of fault and fractures, breaching containment of seal system, etc. Moreover, the other risks associated with CO_2 injection is leakage from abandonment wells, because of cement degradation and casing failure (Raza et al. 2017c). Owing to this, monitoring of storage sites would be essential during and after CO_2 injection.

The aim of monitoring is to ensure that CO_2 stays inside the geological formations, and will not cause any integrity issues upon injection. Any breaching at this state may cause serious environmental issues and contamination of subsurface resources. This monitoring must take place even after stoppage of injection, when pressure builds up and CO_2 plume starts to stabilize, due to physical and mechanical changes induced. There are a number of monitoring techniques at different stages which can be employed to ensure the integrity of the storage reservoir, which include but are not limited to (Winthaegen et al. 2005).

1. Pressure and temperature monitoring practice is a key aspect of monitoring storage sites which is often used by employing different downhole gauges near the injection wellbore.
2. Gravity, electro-magnetic and self-potential geophysical techniques applied at surface or downhole wells.
3. Time lapse (4D) seismic reflection data acquisition can be a valuable tool for monitoring CO_2. In this approach, changes in poroelastic properties of rocks, such as fluid distribution, stress and pressure, taking place in the reservoir during and after injection can be detected.
4. Downhole logging tools such as cement bong logs can be used to verify wellbore integrity and monitor the condition of cement behind the casing.

Summary

Carbon dioxide is known as a greenhouse gas causing global warming and climate change. CO_2 injection for storage in subsurface geologic media is one of the techniques established in the past years as an efficient way to mitigate anthropological CO_2 emission into the atmosphere. This chapter summarized the applications of different technologies proposed to capture and separate CO_2 from different sources, among which Carbon Capture and Storage (CCS) technology is possibly the most feasible option. It was discussed that CO_2 can be effectively stored in subsurface geologic media, but conducting preliminary assessments at basin and reservoir scales for selecting suitable storage sites would be essential. Therefore, a comprehensive characterization of key parameters in CO_2 storage, such as storage capacity, injectivity, trapping mechanisms, and containment, must be carried out to confirm the feasibility and the long-term safety of CCS projects. Monitoring is the last step in a CCS project taking place to ensure that storage integrity is not compromised in a short or long-term during and after injection.

Acknowledgments

The authors would like to acknowledge Curtin University, Malaysia for funding this research through the Curtin Sarawak Research Institute (CSRI) Flagship scheme under the grant number CSRI-6015.

References

Abid, K., Gholami, R., Choate, P. and Nagaratnam, B.H. 2015. A review on cement degradation under CO_2-rich environment of sequestration projects. J. NGSE, Elsevier 27: 1149–1157.

Bachu, S. 2001. Screening and ranking of hydrocarbon reservoirs for CO_2 storage in the Alberta Basin, Canada, US Dept. of Eng.-Nat. Eng. Tech. Lab., National Conference on Carbon Sequestration, pp. 1–60.

Bachu, S., Bonijoly, D., Bradshaw, J., Burruss, R., Holloway, S., Christensen, N.P. et al. 2007. CO_2 storage capacity estimation: Methodology and gaps. Int. J. Greenh. Gas Control 1(4): 430–443.

Bachu, S. 2008. CO_2 storage in geological media: role, means, status and barriers to deployment. Prog. Energy Combustion 34(2): 254–273.

Bennaceur, K., Gielen, D., Kerr, T. and Tam, C. 2008. CO_2 capture and storage: a key carbon abatement option. OECD.

Buhre, B.J.P., Elliott, L.K., Sheng, C.D., Gupta, R.P. and Wall, T.F. 2005. Oxy-fuel combustion technology for coal-fired power generation. Prog. Energy Combust. Sci. 31: 283–307.

Burdyny, T. and Struchtrup, H. 2010. Hydrid membrane/cryogenic separation of oxygen from air for use in the oxy-fuel process. Energy 35: 1884–97.

Carroll, S.A., McNab, W.W. and Torres, S.C. 2011. Experimental study of cement-sandstone/shale-brine-CO_2 interactions. Geochemical Transt. 12(1): 1–19.

Cooper, C. 2009. A technical basis for carbon dioxide storage. Energy Procedia 1(1): 1727–1733.

de Visser, E., Hendricks, C., Barrio, M., Molnvik, M.J., de Koeijer, G., Liljemark, S. et al. 2008. Dynamics CO_2 quality recommendations. Int. J. Greenh. Gas Control 2: 478–84.

Dortmundt, D. and Doshi, K. 1999. Recent developments in CO_2 removal membrane technology. UOP LLC, USA, 1–32.

Fan, S., Wang, Y. and X. Lang. CO_2 capture in form of clathrate hydrate-problem and practice. In: Proceedings of the 7th international conference on gas hydrates (ICGH 2011), UK; July 17–21, 2011.

Göttlicher, G. and Pruschek, R. 1997. Comparison of CO_2 Removal Systems for Fossil Fuelled Power Plant Processes. Energy Convers. Manag., 38, S173–S178. http://dx.doi.org/10.1016/S0196-8904 (96) 00265-8.

Herring, A.L., Andersson, L., Schlüter, S., Sheppard, A. and Wildenschild, D. 2015. Efficiently engineering pore-scale processes: The role of force dominance and topology during nonwetting phase trapping in porous media. Advan. in Water Res. 79(0): 91–102.

Hongjun, Y.A.N.G., Shuanshi, F.A.N., Xuemei, L.A.N.G., Yanhong, W.A.N.G. and Jianghua, N.I.E. 2011. Economic comparison of three gas separation technologies for CO_2 capture from power plant flue gas. Chinese J. Chem. Engineering 19(4): 615–620.

Institute, G.C. 2012. The Global Status of CCS | 2012 Summary Report. 9–10.

Institute, G.C. 2016. The Global Status of CCS | 2016 Summary Report. 1–144.

Ishida, M., Yamamoto, M. and Ohba, T. 2002. Experimental results of chemical looping combustion with NiO/NiAl2O4 particle circulation at 12001C. Energy Convers Manag 43: 1469–78.

IPCC. 2005. IPCC special report on carbon dioxide capture and storage. Prepared by Working Group III of the Intergovernmental Panel on Climate Change, Cambridge, United Kingdom and New York, NY, USA.

Leung, D.Y.C., Caramanna, G. and Maroto-Valer, M.M. 2014. An overview of current status of carbon dioxide capture and storage technologies. Renew. Sustain. Energy Rev. 39: 426–443.

Maroto-Valer, M.M., Lu, Z., Zhang, Y. and Tang, Z. 2008. Sorbents for CO_2 capture from high carbonfly ashes. Waste Manag 28: 2320–8.

Oldenburg, C. and Doughty, C. 2011. Injection, flow, and mixing of CO_2 in porous media with residual gas. Transp. in Porous Media 90(1): 201–218.

Olajire, AA. 2010. CO_2 capture and separation technologies for end-of-pipe application–a review. Energy 35: 2610–28.

Raza, A., Rezaee, R., Gholami, R., Rasouli, V., Bing, C.H., Nagarajan, R. et al. 2015. Injectivity and quantification of capillary trapping for CO_2 storage: A review of influencing parameters. J. Nat. Gas Sci. Eng. 26: 510-517.

Raza, A., Gholami, R., Sarmadivaleh, M., Tarom, N., Rezaee, R., Bing, C.H. et al. 2016a. Integrity analysis of CO_2 storage sites concerning geochemical-geomechanical interactions in saline aquifers. J. Nat. Gas Sci. Eng., Elsevier 36PA: 224–240.

Raza, A., Rezaee, R., Gholami, R., Bing, C.H., Nagarajan, R. and Hamid, M.A. 2016b. A screening criterion for selection of suitable CO_2 storage sites. J. Nat. Gas Sci. Eng., Elsevier 28: 317–327.

Raza, A., Gholami, R., Rezaee, R., Bing, C.H., Nagarajan, R. and Hamid, M.A. 2017a. Assessment of CO_2 residual trapping in depleted reservoirs used for geosequestration. J. Nat. Gas Sci. Eng., Elsevier 43C: 137–155.

Raza, A., Gholami, R., Rezaee, R., Bing, C.H., Nagarajan, R. and Hamid, M.A. 2017b. Preliminary assessment of CO_2 injectivity potential in carbonate storage sites. Petroleum, KeAi 3(1): 144–154.

Raza, A., Gholami, R., Rezaee, R., Bing, C.H., Nagarajan, R. and Hamid, M.A. 2017c. Well selection in depleted oil and gas fields for a safe CO_2 storage practice: a case study from Malaysia. Petroleum, KeAi 3(1): 167–177.

Rochelle, GT. 2012. Thermal degradation of amines for CO_2 capture. Curr. Opin. Chem. Eng. 1–2: 183–90.

Roy, S. 2016. Innovative use of membrane technology in mitigation of ghg emission and energy generation. Procedia Envir. Sciences 35: 474–482.

Songolzadeh, M., Soleimani, M., Takht Ravanchi, M. and Songolzadeh, R. 2014. Carbon dioxide separation from flue gases: a technological review emphasizing reduction in greenhouse gas emissions. The Scientific World Journal.

Tambach, T.J., Koenen, M., Wasch, L.J. and van Bergen, F. 2015. Geochemical evaluation of CO_2 injection and containment in a depleted gas field. Int. J. Greenh. Gas Control. 32(0): 61–80.

Takamura, Y., Narita, S., Aoki, J. and Uchida, S. 2001. Application of high-PSA process for improvement of CO_2 recovery system. Can. J. Chem. Eng. 79: 812–6.

Winthaegen, P., Arts, R. and Schroot, B. 2005. Monitoring subsurface CO_2 storage. Oil & Gas Sci. and Tech. 60(3): 573–582.

2

Membrane Engineering in CO_2 Separations

*Adele Brunetti,** *Enrico Drioli* and *Giuseppe Barbieri*

INTRODUCTION

Carbon dioxide is produced in huge quantities in various sectors; power and hydrogen production, heating systems (for example, in steel and cement industries), natural gas and biogas purification, etc. (Table 2.1) are some examples. CO_2 separation from hydrogen and methane streams have long been used owing to the high value of these streams (Shao et al. 2009, Favre et al. 2009, Basu et al. 2010, Lin et al. 2006, Scholes et al. 2012). Recent constrains and regulations on CO_2 emissions have focused on its separation from flue gas streams where N_2 is the more relevant specie (ca. 80%), whilst CO_2 concentration range 5–20% (Brunetti et al. 2010, Ciferno et al. 2009, Herzog et al. 2001, White 2003, Favre 2007, Merkel et al. 2010, Li et al. 2013, Peters et al. 2011, Daal et al. 2013). Adsorption, absorption, and cryogenic distillation (Tuinier et al. 2011) were the first technologies to be considered suitable for this purpose, however membrane technology is a valid alternative for carbon dioxide separation from the various aforementioned streams. Conventional liquid solvent-based technologies for separating CO_2 are used commercially in the chemical industry (Klara et al. 2007). The U.S. Department of Energy (DOE) estimates that post-combustion capture using conventional solvents will increase the cost of electricity by about 80% and incur a US $68/ton avoided cost for CO_2 (Ku et al. 2011). Corresponding projections for pre-combustion capture are a 30-40% increase in cost of electricity and 32–42 US$/ton avoided cost. Considering that streams containing carbon dioxide coming

Institute on Membrane Technology (ITM-CNR), National Research Council, Cubo 17C, Via Pietro Bucci, Rende CS 87036, Italy.
Emails: e.drioli@itm.cnr.it; g.barbieri@itm.cnr.it
* Corresponding author: a.brunetti@itm.cnr.it

Table 2.1. Typical sources of CO_2 emissions.

	Source	Separation	Feed composition	Temperature and pressure	Ref.
Flue gas streams	Power plants Coal gasification plants Steel factory Cement factory Transportation	CO_2/N_2	5–25% CO_2 65–80% N_2 3–5% O_2 Rest N_2, SO_x, H_2S, H_2O	35–100°C and 1 bar	(Daal et al. 2013)
Natural Gas	Natural gas pipes Sweetening of natural gas, etc.	CO_2/CH_4	1–8% CO_2 70–90% CH_4 0–20%C_2H_6, C_3H_8, C_4H_{10} Rest O_2, N_2, H_2S, Ar, Xe, He	25–30°C and 1.2 bar	(Baudot 2011, Xiao et al. 2009)
Biogas	Various		34–40% CO_2 50–70 %CH_4 Rest N_2, O_2, H_2S, H_2O	25–35°C and 1 bar	(Deng and Hagg 2010)

from power plants or heating systems are waste with no "profit" margin involved in their treatment, a significant separation cost (no less than 20/25 US$/ton) would significantly affect the final cost (e.g., of the electricity) (Brunetti et al. 2010).

Membrane operations are now being explored for CO_2 capture from power plant emissions and other fossil-fuel-based flue gas streams, owing to their interesting engineering and economic advantages over competing separation technologies.

Various materials can be considered suitable for the separation of CO_2 from flue gas or methane streams (Powell et al. 2006, Luis et al. 2012, Ramasubramanian et al. 2011, Park et al. 2007, Jung et al 2010, Calle et al. 2011, Adams et al. 2010, Adams et al. 2011, Falbo et al. 2014, Cersosimo et al. 2015, Brunetti et al. 2016, Falbo et al. 2016, Brunetti et al. 2017), and many advances have been made in the maximization of their mass transport properties. However, questions remain about the scalability and durability of these materials under real conditions. Generally, there is a debate among the scientific community about the fact that it could be more convenient to have a membrane with high permeance and low selectivity, or vice versa (the two limit conditions of Robeson diagram) (Robeson 2008). For all cases, high membrane CO_2 permeance minimizes membrane area requirements, whilst high CO_2/N_2 selectivity improves the CO_2 permeate concentration. The benefits of higher selectivity are accentuated at higher feed-to-permeate pressure ratios, at the expense of increased energy cost. The advantages of higher permeance are most pronounced at lower pressure ratios. The engineering design of a membrane separation unit for the recovery of CO_2 must thus take into account various factors.

This chapter summarizes some of the most significant studies (Brunetti et al. 2010, 2014, 2015) of our group on the engineering approach that in our opinion can be followed for the design of membrane gas separation in CO_2 processing. As performed, this analysis can constitute a useful guide for readers interested in CO_2 separation with membranes and can be considered valid for the design of the performance of a single-stage or multi-stage separation system.

Performance Maps for CO_2 Separation

The engineering design of a membrane separation unit for the recovery of CO_2 has to take into account various factors. In our previous work (Brunetti et al. 2010), we developed a simple tool that uses "maps" to enable analysis of performance and the perspectives of membranes in CO_2 capture.

This tool consists of a dimensionless 1D mathematical model for the multi-species steady-state permeation in no sweep mode and co-current configuration. The results achieved by these simulations are described in terms of general maps of CO_2 permeate concentration versus CO_2 recovery. These are useful for analyzing different design solutions in terms of membrane area and pressure ratio. More details on this tool can be found in Brunetti et al. works (Brunetti et al. 2010, 2014, 2015), where the validation of the model via experimental literature results is also presented.

In the dimensionless form of the equations, the terms θ and ϕ can be distinguished as the permeation number and the feed to permeate pressure ratio, respectively.

$$\theta = \frac{\Pi_i A^{Membrane} p^{Feed}}{x_i^{Feed} Q^{Feed}} \tag{2.1}$$

$$\phi = \frac{p^{Feed}}{p^{Permeate}} \tag{2.2}$$

$$\Pi_i = \frac{i - speciepermeatingflux}{\Delta P_i} \tag{2.3}$$

$$\Delta P_i = P_i^{Feed/Retentate} - P_i^{Permeate} \tag{2.4}$$

The permeation number (2.1) defines a comparison between the two main transport mechanisms involved: permeation through the membrane and convective transport at the membrane inlet. A high permeation number corresponds to a high membrane area and/or permeance (2.3) and, thus, to a high permeation through the membrane with respect to the total flux along the module. The pressure ratio (2.2) is an index of the maximum driving force for permeation. It represents one of the most important and determinant operating parameters affecting the performance of the membrane unit and is the driving force for the separation.

The overall membrane module performance are available in terms of, for instance, final species concentration and total CO_2 recovery in the permeate (2.5) stream, easily calculated considering the value of state variables (species composition, overall flow rate) at the module exit.

$$CO_2\, recovery = \frac{CO_2\, permeate\, flow\, rate}{CO_2\, feed\, flow\, rate} * 100 \tag{2.5}$$

In the maps (Fig. 2.1), all the parametric curves at constant pressure ratio have the same trend: at a higher recovery corresponds a lower CO_2 concentration in the final permeate stream. Then, as the pressure ratio increases the curves shift upward to a higher permeate concentration, at the same CO_2 recovery. At a given permeation number (dashed lines), CO_2 permeate concentration increases as the recovery increases too.

For instance, for a set membrane for which the permeance and selectivity are known and for a known feed stream composition, the pressure ratio and permeation number required to obtain a permeate stream with specific characteristics can be easily identified moving in the map. Figure 2.1 A–D show the maps calculated at different values of membrane selectivity. Each map is developed at various CO_2 feed concentrations for a set value of selectivity, so that it can be utilized to evaluate each membrane gas separation unit constituting the separation system, also in cases of multistage configuration. Considering, for example, a stream containing 10% CO_2 fed to a first membrane unit with a selectivity of 50, at a pressure ratio of 5 and a permeation number of 10, corresponds a recovery of ca. 45% with 30–35% of CO_2 concentration. With these conditions, entering the second diagram corresponding to the membrane with a higher selectivity, the operating conditions necessary to achieve a defined final permeate concentration can be defined. For example, assuming a membrane unit with a selectivity of 100 (Fig. 2.1B), the permeate concentration of

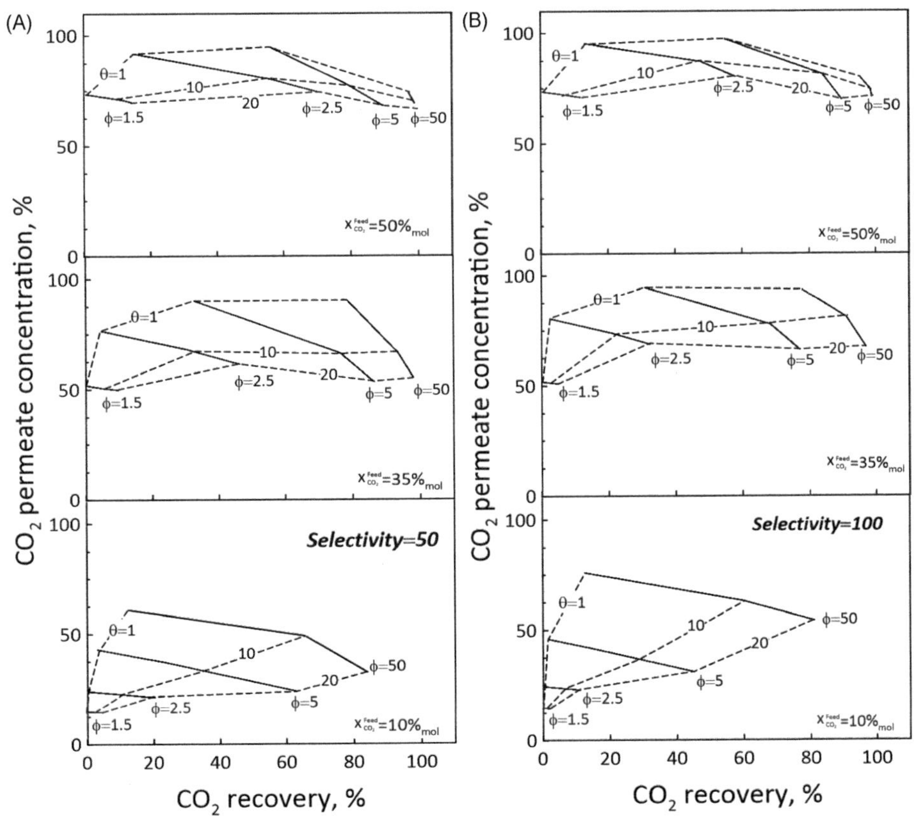

Figure 2.1 contd. ...

... *Figure 2.1 contd.*

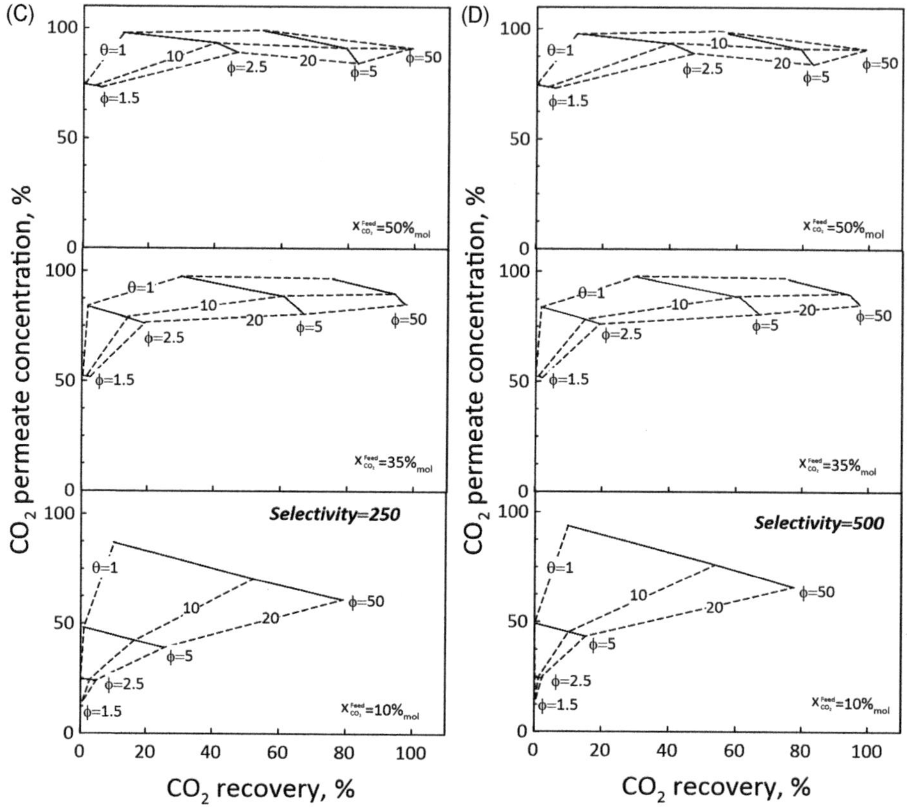

Figure 2.1. Maps of CO_2 permeate concentration as a function of CO_2 recovery at various values of CO_2 feed concentration. Pressure ratio and permeation number is also reported. (A) CO_2/N_2 selectivity = 50; (B) CO_2/N_2 selectivity = 100; (C) CO_2/N_2 selectivity = 250, (D) CO_2/N_2 selectivity=500. Reproduced from [39] with permission from Elsevier.

75% in the second stage can be reached at a pressure ratio of 2.5 but with ca. 20% recovery or at a pressure ratio of 5 with recovery of more than 60%.

Integrated Schemes For CO_2 Separation

The possible application of membranes for CO_2 separation in the treatment of non-valuable streams (e.g., CO_2/N_2, flue gas of a power plant or cement industry) or valuable streams (e.g., CO_2/CH_4, biogas) is strictly related to the characteristics of the stream to be treated (composition, pressure, temperature, etc.) as well as to the target to be achieved.

In the separation of CO_2 from flue gas, the pressure of the stream coming from the chimney of the plant is the very limiting step. The large flows tend to be treated, and the absence of added value in the product to be recovered (CO_2) limits the availability of extra-pressure supplied to the stream by means of compressors. Thus, pressure ratios considered in these calculations are limited (ranging from 1.5–5) when discussing flue gas separation. Different circumstances can be considered for natural gas and biogas separation, where purified methane can be retained in the retentate of the already compressed membrane stream. In this case, the high added value of methane stream, the lower flow rates, and the fact that the purified methane has to be pumped in the pipeline at 40–50 bar, make it affordable to operate with high feed pressure in the membrane unit and, hence, high pressure ratios.

CO_2 separation from flue gas

The aforementioned maps could be the starting point in the *carbon capture and storage* process design. In fact, global economic considerations on the final electricity cost and CO_2 storage technology allow the optimal performance (that is, a point on CO_2 purity vs. recovery plot) to be univocally identified on the maps; the parametric curves crossing on this optimal point provide the corresponding pressure ratio and permeation number.

For instance, 64% CO_2 purity and 61% recovery corresponds to a system having a pressure ratio (ϕ) of 20 and a permeation number (θ) of 0.2. For a given geometry and membrane ($Permeance_{CO_2}$) this pair of parameters can be obtained by means of infinite different couples of operating conditions; looking to the data in Table 2.2 and assuming a 1 m^3(STP) h^{-1} of flue gas per module, in the case of 20 bar of feed pressure, the feed flow rate decreases to 50 or 10%, respectively, when the feed pressure is reduced to 10 or 1 bar. Therefore, to process a given flue gas flow rate, the lower the feed pressure the higher the membrane area. On the contrary, since usually the flue gas disposal takes place at almost atmospheric pressure, a high feed pressure leads to a high compression load to be accounted for in the overall membrane system evaluation.

Table 2.3 reports two different cases, related to 20 (case A) and 1 bar (case B) of pressure ratio. The point "C" of Fig. 2.2 represents both these conditions. In case A, the flue gas feed stream has to be compressed up to 20 bar; assuming the compression load proportional to the flow rate and to the pressure ratio, for 1 m^3(STP) h^{-1} of flue gas this load is 20 a.u. (arbitrary unit).

In case B the feed flow rate to be send in the module is 0.05 m^3(STP) h^{-1} (1/20 of the case A); moreover, a vacuum pump is needed to reduce the permeate down to 0.05 bar. In this case the compression load is 0.12 a.u., being proportional only to the flow rate effectively sucked, that is 6.2 dm^3 (STP) h^{-1} per module $= \dfrac{Q^{Feed} x_{CO_2}^{Feed}}{(CO_2 \text{ permeate purity}) \times (CO_2 \text{ recovery})}$ and to the pressure ratio (always 20). Finally, considering that in case B 20 modules are necessary to process the same flue gas flow rate treated in case A, the conclusion is that passing from case A to case B there is a reduction from 20 to 2.4 a.u. of compression/vacuum load and an increase from 1 to 20 membrane modules (Table 2.3).

A higher CO_2/N_2 ideal selectivity leads to different maps (the curves move up Fig. 2.3).

Figures 2.4, 2.5 and 2.6 show the effect of ideal selectivity on the membrane module performance at pressure ratios of 10, 20 and 50, respectively. In Fig. 2.4 the

Table 2.2. Different possible operating conditions for feed ratio of 20 and permeation number of 0.2. Reproduced from [6] with permission from Elsevier.

P^{Feed}, bar	$Q_{CO2}{}^{Feed}$/, m^3 (STP) h^{-1}	$Q_{CO2}{}^{Feed}/(Q_{CO2}{}^{Feed}$ @ 20bar), %
20	5 Permeance$_{CO_2}$ A^{Membrane}	100%
10	2.5 Permeance$_{CO_2}$ A^{Membrane}	50%
1	0.25 Permeance$_{CO_2}$ A^{Membrane}	10%

Table 2.3. Comparison of compression/vacuum load and membrane area for two different sets of operating condition. Reproduced from [6] with permission from Elsevier.

	Case A	Case B
ϕ, –	20	
θ, –	20	
CO_2 permeate purity, %	64	
CO_2 recovery, %	61	
P^{Feed}, bar	20	1
$P^{Permeate}$, bar	1	0.05
Q^{Feed}, m^3 (STP) h^{-1}	1	0.05
Compression load, a.u	20	2.4
Membrane modules, -	1	20

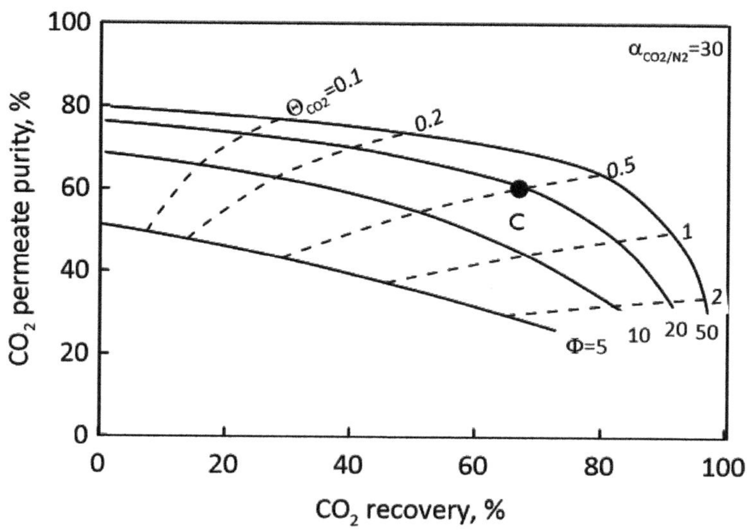

Figure 2.2. CO_2 permeate purity versus CO_2 recovery for different pressure ratios (continuous lines) and permeation numbers (dashed lines). Feed CO_2 composition = 13%. CO_2/N_2 ideal selectivity = 30. Reproduced from [6] with permission from Elsevier.

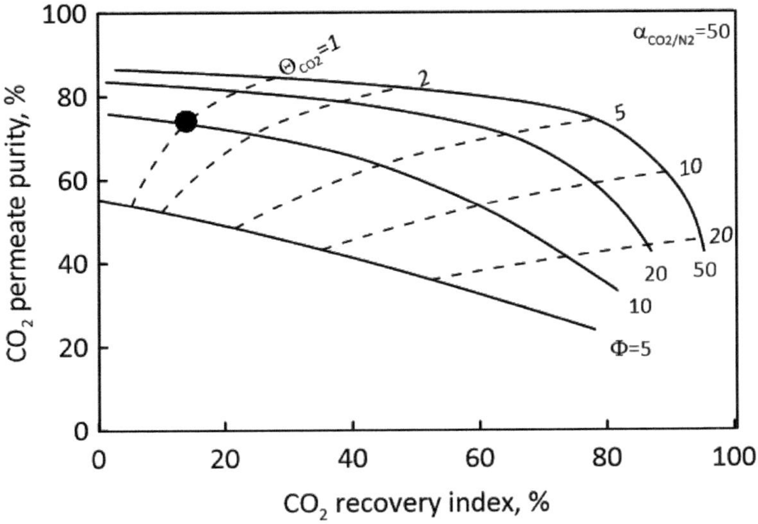

Figure 2.3. CO_2 permeate purity versus recovery index for different ϕ (continuous lines) and different θ (dashed lines). Feed CO_2 composition = 13%. CO_2/N_2 ideal selectivity = 50. Reproduced from [6] with permission from Elsevier.

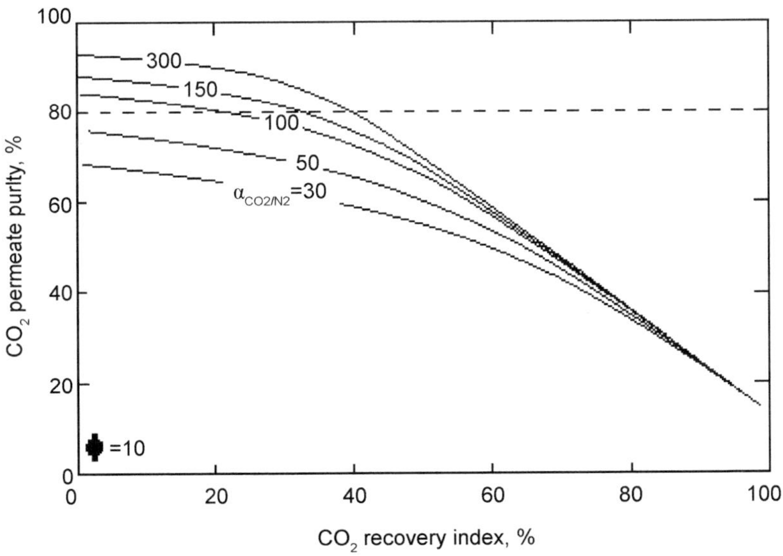

Figure 2.4. CO_2 permeate purity versus recovery index for pressure ratio 10 at different CO_2/N_2 ideal selectivity from 30 to 300. Reproduced from [6] with permission from Elsevier.

CO_2 permeate purity as a function of recovery was calculated for a pressure ratio of 10. In the case of $\alpha_{CO2/N2}$ = 30, the CO_2 purity reduces progressively from 69 (its maximum value, when recovery is close to 0%) to 13% (the feed composition, when

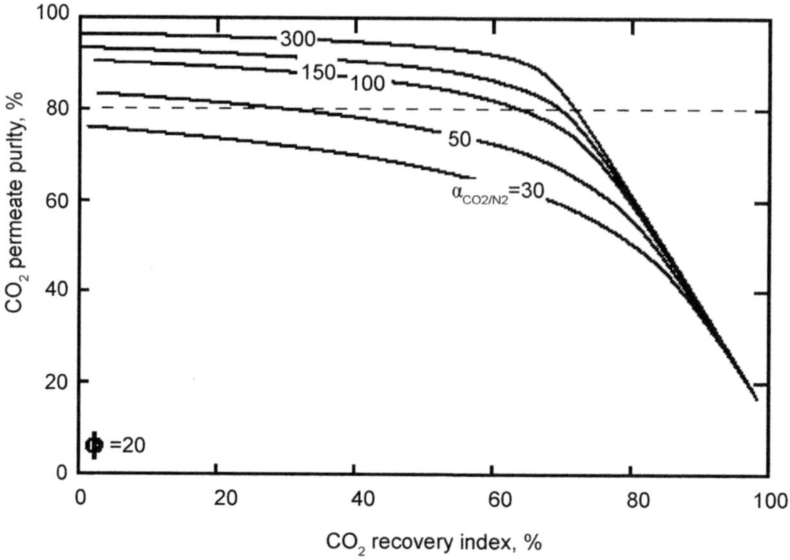

Figure 2.5. CO_2 permeate purity versus recovery index for pressure ratio 20 at different CO_2/N_2 ideal selectivity from 30 to 300. Reproduced from [6] with permission from Elsevier.

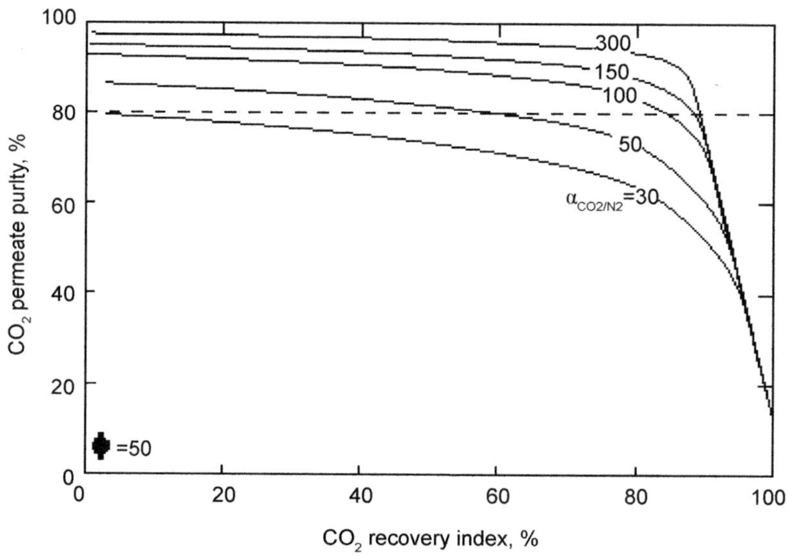

Figure 2.6. CO_2 permeate purity versus recovery index for pressure ratio 50 at different CO_2/N_2 ideal selectivity from 30 to 300. Reproduced from [6] with permission from Elsevier.

the recovery is total). When ideal selectivity of 50, 100, 150 and 300 are considered, the maximum achievable purity increases up to 76, 84, 88 and 93%, respectively. On the other hand, at a high recovery all the curves at different selectivity approach asymptotically to a limit trend, with the permeate converging to the feed composition (13% of CO_2).

Same simulations are shown in Figs. 2.5 and 2.6 for pressure ratio = 20 and 50 where all the curves at different selectivity are shifted towards higher purity and recovery. At a selectivity = 100, the maximum achievable purity increases from 84 (pressure ratio = 10) to 91 and 94% for the pressure ratios 20 and 50, respectively. At a higher pressure ratio also the maximum achievable CO_2 recovery is higher and for selectivity 100 it exceed the 60 and 80% for pressure ratio = 20 and 50, respectively. In general, at low CO_2 recovery, the permeate stream purity is strongly affected by the membrane selectivity: a high selectivity allows high CO_2 purity to be achieved. At a high CO_2 recovery, on the contrary, the membrane selectivity does not affect at all the membrane module performance (the asymptotical approach of all the curves), the permeate purity being controlled only by the pressure ratio: a high pressure ratio allows to produce a permeate stream highly concentrated in CO_2.

Bio-methane production from biogas

The biogas can be used in a wide range of applications (Poschl et al. 2010), for example to co-generate thermal energy or it can be burned to generate heat energy in boilers. It is also important as a direct fuel for automotive applications or in reforming processes to generate hydrogen to be further supplied to fuel cells (Herle et al. 2004, Papadias et al. 2012, Ryckebosch et al. 2011). However, the presence of gases like CO_2 and H_2S strongly lowers the fuel calorific value, reduces the possibility to compress and transport over long distances, due to the corrosive nature of these gases. In addition, the presence of fouling traces including, for examples, siloxanes can induce the formation of fouling in engines and turbines. The biogas upgrading is currently one of the most studied options in biogas treatment leading to the production of bio-methane that can be directly supplied to natural gas grids. As case studies to demonstrate the suitability of membranes for biogas separation, in an our previous work (Brunetti et al. 2015) attention was paid to the use of Matrimid 5218 and Hyflon membranes in multistage systems. Using previously developed performance maps (Brunetti et al. 2014), membrane-integrated systems were analyzed identifying suitable operating conditions and proposing possible process schemes.

Single-stage separation systems

As aforementioned, two different commercial types of membranes are compared, the first is the Hyflon AD60X exhibiting high permeability but low CO_2/CH_4 selectivity and the Matrimid 5218 offering lower permeability but greater selectivity. The performance of a single stage separation unit is analyzed using both membranes, respectively. The targets for delivering CH_4 as a product are (a) gas purity higher than 95% and (b) recovery higher than 90%. As it can be seen in Fig. 2.7, when Hyflon membrane used in the single stage system does not allow to get both targets of purity and recovery. In particular, a purity of methane-rich stream close to 95% and with a CO_2 content below 2% can be achieved with very high values of permeation number, or, in other words, high membrane area (for fixed membrane type, feed compositions and flow rate) and pressure ratio (ca. 40) but with a low CH_4 recovery (ca. 30%).

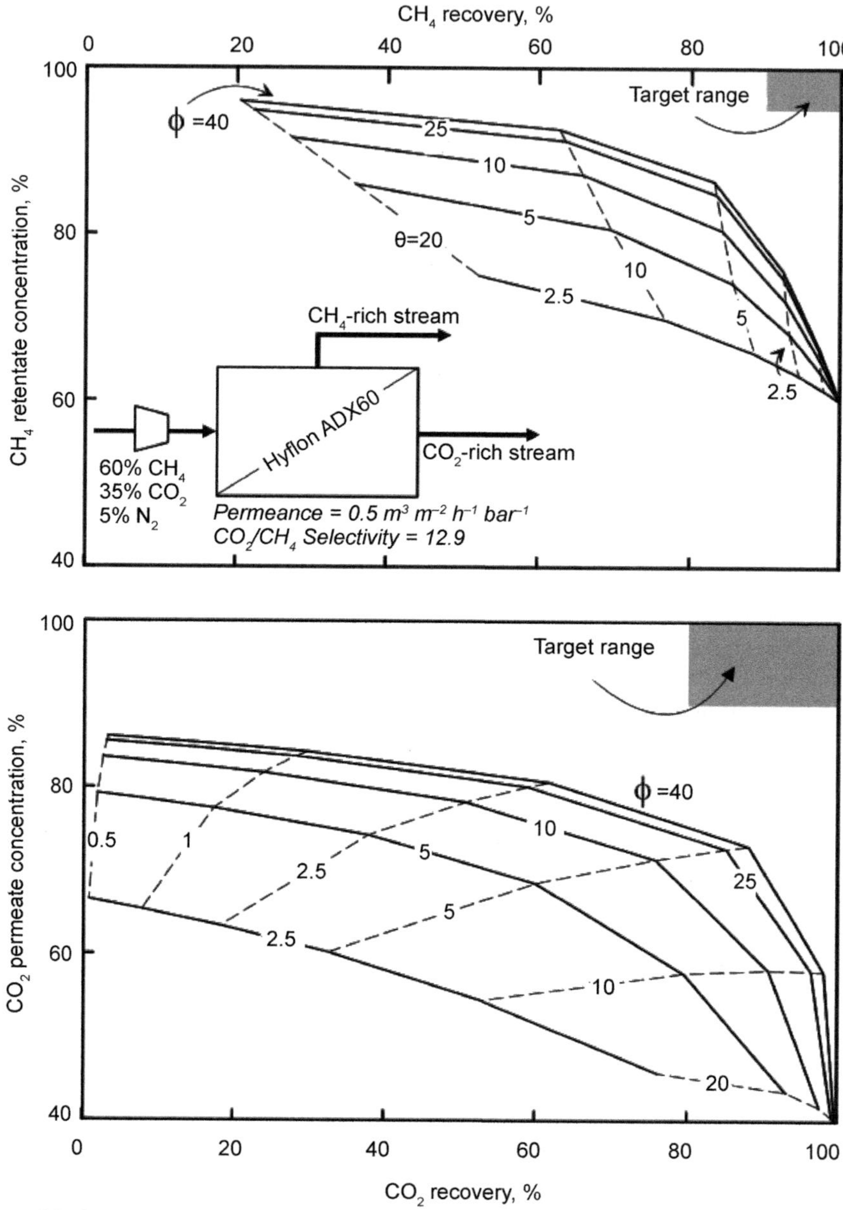

Figure 2.7. Stage 1—Maps of CH_4 and CO_2 concentration in retentate and permeate streams, respectively, as a function of correspondent recovery at various values of pressure ratio and permeation number. Membrane: Hyflon ADX60. CO_2/CH_4 selectivity = 12.9. Reproduced from [38] with permission from Elsevier.

An analogous approach is used to analyze the performance of a single stage operation considering other two types of membranes, i.e., Matrimid 5218 both as it is and with improved higher selectivity.

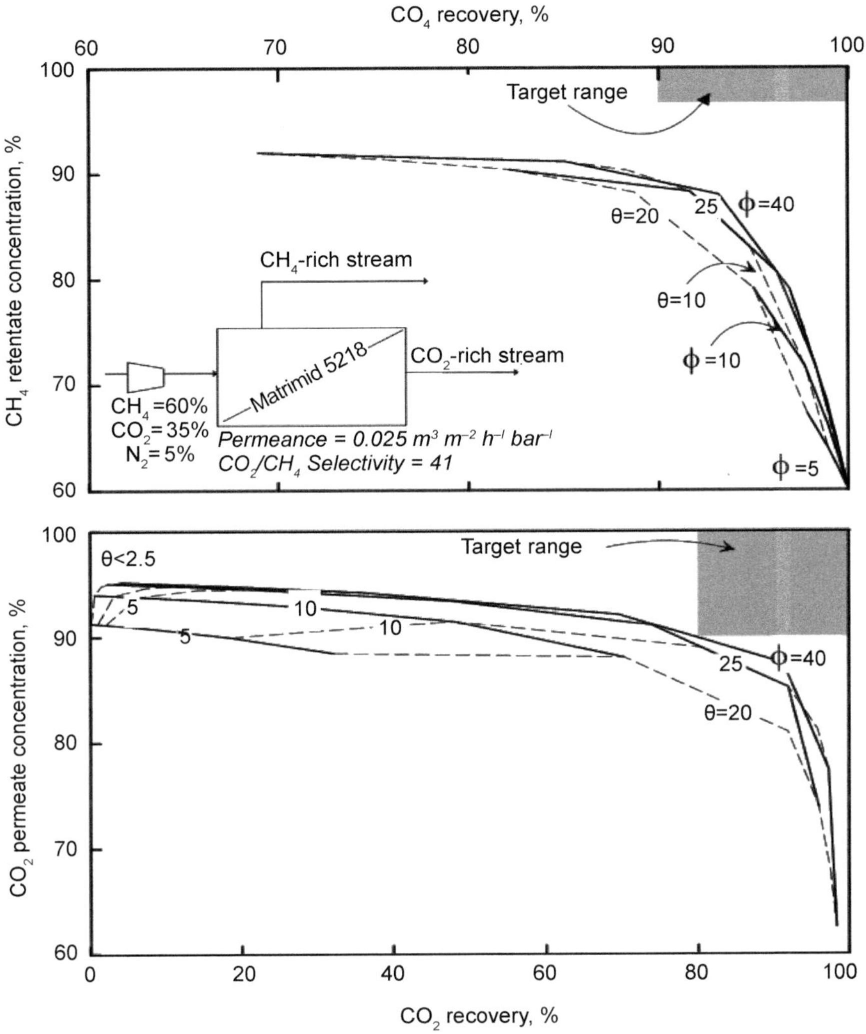

Figure 2.8. Stage 1—Maps of CH_4 and CO_2 concentration in retentate and permeate streams, respectively, as a function of correspondent recovery at various values of pressure ratio and permeation number. Membrane: Matrimid 5218 (CO_2/CH_4 selectivity = 41). Reproduced from [38] with permission from Elsevier.

Figure 2.8 and Fig. 2.9 show the performance maps for the first gas separation stage with Matrimid 5218 with selectivity of 41 and 100, respectively. Also with these membranes, the targets for delivering CH_4 as a product cannot be reached by a single stage.

Matrimid 5218 membrane with improved selectivity (Fig. 2.9) does not allow the achievement of a high-purity CH_4-rich stream so highly pure. However, CO_2 permeate stream can get the targets for storage with interesting values of recovery (80% ca.) for a pressure ratio of 40 and a permeation number of 10. Based on the

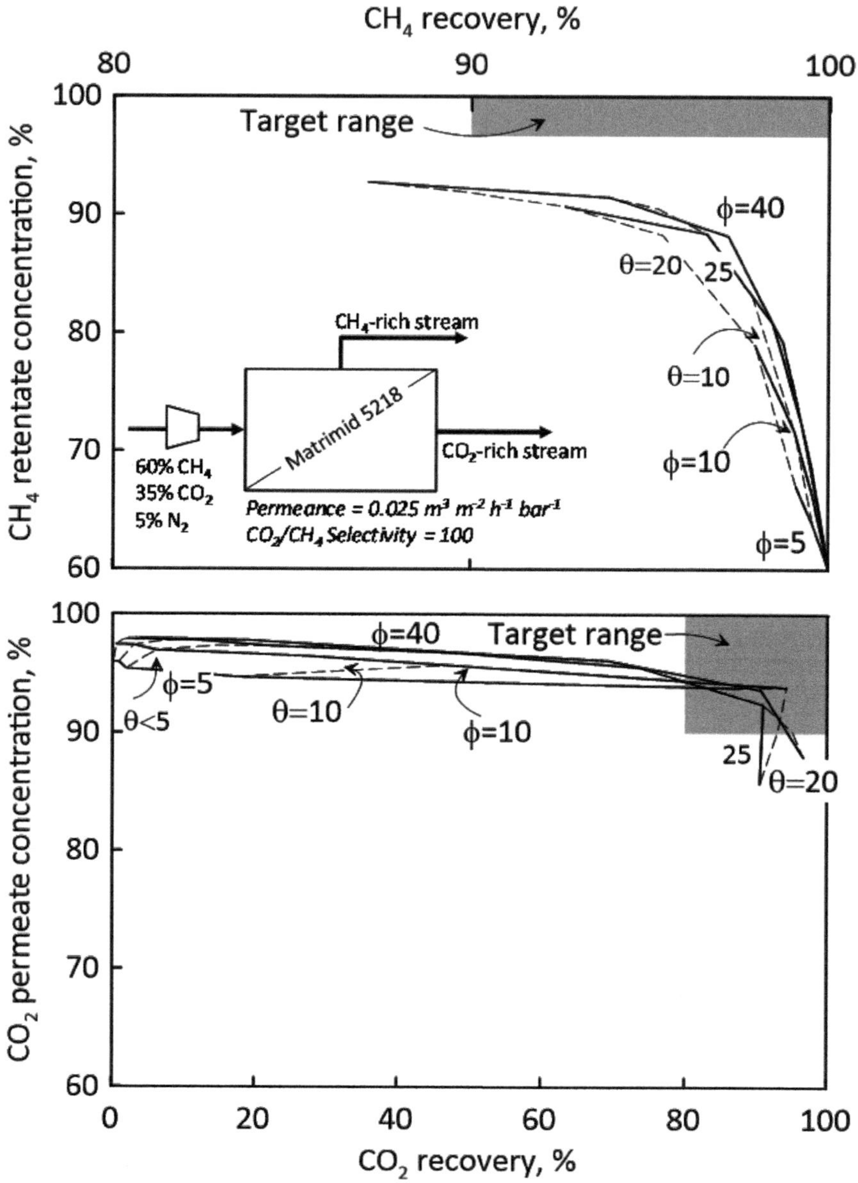

Figure 2.9. Stage 1—Maps of CH_4 and CO_2 concentration in retentate and permeate streams, respectively, as a function of correspondent recovery at various values of pressure ratio and permeation number. Membrane: Matrimid 5218 (CO_2/CH_4 selectivity = 100). Reproduced from [38] with permission from Elsevier.

aforementioned results, Hyflon membranes do not allow to obtain the required purity therefore can be used for as first stage of a multi-stage membrane system for concentrating the retentate and permeate streams.

Double-stage Separation Schemes

Figure 2.10 shows a possible scheme operating for treatment and purification of the CH_4 rich-stream, in which the first stage consists of Hyflon membranes and the successive stage is constituted, instead, by the high-selective Matrimid 5218 membrane in order to obtain highly pure streams.

The addition of the second separation stage allows the selection of a permeation number at the first stage, which would imply a good compromise between the retentate and permeate stream characteristics. This in the logic of adding also a third stage for the treatment of the permeate. A high recovery of the reference species and a concentration sufficiently high to assure a satisfactory permeation driving force in the other membrane stage are well desired for both retentate and permeate stream. Consequently, a permeation number equal to 5 is chosen for the first stage (Fig. 2.11) along with a pressure ratio of 40, which correspond to the streams characteristics reported in Table 2.4. Figure 2.11 shows the strategy to be used in choosing the operating conditions for the other membrane separation stages. The blue curves are the performance of the first stage for a pressure ratio of 40. The red performance curve refers to the second stage for the treatment of the retentate stream at the same pressure ratio.

For a permeation number of 5, two arrows identify two possible performance targets in the second stage, characterized by a certain value of permeation number. The two options selected consider a Permeation Number of 100 and 50, respectively. The retentate and permeate characteristics obtained in these conditions are summarized in Table 2.5. A high permeation number, firstly indicates a greater CO_2 permeation and, thus, a greater recovery. However, also the permeation of the less permeable species is promoted and, thus, CO_2 concentration in the permeate decreases, as well as it does the CH_4 recovery in the retentate.

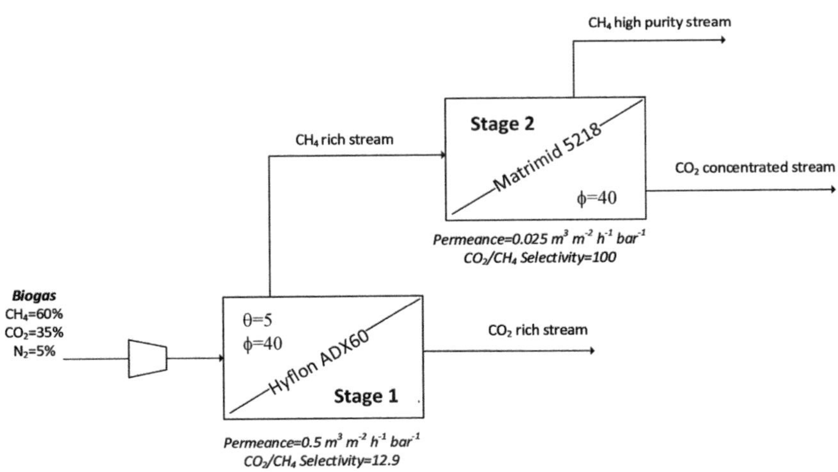

Figure 2.10. Double stage module configuration. Reproduced from [38] with permission from Elsevier.

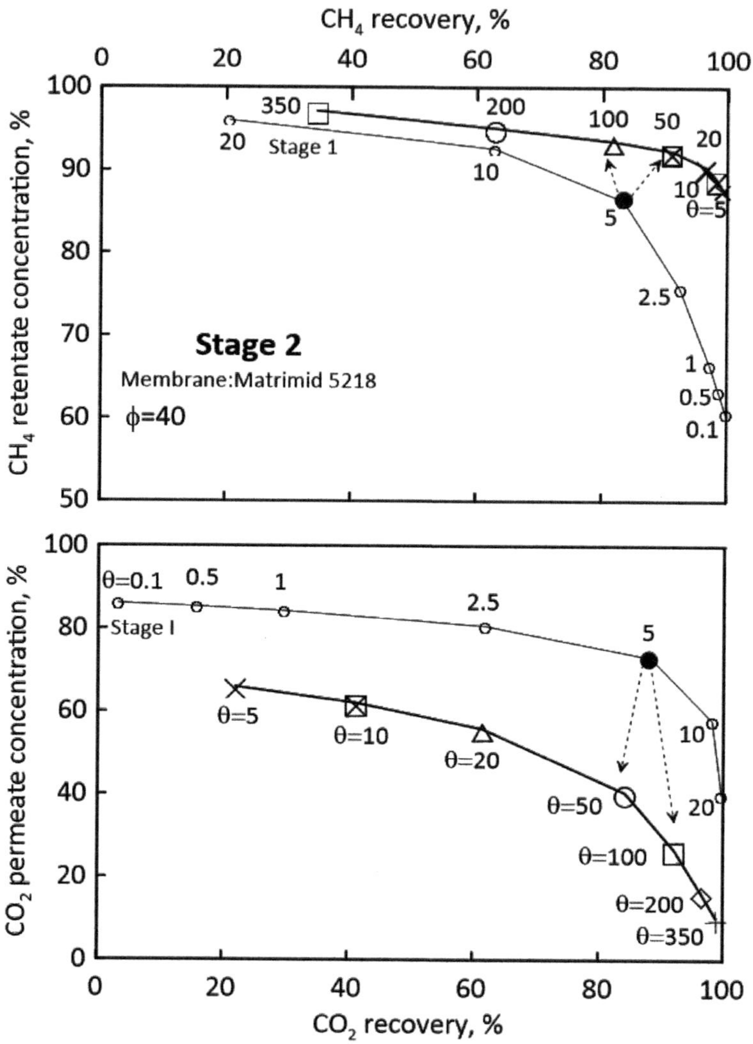

Figure 2.11. Stage 1 and Stage 2: Maps of CH_4 and CO_2 concentration in retentate and permeate streams, respectively, as a function of correspondent recovery at various values of permeation number. Pressure ratio = 40. Reproduced from [38] with permission from Elsevier.

Table 2.4. Retentate and permeate streams characteristics at the first stage. Permeation number = 5; Pressure ratio = 40. Reproduced from [38] with permission from Elsevier.

	Retentate		Permeate	
	CH_4	CO_2	CH_4	CO_2
Concentration, %	86.41	7.15	23.94	73.02
Recovery, %	83.1	-	-	88.2
Content, %	-	7.5	-	-

Table 2.5. Characteristics of retentate and permeate streams at the second stage. Reproduced from [38] with permission from Elsevier.

	Case 1	Case 2
ϕ, –	40	40
θ, –	100	50
Retentate		
CH_4 Purity, %	93.5	92.4
CH_4 Recovery, %	81.7	91.0
CO_2 Content, %	0.8	1.3
Permeate		
CO_2 Purity, %	26.8	40.4
CO_2 Recovery, %	92.1	84.2

Multi-stage separation scheme

Looking at the results achieved with a two-stage configuration, a third stage is added where the permeate streams of first and second stages are fed after mixing. As for the other separation stages, also this one is operated at a pressure ratio of 40 (Fig. 2.12). The feed stream of the third stage consists of a flow rate corresponding to the sum of the permeate flow rates of first and second stage and a composition coming out from the mixing of the two streams. Consequently, changing the operating parameters of the second stage, also the characteristics of the feed stream of the third stage are affected. Indeed, in Fig. 2.12, each map shows two curves, each one corresponding to the case of operation at the stage II (Table 2.5). Case 1 corresponds to high concentration of CH_4 in the retentate (93.5%) but low CO_2 concentration in permeate (26.8%). In Case 2, the CO_2 concentration in permeate was higher (40.4%) than that in Case 1. As a result, the performances at stage III shows a greater purity of the methane-rich stream operating in the conditions of Case I and a greater purity of CO_2 permeate stream when the conditions of Case II are considered. The outlet conditions at each stage, depending on the permeation number chosen for each one, in terms of purity and recovery of CH_4 and CO_2 in retentate and permeate stream, respectively are summarized in Table 2.6. It results do not make it possible to reach at the same time the target of purity and that of the recovery for CO_2 and CH_4 in any of the schemes considered. Even though the molar fraction of CO_2 in retentate stream is lower than 2% in all schemes except that in Scheme 3, nevertheless the target purity required for CH_4 stream can be achieved only under conditions of Scheme 2, although a recovery of 70% ca. is obtained.

Concerning CO_2, the stream targets are well fitted by both Scheme 1 and Scheme 3. In particular, the first restitutes a higher recovery (97.5%) with a purity close to target, whereas Scheme 3 allows achieving a very pure stream (95.9%) with a slightly lower recovery.

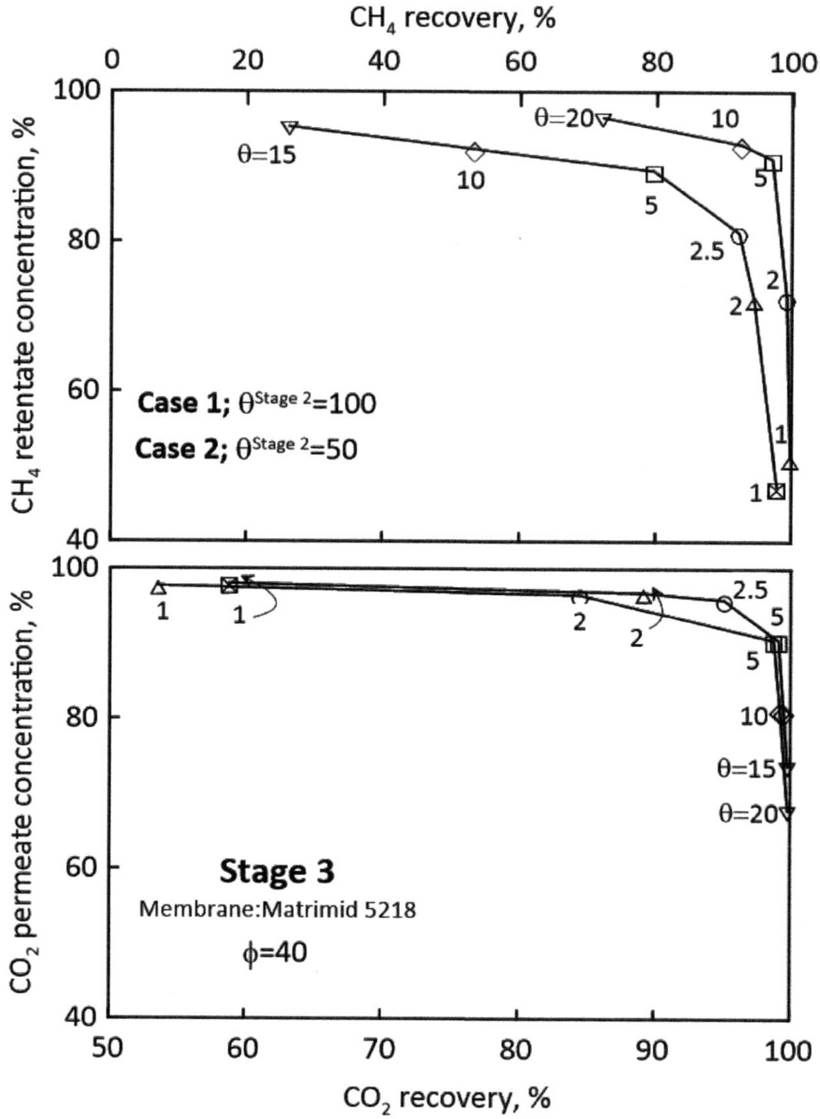

Figure 2.12. Stage 3—Maps of CH_4 and CO_2 concentration in retentate and permeate streams, respectively, as a function of correspondent recovery at various values of permeation number. Pressure ratio = 40. Reproduced from [38] with permission from Elsevier.

Table 2.6. Performances of membrane stages for the different process schemes. Reproduced from [38] with permission from Elsevier.

	Scheme 1			Scheme 2			Scheme 3			Scheme 4			
Stage	I	II	III	I	II	III	I	II	III	I	II	III	
Permeation number	5	100	5	5	100	20	5	50	2.5	5	50	15	
Outlet conditions													Targets
CH_4 purity, %	86.4	93.5	94.1 (1.3%CO_2)	86.4	93.5	95.1 (0.77%CO_2)	86.4	92.4	89.4 (3.6%CO_2)	86.4	92.4	92.5 (1.4%CO_2)	> 95 (CO_2 < 2%)
CH_4 recovery, %	83.1	81.7	89.7	83.1	81.7	70.1	83.1	91.0	92.8	83.1	91.0	77.1	> 90
CO_2 purity, %	73.0	26.8	90.5	73.0	26.8	67.8	73.0	40.4	95.9	73.0	40.4	73.7	> 90
CO_2 recovery, %	88.2	92.1	97.2	88.2	92.1	98.2	88.2	84.2	92.4	88.2	84.2	97.1	> 80

Membranes for CO_2 Separation

Membrane modules

The use of membranes on a technical scale usually requires large membrane areas. The smallest unit into which the membrane area is packed is called "module" and is the central part of membrane installation (Scholz et al. 2011) .

Figure 2.13 shows a feed stream entering the module; it is characterized by its own temperature, pressure, composition and flow rate. Because the membrane has the ability to transport one component faster than another, both the composition and the flow rate inside the module will change along the module length. The feed stream is separated during the passage through the module into a permeate stream and a retentate stream. The permeate stream is the fraction of the feed stream passed through the membrane whereas the retentate stream is the fraction retained. The aspects to take into account in the membrane module design are various:

- Good mechanical, thermal and chemical stability
- Good flow distribution (no dead zones, no by-pass)
- High packing density
- Low pressure drop
- Possibility of cleaning
- Ease of maintenance and operation
- Cheap manufacturing
- Compactness of the system scale
- Possibility of membrane replacement

The major module configurations are: plate-and-frame, spiral wound, and hollow fiber.

In a *plate-and-frame module* (Fig. 2.14a) two membranes are placed in a sandwich-like configuration with their feed sides facing each other. A spacer is placed between each feed and permeate compartment. Several sets of two membranes constitute a stack. The packing density (membrane surface per module volume) of such modules is low and about 100–400 m^2/m^3.

Spiral-wound modules (Fig. 2.14b), flat membrane envelopes are wrapped around a central tube. The feed, distributed by a central tube, passes along the length of the module and the permeate passes into a membrane envelope and then goes out *via* the collecting tubes. Both the feed and permeate are transported through the module in fluid-conductive spacer. Modern modules tend to contain multiple membranes that are all attached to the same central tube. Currently, the spiral wound contains around 1–2 m of rolled sheets, for 20–40 m^2 of membrane area.

Hollow-fiber modules (Fig. 2.14c) contain a large number of membrane fibers housed in a shell resulting in a tubes-and-shell configuration. The free ends of the fibers are potted with agents such as epoxy resins, polyurethanes, silicone rubber, thermoplastics, or inorganic cements. Feed can be introduced on either the fiber or

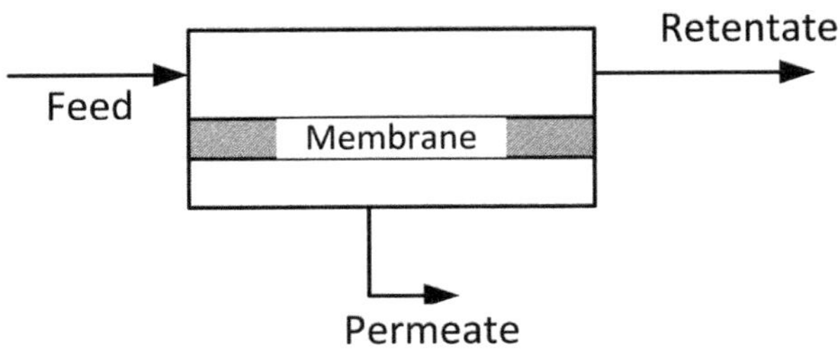

Figure 2.13. Schematic drawing of a membrane module.

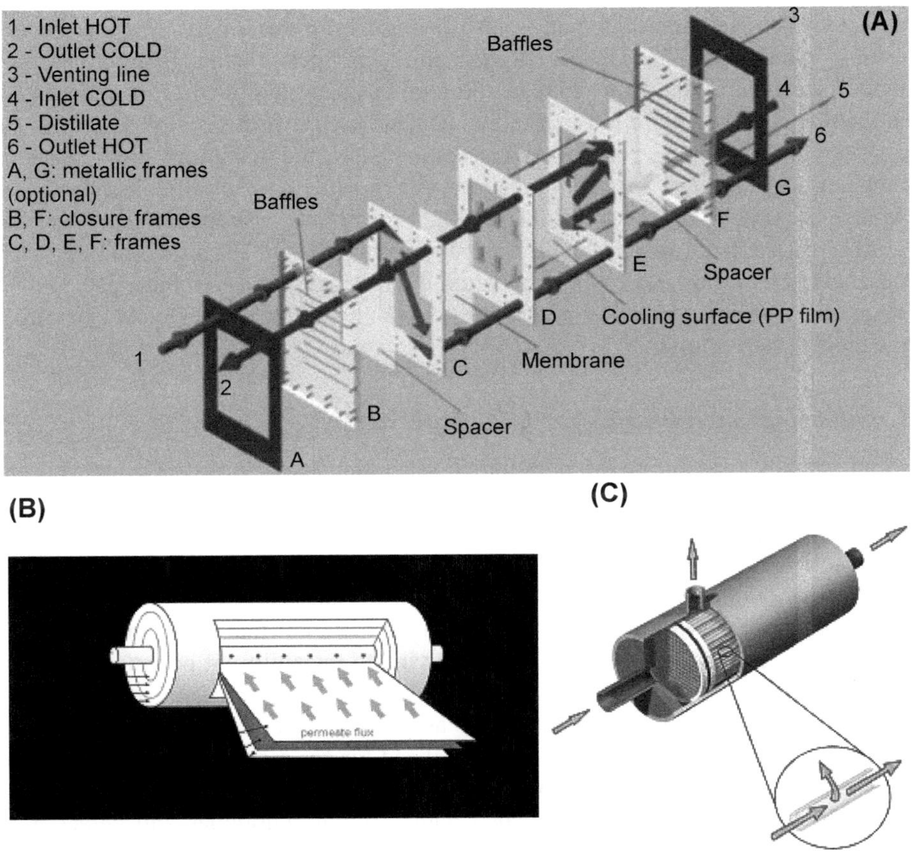

Figure 2.14. Schematic drawing of membrane module configurations: (A) plate and frame (Reprinted with permission of Elsevier from (Cipollina et al. 2012); (B) spiral wound and (C) hollow fiber (https://en.wikipedia.org/wiki/Membrane_technology. CC0-1.0 public domain (accessed 21/07/2017).

shell side. Permeate is usually withdrawn in a co-current or counter-current manner, with the latter being generally more effective. Hollow fibers are the cheapest modules on a per-square-meter basis (with the highest membrane area to module volume ratio); however, the preparation/fabrication of very thin selective layers in hollow-fiber form is harder than in flat sheet configuration. This implies that the permeance of hollow fibers is generally lower than that of a flat sheet membrane prepared with the same material; therefore, more membrane area is required for achieving the same separation. Hollow fiber modules also require more pre-treatments of the feed than is usually required by spiral wound modules for removing particles, oil residue and other fouling components. These factors strongly affect the final cost of the hollow fiber module systems. Therefore, currently, spiral wound modules are employed in several separations (e.g., in natural gas processing) particularly for those separations which cannot support the costs associated with the required pre-treatments of hollow fiber modules.

Spiral-wound modules are also used where pressure drop has to be considered and when counter current flow is not needed to maximize separation efficiency. The choice of the membrane module is also determined by economic considerations, even if in industrial plants, especially in refinery and petrochemical operations, the module costs are only 10 to 25% of the total costs, so that significant reductions in membrane costs might not markedly change the cost of the complete whole. In a hollow fiber spinning plant operating continuously, the membrane costs are in a range of US\$ 2 to 5 m^{-2} of membrane area. An equivalent of spiral-wound modules would cost US\$ 10 to 100 m^{-2} (Baker 2002). A membrane gas separation production process can be built, assembling the membrane modules in several configurations, depending on the particular type of separation. The single-stage, double-stage, multi-stage and in series, parallel or combined flow-configuration with recycle constitute the main design solutions.

Commercial membranes for CO_2 separation

Commercially available membranes for gas separation applications are usually based on polymeric materials forming a dense ultra-thin layer as either asymmetric or composite structures (Spillman 1995).

Table 2.7 summarizes some of the most important commercially available membranes, companies and principal membrane materials (Scholes et al. 2017). These membranes are based on a few polymeric materials that have dominated the industry for the past few decades, mainly owing to the ability of polymers to form low-cost membranes with stable thin active layers that can be processed into modules.

Inorganic and carbon membranes, as well as facilitated transport based membranes, have yet to be fully commercialized for gas separation applications, as they are generally considered more expensive than polymeric membranes, owing to the high fabrication costs associated with forming continuous and defect free membranes, as well as handling issues around their potential brittleness and adhesion to the module (Scholes et al. 2014).

In recent years, the price of gas separation membranes has been settled toUS\$ ~ 50 m^{-2}, mainly associated with the polymeric materials and fabrication method. New membranes are based on more expensive custom made polymers, and while these have improved separation properties they often cost as much as US\$ 1000 to US\$ 10,000 per kg to synthesis. This restricts them to composite membrane structures, based on an ultrathin layer of the polymer to keep the amount used to less than 1 g per m^2, while an inexpensive polymer is used for the porous support layer. The membrane configuration also affects the price. Hollow fiber production costs

Table 2.7. Most important commercially available membranes for CO_2 separation (Scholes et al. 2017).

Membrane	Supplier	Material
Cynara	Cameron	Cellulose acetate
Prism	Air Products	Polysulfone
Medal	Air Liquide	Polyimide/polyaramide
	Generon	Polycarbonate
Separex	UOP	Cellulose acetate
IMS	Praxair	Polyimide
Grace	Kvaerner	Cellulose acetate
UBE	Ube Industries	Polyimide
Polaris	MTR	

are estimated to between US$2 to 5 m^{-2} compared to spiral-would production, in the range of US$ 10 – 100 m^{-2}. For natural gas sweetening, the membrane itself may be less than 5% of the total capital cost (Spillman 1995). This is because the cost of the membranes used is a small fraction of the final capital investment because of the cost associated with the pressure vessels, expensive controls and instrumentation as well as piping, flanges and valves. Baker and Lokhandwala (2008) estimated that these additional costs lead to a membrane skid costing US$ 500 m^{-2}, which is an order of magnitude greater than the estimated membrane price, and demonstrates that for high pressure applications, membrane price is a small variable in the overall cost. For low pressure applications, such as post-combustion carbon capture, these additional expenses are not necessary because there is no need for pressure vessels and extensive instrumentation for pressure control. The membrane module can also be based on fiberglass wrap or plastic comparable to those used in reverse osmosis membranes units, rather than steel. This will yield membrane module costs for low pressure carbon capture comparable to those of reverse osmosis at US$ 30–50 m^{-2}. For non-polymeric gas separation membranes, techno-economic studies on membrane prices are highly speculative, because none have been fabricated on a large enough scale (Scholes 2017).

Conclusions and Final Remarks

CO_2 capture for reducing greenhouse emissions is today one of the separations that more than other attracts the attention on the use of membranes. The modelling and simulation leading to the design of new integrated membrane processes can significantly promote the development of this technology on industrial scale.

The possibility of using a membrane unit in the separation of a gaseous stream is strictly connected to three main factors: the composition of the feed, the available operating conditions, and the separation properties of the membrane chosen for the specific application. This chapter systematically discusses the performance maps and their practical use in the preliminary design of a membrane separation system for CO_2 capture, taking into account a wide range of feed and operation conditions, as well as membranes and membrane module separation properties. Together with the feed conditions, one variable significantly affecting the performance of the membrane module is the feed/permeate pressure ratio. The permeation number is a determining parameter for the module performance. For a set feed flow rate, a set membrane type, and defined pressure ratio, a low permeation number indicates low recovery and high permeate concentration, and vice versa. The low CO_2 concentration in the feed, as it is in flue gas streams, does not allow high permeate concentration streams, even when increasing the pressure ratio, and thus more separation stages are necessary. The effect of the selectivity on the performance of the membrane module is negligible at low pressure ratios but becomes important as this value is increased. For a high value of selectivity, doubling of the pressure ratio implies a recovery 2–3 times greater and improvements in the CO_2 permeate concentration.

In general, membrane engineering plays a fundamental role in the integration of the units in a single plant and, at the same time, in the definition of the knowledge necessary to drive the process by maximizing the gains both in terms of efficiency and plant size reduction. Therefore, it can make a fundamental contribution to the sustainable industrial development for CO_2 capture by introducing a new design philosophy based on the principles of Process Intensification Strategy, which leads to the development and the re-designing of more compact and efficient new processes.

References

Adams, R., Carson, C., Ward, J., Tannenbaum, R. and Koros, W. 2010. Metal organic framework mixed matrix membranes for gas separations. Micr. Mes. Mat. 131(1): 13–20.

Adams, R.T., Lee, J.S., Bae, T.-H., Ward, J.K., Johnson, J.R., Jones, C.W. et al. 2011. CO_2–CH_4 permeation in high zeolite 4A loading mixed matrix membranes. J. Mem. Sci. 367: 197–203.

Baker, R.W. 2002. Future directions of membrane gas separation technology. Ind. Eng. Chem. Res. 41: 1393–1404.

Baker, R.W. and Lokhandwala, K. 2008. Natural gas processing with membranes: an overview. Ind. Eng. Chem. Res. 47: 2109–2121.

Basu, S., Khan, A.L., Cano-Odena, A., Liu, C. and Vankelecom, I.F.J. 2010. Membrane-based technologies for biogas separations. Chem. Soc. Rev. 39: 750–768.

Baudot, A. 2011. Gas/vapour permeation applications in the hydrocarbon-processing industry. *In*: Drioli, E. and Barbieri, G. (eds.). Membrane Engineering frovthe treatment of gases, The Royal Society of Chemistry, Cambridge, The United Kingdom, 2011, pages 150–195, ISBN 978-1-84973-239-0.

Brunetti, A., Scura, F., Barbieri, G. and Drioli, E. 2010. Membrane technologies for CO_2 separation. J. Mem. Sci. 359: 115–125.

Brunetti, A., Drioli, E., Lee, Y.M. and Barbieri, G. 2014. Engineering evaluation of CO_2 separation by membrane gas separation systems. J. Membr. Sci. 454: 305–315. https://doi.org/10.1016/j.memsci.2013.12.037.

Brunetti, A., Sun, Y., Caravella, A., Drioli, E. and Barbieri, G. 2015. Process intensification for greenhouse gas separation from biogas: more efficient process schemes based on membrane-integrated systems. Int. J. Greenhouse Gas Control 35: 18–29.

Brunetti, A., Cersosimo, M., Dong, G., Woo, K.T., Lee J., Kim, J.S. et al. 2016. *In situ* restoring of aged thermally rearranged gas separation membranes. J. Membr. Sci. 520: 671–678.

Brunetti, A., Cersosimo, M., Kim, J.S., Dong, G., Fontananova, E., Lee, Y.M. et al. 2017. Thermally rearranged mixed matrix membranes for CO_2 separation: An aging study. Int. J. Greenhouse Gas Control 61: 16–26.

Calle, M. and Lee, Y.M. 2011. Thermally rearranged (TR) poly(ether-benzoxazole) membranes for gas separation. Macromol. 44: 1156–1165.

Cersosimo, M., Brunetti, A., Fiorino, F., Drioli, E., Dong, G., Woo, K.T. et al. 2015. Separation of CO_2 from humidified ternary gas mixtures using thermally rearranged polymeric membranes. J. Membr. Sci. 2015(492): 257–262.

Ciferno, J.P., Fout, T.E., Jones, A.P. and Murphy, J.T. 2009. Capturing carbon from existing coal-fired power plant. Chem. Eng. Prog. 33–41.

Cipollina, A., Di Sparti, M.G., Tamburini, A. and Micale, G. 2012. Development of a Membrane Distillation module for solar energy seawater desalination. Chem. Eng Res. Des. 90: 2101–2121.

Daal, L., Claassen, L., Bruns, R., Schallert, B., Barbieri, G., Brunetti, A. et al. 2013. Field tests of carbon dioxide removal from flue gasses using polymer membranes. VGB Powertech 6: 78–84.

Deng, L. and Hagg, M.-B. 2010. Techno-economic evaluation of biogas upgrading process using CO_2 facilitated transport membrane. Int. J. Greenhouse Gas Control 4: 638–646.

Falbo, F., Tasselli, F., Brunetti, A., Drioli, E. and Barbieri, G. 2014. Polyimide hollow fibre membranes for CO_2 separation from wet gas mixtures. Brazilian Journal of Chemical Engineering 31: 1023–1034.

Falbo, F., Brunetti, A., Barbieri, G., Drioli, E. and Tasselli, F. 2016. CO_2/CH_4 separation by means of Matrimid hollow fibre membranes. App. Petr. Res. J. 6, 4, 439–450.

Favre, E. 2007. Carbon dioxide recovery from post-combustion processes: Can gas permeation membranes compete with absorption? J. Mem. Sci. 294: 50–59.

Favre, E., Bounaceur, R. and Roizard, D. 2009. Biogas, membranes and carbon dioxide capture. J. Mem. Sci. 328: 11–14.

He, X. and Hägg, M.-B. 2004. Membranes for environmentally friendly energy processes. Memb. 2(4): 706–726.

Herle, J.V., Membrez, Y. and Bucheli, O. 2004. Biogas as a fuel source for SOFC co-generators. J. Power Sou. 127: 300–12.

Herzog, H. 2001. What future for carbon capture and sequestration? Env. Sci. Tech. 35(7): 148–153.

Jung, C.H., Lee, J.E., Han, S.H., Park, H.B. and Lee, Y.M. 2010. Highly permeable and selective poly(benzoxazole-co-imide) membranes for gas separation. J. Mem. Sci. 350: 301–309.

Klara. J.M. 2007. Cost and performance baseline for fossil Energy plants. Volume 1: Bituminuous coal and natural gas to electricity final report, DOE/NETL-2007-1281, (2007) http://www.netl.doe.gov/energy-analyses/pubs/Bituminuous%20Baseline_Final%20Report.pdf.

Ku, A.Y., Kulkarni, P., Shisler, R. and Wei, W. 2011. Membrane performance requirements for carbon dioxide capture using hydrogen-selective membranes in integrated gasification combined cycle (IGCC) power plants. J. Mem. Sci. 367: 232–239.

Li, B., Duan, Y., Luebke, D. and Morreale, B. 2013. Advances in CO_2 capture technology: A patent review, App. En. 102: 1439–1447.

Lin, H., Van Wagner, E., Raharjo, R., Freeman, B.D. and Roman, I. 2006. High-performance polymer membranes for natural-gas sweetening. Adv. Mater. 18: 39–44.

Luis, P., Van Gerven, T. and Van der Bruggen, B. 2012. Recent developments in membrane-based technologies for CO_2 capture. Progr. En. Com. Sci. 38(3): 419–448.

Merkel, T.C., Lin, H., Wei, X. and Baker, R. 2010. Power plant post-combustion carbon dioxide capture: An opportunity for membranes. J. Mem. Sci. 359, 1–2, 1 (2010) 126–139.

Papadias, D., Ahmed, S. and Kumar, R. 2012. Fuel quality issues with biogas energy e an economic analysis for a stationary fuel cell system. Energy 44: 257–77.

Park, H.B., Jung, C.H., Lee, Y.M., Hill, A.J., Pas, S.J., Mudie, S.T. et al. 2007. Polymers with Cavities Tuned for fast and Selective Transport of Small Molecules and Ions. Sci. 318: 254–25.

Peters, L., Hussain, A., Follmann, M., Melin, T. and Hägg, M.B. 2011. CO_2 removal from natural gas by employing amine absorption and membrane technology—A technical and economical analysis. Chem. Eng. J. 172: 952–960.

Pidwirny, M. 2006. Atmospheric Composition. Fundamentals of Physical Geography, 2nd Edition. Date Viewed. http://www.physicalgeography.net/fundamentals/7a.html.

Poschl, M., Ward, S. and Owende, P. 2010. Evaluation of energy efficiency of various biogas production and utilization pathways. Appl. Energy 87: 3305–21.

Powell, C.E. and Qiao, G.G. 2006. Polymeric CO_2/N_2 gas separation membranes for the capture of carbon dioxide from power plant flue gases. J. Mem. Sci. 279: 1–49.

Ramasubramanian, K., W.S. Ho. 2011. Recent developments on membranes for post-combustion carbon capture. Curr. Op. Chem. Eng. 1(1): 47–54.

Robeson, L.M. 2008. A revised upper bound. J. Mem. Sci. 320: 390–400.

Ryckebosch, E., Drouillon, M. and Vervaeren, H. 2011. Techniques for transformation of biogas to biomethane. Biomass and Bioenergy 35: 1633–45.

Scholes, C.A., Bacus, J., Chen, G.Q., Tao, W.X., Li, G., Qader, A. et al. 2012. Pilot plant performance of rubbery polymeric membranes for carbon dioxide separation from syngas. J. Mem. Sci. 389: 470–477.

Scholes, C.A. 2017. Cost competitive membrane processes for carbon capture, chapter 8. pp. 216–241. *In*: Drioli, E., Barbieri, G. and Brunetti, A. (eds.). Membrane Engineering for the Treatment of Gases. Volume 1 Gas-separation issues with membranes, 2017. The Royal Society of Chemistry, Cambridge, The United Kingdom.

Scholes, C.A., Motuzas, J., Smart, S. and Kentish, S.E. 2014. Membrane adhesives. Ind. Eng. Chem. Res. 53: 9523–9533.

Scholz, M., Wessling, M. and Balster, J. 2011. *In*: Drioli, E. and Barbieri, G. (eds.). Membrane Engineering for the Treatment of Gases. The Royal Society of Chemistry, Cambridge, The United Kingdom.

Shao, L., Low, B.T., Chung, T.-S. and Grenberg, A.R. 2009. Polymeric membranes for the hydrogen economy: Contemporary approaches and prospects for the future. J. Mem. Sci. 327(1-2): 18–31.

Spillman, R. 1995. Economics in membrane gas separation processes. pp. 589–667. *In*: Noble, R.D. and Stern, S.A. (eds.). Membrane Separations Technology. Elsevier Science, Eastbourne.

Tuinier, M.J., Hamers, H.P. and van Sint Annaland, M. 2011. Techno-economic evaluation of cryogenic CO2 capture—A comparison with absorption and membrane technology. Int. J. Greenhouse Gas Contr. 5: 1559–1565.

White, C.M. 2003. Separation and capture of CO_2 from large stationary sources and sequestration in geological formations. J. Air Waste Management Association 53: 645–715.

Xiao, Y., Low, B.-T., Hosseini, S.S., Chung, T.S. and Paul, D.R. 2009. The strategies of molecular architecture and modification of polyimide-based membranes for CO_2 removal from natural gas—A review. Progress in Polymer Science 34: 561–580.

3

Polymeric Membrane for Flue Gas Separation and Other Minor Components in Carbon Dioxide Capture

A.L. Ahmad,[1,*] *Y.O. Salaudeen*[1] and *Z.A. Jawad*[2]

INTRODUCTION TO POLYMER

A polymer is a large molecule with long a chain composed of a lot of small unities, called monomers. A polymer is consequently characterized by a high molecular weight. The various properties of a polymer do not only originate from the chemical nature of the repetitive pieces but also from their average molecular weight or from the conformation of the chain.

The polymer is called **homopolymer** as in row 1, if all the repetitive units are the same. If the repetitive units are made of two or three different monomers, it is called **copolymer**. Due to differences in the mode in which various unities are linked together, polymers have different properties as well. It is an **alternating copolymer** as in row 2 if the polymer is made of A and B units in regular alternating sequence. But if without any order, it is termed **random copolymer** as in row 3. A **block copolymer** is made of two or more homopolymer subunits interconnected by coordinate bonds as in row 4. The polymer is **grafted** as in row 5 if the key chain is made of A monomers and several side chains of B are linked to the first one, as all this term are in Fig. 1 (Iupac 1996).

[1] Universiti Sains Malaysia, School of Chemical Engineering, Engineering Campus 14300, Nibong Tebal, Pulau Pinang, Malaysia.
Email: yuslah9@gmail.com

[2] School of Engineering and Science, Department of Chemical Engineering, Curtin University Malaysia, CDT 250, Miri 98009, Sarawak, Malaysia.
Email: zeinab.aj@curtin.edu.my

* Corresponding auhor: chlatif@usm.my

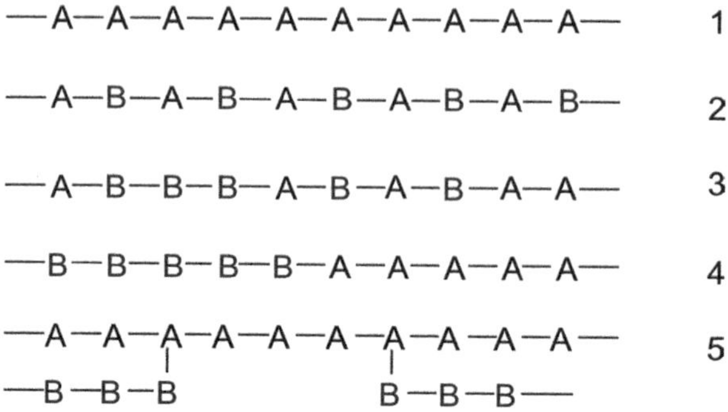

Figure 3.1. Different structure of polymers.

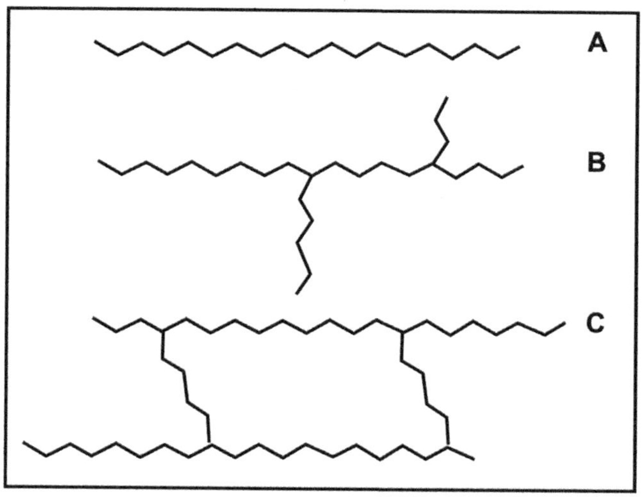

Figure 3.2. Molecular structure of polymer.

Figure 3.2 is based on molecular structure classification; which explains that polymers can be **linear (A)**, **branched (B)** or **cross-linked (C).** The interlinking has a portentous effect on the physical, heat conductivity and thermo-mechanical properties of a polymer: it can lose its solubility and hence become resistant to chemical reagent. Chemical or physical crosslinking can be used (Mulder 1996).

State of a Polymer

Amorphous and **crystallinity** can be used to generalize the two phases of a polymer. Such a polymer is then classified based on its degree of crystallinity and termed **semi crystalline** polymer. There is the possibility of producing only amorphous polymer but a complete crystalline polymer is rarely found. Lack of order at the molecular level denotes the amorphous state. If the polymer is a very regular one and has a symmetrical structure and their chains are orderly linked hence the structure becomes crystalline. Glass transition temperature T_g is what mainly determines the state of the polymer (except for crystalline polymers for which the melting temperature T_m

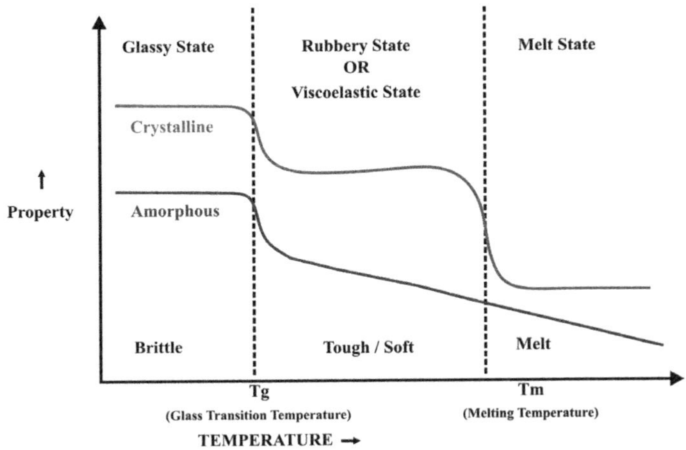

Figure 3.3. Thermal transitions in polymer.

is more important) because glass transition reflects the reaction of macromolecular chains that are arranged in a disordered fashion (amorphous phase) when the material is heated. As a result of further movement of chains, the material character changes from glassy state to a rubbery one. Meanwhile, the melting temperature is destruction of ordered chains in crystallites (crystalline phase). This temperature demotes the border between two different states of the polymer: **glassy** for $T < T_g$ and **rubbery** for $T > T_g$. This modification can be noticed by ensuing various properties, as for example tensile modulus E, when raising the temperature (Mulder 1996). The tensile modulus is the force applied to a given area to obtain a given deformation and it is obtain from the thermo gravimetric curves as in Fig. 3.3.

The polymer is defined as a glassy polymer when the temperature is below glass transition temperature (T_g). It is hard and brittle which is related to the limited chain mobility. The movement of vibration is only allowed due to the intermolecular forces between the chains. The selectivity of the glassy polymer is very good but the penetrant diffusion is low (Collins and Ramirez 1979). However, permeability coefficients of glassy usually decrease with enhancing penetrant size in the absence of plasticization (Freeman and Pinnau 1999). Hence, glassy polymer is generally acceptable by the industries as a result of high mechanical strength and good separation properties. The species with low molecular diameter often permeate faster in glassy polymers and the differences in molecular dimension result into selectivity (Bernardo et al. 2009).

Meanwhile, the polymer is in a rubbery state when it is above the glass transition temperature. The tough and flexible polymer in a rubbery state can be related to the free chain motion. This polymer has the capacity to swiftly respond to external stresses compared to glassy polymers. Thus, there would be an adjustment in the new equilibrium state that form when there is a change in temperature (Collins and Ramirez 1979). The selectivity of rubbery polymers is largely determined by differences in the condensability of the gas species and most often present high permeabilities especially when applied to separate an organic vapour from nitrogen, preferentially used to permeate organic molecules. A branch in organic chemistry, poly(organosiloxanes) has been widely studied by the researcher due the vast utility of polydimethylsiloxane (PDMS) that possess high permeability coefficients which is as a result of its free-volume, and high selectivity for condensable gases (Bernardo et al. 2009). Silicone rubber has an enormous permeability and adequate vapour/inert gas selectivity for most applications; composite membranes of silicone rubber are majorly utilized in all the installed vapour separation systems (Baker et al. 1998). Both glassy and rubbery polymers are used to produce polymeric membranes. Glassy polymers have mechanical properties that permit them to be used in a self-supported structure like flat sheet while rubbery material has a proper substrate because of their softness.

Polymeric Membrane

The most important advantages of polymeric membrane over the other established process for separation such as cryogenic distillation, pressure (and vacuum) swing adsorption, and chemical absorption process, are low capital and operating costs, ease of operation, and low energy requirements. Hence, using polymeric membrane for separation of gas has attained great importance in the manufacturing setting as well as signifies a potent answer for CO_2 capture from the gaseous streams (Hu et al. 2006). Research is on course for development membranes using various materials with suitable properties for CO_2 separation. Characteristics such as proper separation properties, constancy under real conditions, thermo-mechanical and chemical resistivity in the presence of unfavourable environmental conditions, reproducing at a very high scale, easy handling, etc., are the major requirement of the features of the membranes to be suitable for industrial applications (Falbo et al. 2012). The selectivity and permeability is the major player that determines the efficiency of a polymer membrane-based gas separation process. The efficient separation of a membrane is as a result of higher selectivity of the membrane with high recovery as well as low power cost whereas; polymer membrane that produces higher permeability leads to higher productivity and low capital cost. However, the most economical gas separation process would be achieved if the membrane concurrently possess a high value of selectivity and permeability (Ismail and Yean 2003). Although the main obstacle of a polymeric membrane is its very low thermal stability these membranes may be plasticized with influence of CO_2 in membrane while their selectivity and permeability is usually high for CO_2 separation. Hence, for the membrane to be useful in post-combustion capture, the heat of flue gas must first be cooled down to 313–333 K (Favre 2011). The gas permeability of polymer is rule by the following physicochemical characteristics:

- The correlation of the glass transition temperature, T_g with the mobility of the polymer.
- The corresponding polymer free volume and the intersegmental spacing.
- The interaction between the polymer and the penetrant gas that is correlated with the gas solubility (Huang et al. 2006).

The transports of gases in a nonporous film of the selective layer polymeric membranes are generally by the solution-diffusion mechanism. Examples of polymeric membranes used for separation of CO_2 include polyetherimides, polysulphones, polyacetylenes, polyaniline, polyarylates, polyarylene ethers, polycarbonates, polyimides, polyethylene oxide, polypyrroles, polyphenylene oxides, and amino groups such as polyethyleneimine blends, polymethacrylates (Doshi and Dortmundt 1999). Selective polymeric membranes are categorized namely into two: glassy and rubbery. The high gas selectivity and good mechanical properties is often referred as a glassy polymeric membrane for CO_2 separation. Contrariwise, the flexible and soft with a low selectivity but a high permeability is often as rubbery polymeric membranes, whereas glassy polymers exhibit a high selectivity but low permeability (Scholes et al. 2009). The minute differences in molecular dimensions usually determine the efficiency of a glassy polymer to separate molecules. They are appropriate for CO_2 separation because of their distinctive possessed size and shape than that of the rubbery polymers. Henry's law can be best described the sorption of gases in rubbery polymers while the sorption in glassy polymers can be described by complex sorption isotherms concomitant to the unrelaxed volume of the matrix when the materials are quenched below their glass transition temperature (Baker 2004). Moreover, the glassy polymer provides mechanical and structural support for the rubbery matrix (Scholes et al. 2009). The commercial polymeric membranes generally possess a structure which is not uniform with a very thin selective layer supported by a thicker porous sub-layer. The higher gas fluxes through the membrane would be as a result of thin selective layer, while the structural integrity of the membrane would be as a result of thick support layer. Recently, reports have been made by researchers that combining selective polymers would produce membranes with improved CO_2/N_2 permeances and selectivities. In the case of blended polymer membranes, the incorporation of a glassy polymer into a rubbery polymer matrix to combine the high selectivity of the glass polymer with the high diffusivity of the rubbery polymer (Ghalei and Semsarzadeh 2007). The gas permeation is ruled by the solution–diffusion mechanism and this mechanism is usually used to describe the polymeric membranes which are generally nonporous, and the ability of gases to diffuse and be soluble within the polymer matrix is determined by the solution-diffusion as the term implies (Scholes et al. 2009). The gas-membrane interface will attract the gaseous molecules to pass through from the feed side where they dissolve and penetrate across the membrane interface by random molecular diffusion, followed by desorption and diffusion into the permeate bulk stream. The occurrence of the unceasing appearance and disappearance of the thermal motion causes the diffusion between the free-volume elements of 0.2 and 0.5 nm in size and the polymer chains. The inherent slow diffusion of gas through the nonporous polymeric structures allow the polymeric membranes to exhibit low CO_2 permeance with moderate CO_2/N_2 selectivity (Iarikov and Oyama 2011).

Membranes can be categorically classified depending on flux density and selectivity as follows: (a) porous, (b) non-porous and (c) asymmetric.

A porous membrane can be recognized with some feature such as its rigidity, highly voided structure with randomly distributed interconnected pores. Separation of materials by porous membranes helps in determining membrane properties like the molecular size of the membrane polymer and the permeate character. The structure and function of a porous membrane is very identical to a conventional filter. Generally, the microporous membranes can be used to effectively separate the substantial differentiation of the molecules in size. The featured of the microporous membranes are as follows such as the membrane porosity ϵ, average pore diameter d, and the tortuosity of the membrane (Abedini and Nezhadmoghadam 2010). Solution casting, sintering, stretching, phase separation and track etching are the methods that can be easily adopted to prepare the porous polymeric membranes. The characteristics features of the materials and the process conditions undergone will determine the final morphology of the membrane produced (Abedini and Nezhadmoghadam 2010).

An imperative feature of non-porous/dense membrane is that, the permeants of similar size may be separated if their solubility in the membrane considerably varies. The rates at which gases are transported are usually low for non-porous/dense membranes while they are highly selective for the separation of gases from their mixtures. Melt extrusion is one of the methods usually adopted for the preparation of dense membranes where a melt is anticipated as a solution in which the polymer is both a solute and solvent. In solution-casting technique, dense membranes are cast from polymer solutions prepared by dissolution of a polymer in a solvent container to form a film, this is followed by complete evaporation of the solvent after casting either by immersing into a non-solvent or by allow to dry at room temperature (Pandey and Chauhan 2001).

Asymmetric membrane is made up of a thin, dense selective skin which is usually at the upper layer and the other a thick, porous matrix which is at the sub-layer to provide a physical support for the thin upper layer (Pandey and Chauhan 2001).

Polymeric Membrane Fabrication Process

Today, virtually all the membranes for separation of gases are made by processes depending on the concept of phase inversion because it is the only commercially viable way known for making thin (i.e., of the order of 100 nm or less), defect-free membranes at large enough surface areas to be useful for practical applications. The phase inversion process, invented by Loeb and Sourirajan to make cellulose acetate desalination membranes (Loeb 1962), can be utilized to produce asymmetric membranes with very thin, dense films on a porous substrate and delivered a practical route to prepare high flux membranes. This method remains the primary method by which commercial gas separation membranes are prepared and has permitted membranes to be prepared in adequately large area to process, for example, from 50 to 700 million standard cubic feet of natural gas per daily basis in particular locations (Sanders et al. 2013). To prepare membranes through phase inversion, a water-insoluble polymer is dissolved in a water miscible, high boiling solvent or mixture of solvents; the resulting solution is usually referred to as a polymer dope. This dope is then cast using a doctor blade onto a porous backing material, for flat sheet, spiral wound membranes, or extruded through a hollow fibrespinneret to prepare hollow fibres (Baker 2004). Afterpassing via a short air gap to allow solvent evaporation from the surface of the nascent membrane to initiate formation of a thin, dense layer on the surface, the cast or extruded polymer dope is immersed in a non-solvent (usually water) bath. In this bath, solvent exchange and polymer coagulation produce a porous substrate with a thin, dense skin on top of it, in the case of spiral wound membranes, or on the outside of hollow fibres. Usually, the polymer dope

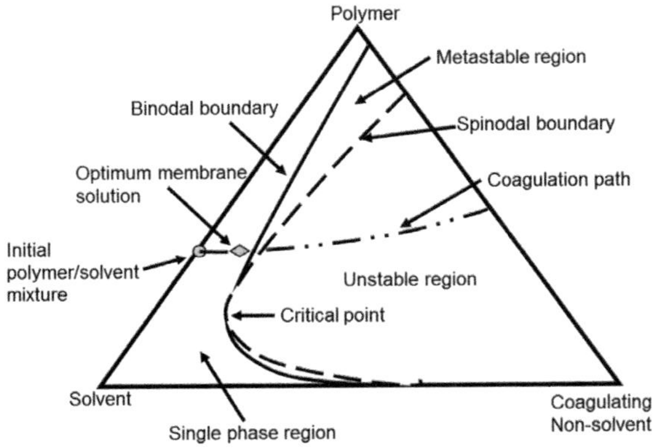

Figure 3.4. Qualitative illustration of a phase diagram for a polymer/solvent/non-solvent system. The path followed during formation of a phase inversion membrane is illustrated by the line labelled "Coagulation path" (Chen et al. 2011).

concentration would be in the range of 10–35 wt. %, and small amounts of a water soluble polymer (e.g., poly (vinyl pyrrolidone) or poly (ethylene oxide) oligomers) could be added to augment the porous structure (Baker 2004). To reduce macrovoid formation, water, alcohols or other additives are included in the original polymer dope, to make sure the dope is very close to the phase separation boundary before being cast or extruded (Baker 2004).

An example of a ternary phase diagram for polymer, solvent(s), and a coagulating non-solvent is display in Fig. 3.4. The path followed by the polymer during phase inversion is labelled "Coagulation path" in this figure. Two-phase separation processes can occur, as noted in the literature (Robeson 2007). Spinodal decomposition will occur if the membrane formation path goes through the critical point or rapidly into the unstable region. Nucleation and growth can occur in the metastable region in addition to the unstable region (Baker 2004). As the phase separation process continues, the initial features of spinodal decomposition can become less noticeable due to structure coalescence. The thermodynamic interpretations of nucleation and growth versus spinodal decomposition and information directly related to membrane formation are available elsewhere (Kesting 1985). Many commercial gas separation systems are based on hollow fibres produced by this phase inversion process (Berichte 1993). During spinning, the bore of the hollow fibre usually has a bore fluid to provide a recompensing pressure to maintain the hollow interior, coagulate the spin dope, and stabilize the forming fibre; gases such as N_2 and liquids such as H_2O or aqueous based liquids have been recounted as bore fluids (Berichte 1993). After the phase inversion process, the resultant membrane is usually stored in water to further remove the solvent before drying and following post-treatment to block any defects in the membrane surface (Henis and Tripodi 1981). A major concern for membranes prepared via phase inversion is the elimination of pinholes in the dense layer. It is difficult to produce pinhole-free membranes, so repair techniques usually are employed to cover any defects to achieve efficient dense layer thicknesses of 100 nm or less. Henis and Tripodi 1981, reported that pinhole defects in asymmetric membranes could be fixed by coating a slightly defective phase inversion membrane with a thin layer of a highly permeable (but relatively non-selective) polymer such

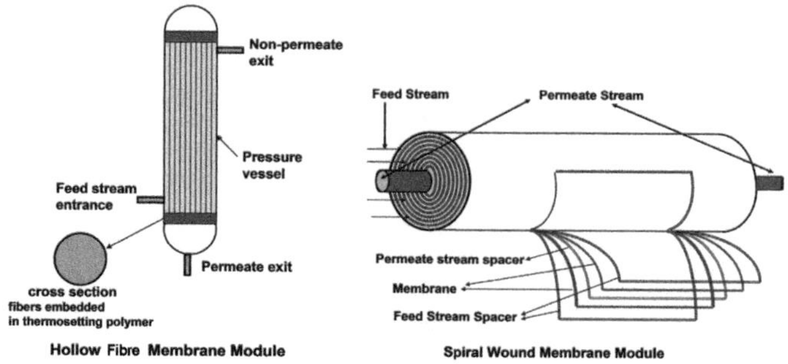

Figure 3.5. Modular constructions employed for gas separation processes (Robeson 2012).

as silicone rubber (i.e., poly (dimethylsiloxane)) (Henis and Tripodi 1981). This so-called "blocking" step basically removes non-selective pore flow through defects with little alteration in the intrinsic flux of the dense layer. Typically, flat sheet membranes would be used to prepare spiral wound membrane modules, and hollow fibres would be assembled into a hollow fibre module (Baker 2002).

These two configurations of membranes are exhibited schematically in Fig. 3.5. The objective of putting membranes into such configurations is to exploit the amount of membrane surface area that can be accommodated in a given volume. Higher surface to volume ratio assemblies of membranes lessen the cost of pressure vessels, etc. Typically, the configuration of hollow fibre bundle contains the fibre thread on the order of 105 hollow fibres which are firmly packed and both ends entrenched in a temperature regulating polymer epoxy (Coker and Freeman 1998). A bundle of the hollow fibre membrane is then carefully packed in a polymeric or metal pressure vessel; this is determined by the pressure it was projected to encounter during operation. The application would determine if a feed gas with high pressure can be applied into the shell side of a hollow fibre module. Generally, to maximize the existing driving force for mass transfer, hollow fibre modules are designed to operate as close to counter-current flow as possible. However, due to flow mal-distribution in the fibre bundles or other factors, gas flow via the shell is classically not in perfect counter-current flow (Yampolskii and Freeman 2010). The spiral wound membrane module configuration involves alternating layers of flat sheet non uniform membranes with the parts which are porous in between the membrane sheets (Baker 2004). The permeate and feed stream travel via alternate layers in a spiral wound module. Due to cost of production, higher membrane surface area to module volume and generally easier production methods (Yuri Yampolskii 2011). the hollow fibre module is much more common than spiral wound systems for gas separation (Baker 2002).

Transport Mechanism of Polymeric Membrane

Adolph Fick's laws of diffusion determined the major scientific context for mass transfer across nonporous membranes, and with Fourier's law of heat conduction and Ohm's law of electrical conduction, the analogue was derived (Fick 1995). Meanwhile, the principles of solution-diffusion mechanism was first publicized by Sir Thomas Graham in 1866 at a seminal, setting a pace in the understanding of

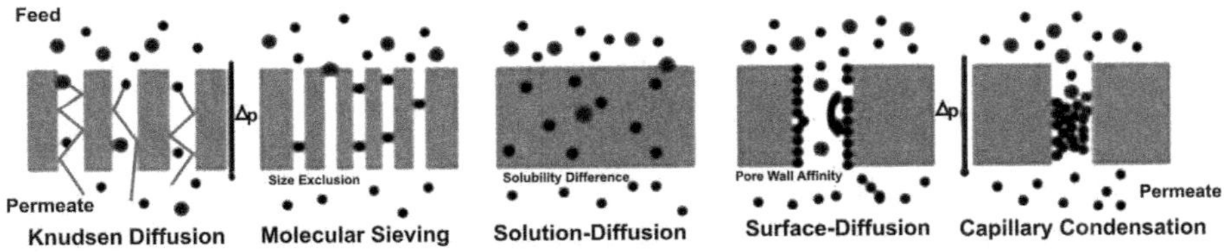

Figure 3.6. Schematic of membrane gas separation mechanisms (Scholes et al. 2009).

transport of gases in all nonporous polymeric membranes used for gas separation. (Graham 1866). The explanation of the mechanism was that, the transportation of gas is exposed to high gas pressure via a nonporous membrane or polymer film by dissolving into the surface of the membrane, diffusing through the polymer films, and exposed to low pressure by desorbing from the surface of the membrane (Wijmans and Baker 1995). The rate-regulating step for gas permeation in all the polymer membranes is the diffusion via the polymer which is known as the intermediate step. This rate-regulating step in diffusion means the entering and exit of transient gaps (free volume elements) of the local scale segmental dynamics of the polymer chains in the polymer; the gas molecules perform Brownian motion (i.e., diffusion) through these free volume components. In achieving the diffusion of small gas molecules through polymers two critical variables are crucial: namely; the packing of polymers and the local segmental motions of polymer chains (Sanders et al. 2013).

Two ways can be used to represent the transport mechanism in membrane: (1) Porous membrane (2) Non-porous/ Dense membrane.

Macropores (> 500 Å), mesopores (500–20 Å) or micropores (< 20 Å) can be used to categorize the porous membrane in accordance to their pore size. Gas transport occurs through a host of different mechanism across these pores (Javaid 2005). These mechanisms were briefly discussed below and clearly illustrated in Fig. 3.6.

(a) *Knudsen diffusion*: In Knudsen diffusion, the transport occurs as a result of smaller pore size than the mean free path. The frequency of the molecules colliding with the pore wall dominated more than the inter-molecular collision. This mechanism is often experienced in macroporous and mesoporous membrane.

(b) *Molecular sieving*: the diffusion occurs mainly through inter-molecular collision while the gas molecule mean free path is smaller than the available pore size. The composition gradient provides the driving force for molecular diffusion.

(c) *Surface diffusion*: the membrane surface has a strong affinity and absorbs gases along the pore wall exists. Hence the separation that occurs is as a result of the differences in the amount of adsorption of the permeating species

(d) *Capillary condensation*: often takes place in a mesoporous and small macropores at a critical relative pressure, this relative pressure may be ascertained by the kelvin equation in which the pores are completely filled by the condensed gas. It is a form of flow of gases on the membrane surface and consequent condensation of a constituent condensable gas (Javaid 2005).

Transport mechanism in non-porous/dense membrane: Gas permeation can be best described as the solubility and diffusion of certain gases within the membrane through the dense polymer matrix where the solution-diffusion mechanism occurs in non-porous polymeric membranes. Thus, the physio-chemical bonding between the certain gases and the polymer determine separation of the gas (Scholes et al. 2009). The permeability of gas through the membrane will be determined by the solubility and diffusivity in the membrane matrix. Hence, the possible efficient separation will be determined by the nature of the gas and the polymer, and this will consequently affect the permeability of the membrane. The product of a thermodynamic factor known as the solubility coefficient S, and a kinetic parameter known as the diffusion coefficient D, is denoted as permeability coefficient, P.

$$\boldsymbol{P = SD} \tag{1}$$

Where P represents the permeability coefficient in cm^3 (STP) cm cm^{-2} s^{-1} $cmHg^{-1}$, or in Barrer (10^{-10} cm^3 (STP) cm cm^{-2} s^{-1} $cmHg^{-1}$). D is the mobility of the gas through the membrane which denotes the diffusion coefficient in cm^2 s^{-1}, and S represents the solubility coefficient in cm^3 (STP) cm^{-3} $cmHg^{-1}$. The permeability is related to the gas permeation rate through the membrane, or flux (Q), the surface area of the membrane (A), the thickness of the membrane (l) and the pressure difference across the membrane (Δp), which is the driving force for separation.

$$\left(\frac{P}{l}\right) = \frac{Q}{A\Delta p} \tag{2}$$

The ideal selectivity (a) of one gas, A, over another gas, B, is defined as:

$$\alpha = \frac{P_A}{P_B} \tag{3}$$

The classification of the polymeric membranes majorly depend on their operating temperature in relation to the glass transition temperature (Tg) of the polymer, and this can be classified as either rubbery or glassy polymer (Alentiev et al. 1997). Rubbery polymers operate above the glass transition temperature in thermodynamic equilibrium and can easily rearrange itself on a considerable time scale. Thus, gas solubility within the rubbery polymer matrix is in line with Henry's Law and directly proportional to the partial pressure (Alentiev et al. 1997):

$$C_D = K_D P \tag{4}$$

C_D Means the concentration of gas in the polymer matrix and is directly proportional to the Henry's Law constant (K_D).

The glassy polymer membranes operate below the glass transition temperature which means the membrane never reaches thermodynamic equilibrium; hence, the polymer rearrangement takes a very long time. Therefore, the provision of the excess free volume is as a result of improper packing of the polymer chains which is in form of microscopic voids (Alentiev et al. 1997). The solubility will be increased by the absorption of gases within these voids. Hence, equation (5) can be used to describe the overall concentration of absorbed gas within a glassy membrane (C):

$$C = C_D + C_H \tag{5}$$

where C_H is the standard Langmuir adsorption relationship.

Polymer Structure

The last three decades has been tedious years for researchers to make polymer membranes an established practical industrial process for gas separations. Within this time so many polymers have been documented as common gas separation membranes. The details of transport properties of CO_2 and N_2 gas of several polymers are the objective of this section. Recently, focus has been shifted to the glassy section or block of the polymers in the scientific literature. The samples of the polymers used in the construction of gas separation membranes include polyacetylenes (Stern 1994), polyaniline (Anderson et al. 1991), poly(arylene ether)s (Xu 2002), polyarylates (Paul 1995b), polycarbonates (Aguilar-Vega and Paul 1993), polyetherimides (Shamsabadi et al. 2013), poly(ethylene oxide) (Lin and Freeman 2004), polyimides (Stern et al 1993), poly(phenylene oxide)s (Hamad et al 2002), poly(pyrrolone)s (Zimmerman and Koros 1999), and polysulphones (Kim et al. 2001). Some of these polymers are illustrated in Fig. 3.7. Meanwhile, the coverage of the details of some polymers at the expense of others is as a result of an extensive body of literature on gas separation membranes.

The Robeson values of carbon dioxide permeability versus carbon dioxide/nitrogen selectivity were used to display a numerous number of polymers as in Fig.3.8. The exhibited polymers in Fig. 3.8 were prudently selected and all have been measured at 35°C and at 10 atm for the carbon dioxide permeability. The lower pressure used relative to many other polymers gives a clear upper bound for the poly (imide ethylene oxide) and cross-linked poly (ethylene oxide)s observed, this can appear to give both very high permeabilities and selectivities. Hence, these polymers were not presented in Fig. 3.8. Moreover, the results presented here are from pure gas measurements. Different values would be achieved from the test on a mixed gas. Pure gas measurements data are considerably more available in the text; though mixed gas data would give results more applicable for industrial use. The permeability of carbon dioxide often strongly relies on the carbon dioxide partial pressure, whereas the nitrogen permeability is significantly less dependent. The permeability might drop when pressure increases. The permeability will reach a minimum if the membrane is exposed to plasticisation, and then rise abruptly with increased pressure (Powell and Qiao 2006). Preparation such as thermal annealing of the membrane would considerably alter the permeabilities; this is mostly prominent in glassy polymers. This presents complications in relating results which in turn decreases the value of gas permeation tests as a device responsible for structure/activity relationships. For instance, various laboratories has been used to determine the carbon dioxide permeability of polyimide 6FDA–durene with values ranging from 400 to 456 barrer were achieved (Lin et al. 2000). The permeability drops from 400 to 230 barrier upon thermally annealing a 6FDA–durene membrane for the former at 100°C and the latter at 180°C for 7 hours each (Azeez et al. 2013).

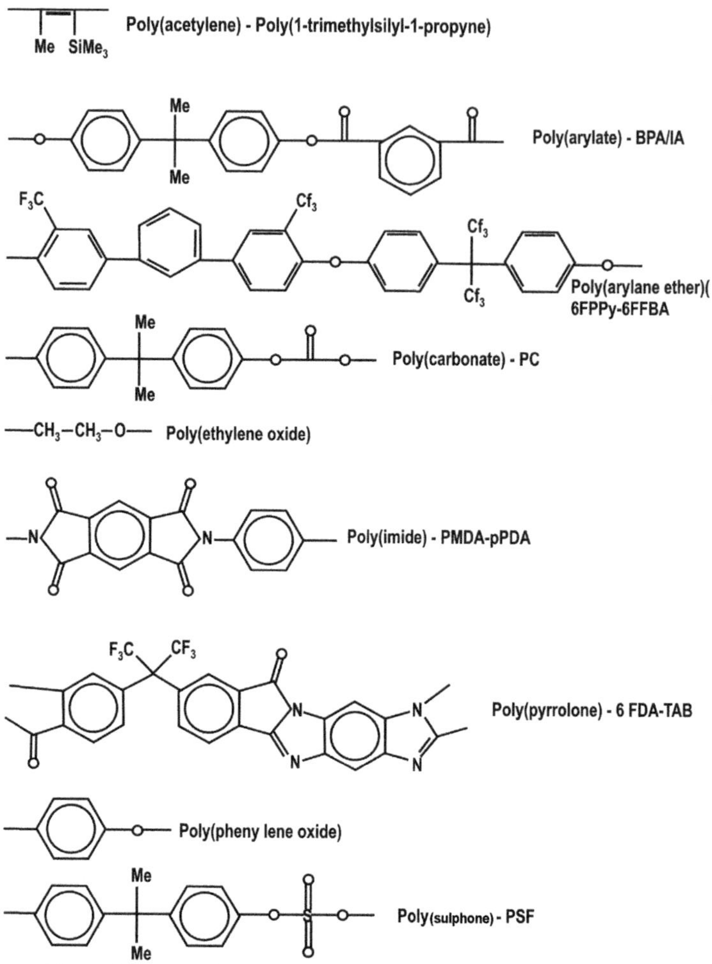

Figure 3.7. Examples of polymeric structures (Powell and Qiao 2006).

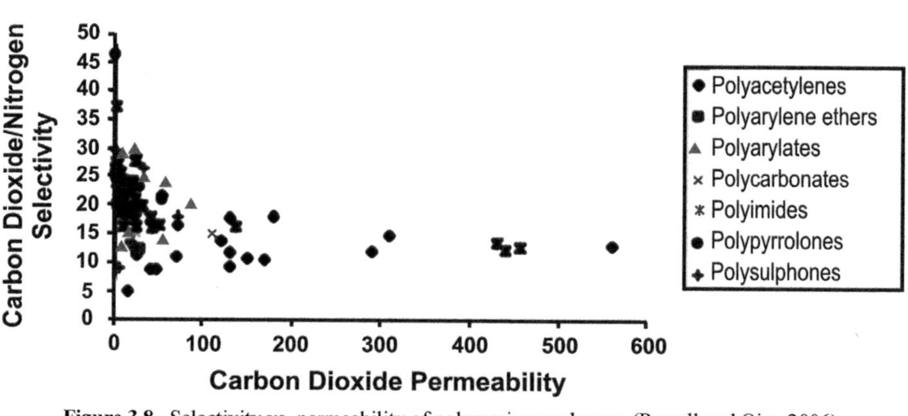

Figure 3.8. Selectivity vs. permeability of polymeric membranes (Powell and Qiao 2006).

Polysulphone

The satisfactory performance and low cost of polysulphones allows for tremendous researches on the gas transport properties of this polymer. The commercialized membrane prepared from polysulphones has already been applied in industrial gas separation processes (Baker 2002). Condensation reaction between a bisphenol and a di-halogenated di-phenylsulphone are largely used for the preparations of polysulphones. Bisphenol A is mostly used to produce the widely studied Polysulfone; PSF. Figure 3.9 displayed the available two paths for the formation of PSF. The structures of other polysulphones are related to this polymer. A lengthen study of the gas transport features of PSF was performed by Paul and co-workers (Barbari et al. 1988). The chemical and thermally durable properties of polysulphones considered it

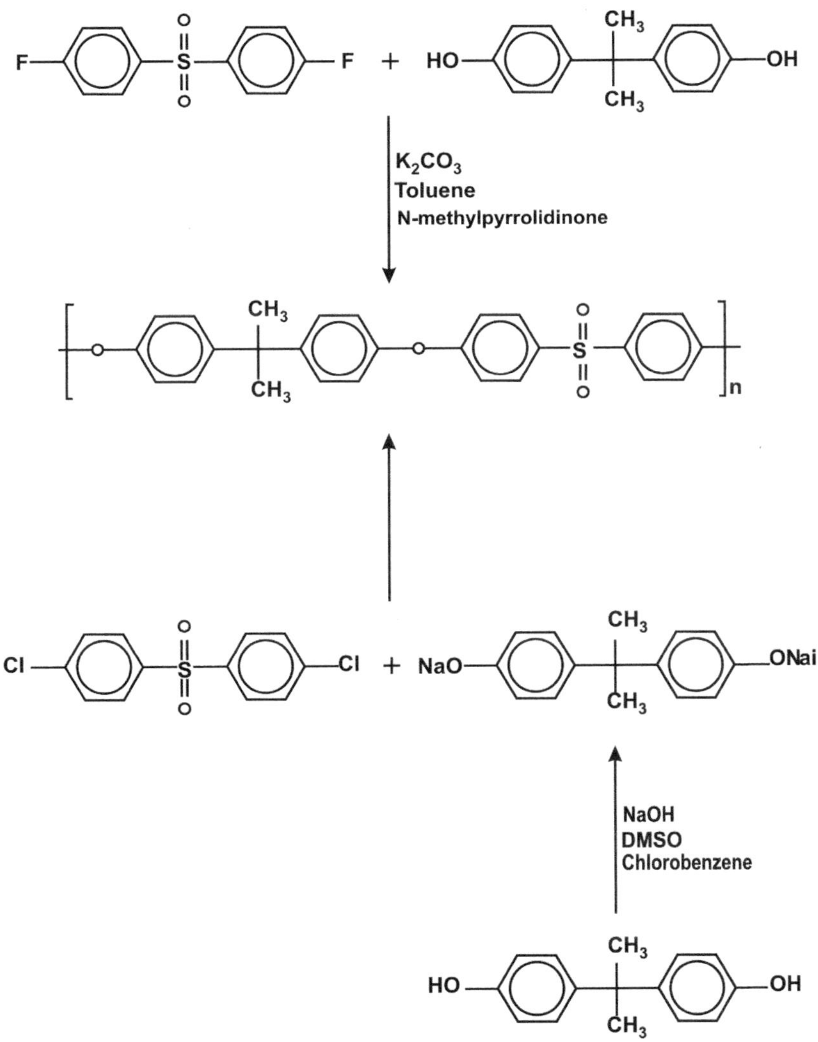

Figure 3.9. Synthesis of Polysulphone (Powell and Qiao 2006).

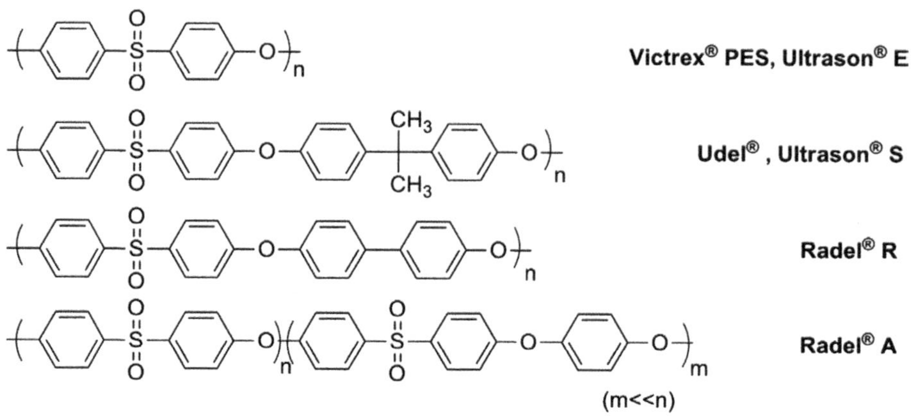

Figure 3.10. Chemical structures of several polysulphones (Sanders et al. 2013).

well recognized among the thermoplastic polymers known, and they are characterized by diphenylene sulphone repeat units (-Ar-SO_2-Ar′-) (Guo and McGrath 2012). The backbone rigidity was generated by repeating units of phenylene rings, an electronic attraction of resonating electron systems between adjacent molecules and steric prevention from rotation within the molecule. The high degree of immobility of the molecules is because of these properties, resulting in high rigidity (high Tg), dimensional stability, good creep resistance, high strength, and high heat deflection temperature (McGrail 1996). Varieties of PSF derivatives have been synthesized by substituting bisphenol A with a different diol. These display a wide range of carbon dioxide permeabilities and selectivities. The transport of different adamantine-based polysulphone membranes was investigated by Pixton and Paul (Paul 1995a). The basic repeat units of several commercially available polysulphones (Vitrex® PES, Udel® PSF and Radel®R) are demonstrated in Fig. 3.10. A widespread range of polysulphones can be prepared through nucleophilic aromatic (S_NAr) polycondensation of an aromatic dihydroxy compound with a bis-(halophenyl) sulphone (Guo and McGrath 2012). Polysulphones are an vital commercial membrane material for gas separations due to their outstanding mechanical properties, fairly good chemical resistance, a wide operating temperature range, and easy production of membranes in a wide selection of configurations and modules (Guo and McGrath 2012). Extensive studies has been made through the gas transport properties of commercial PSF and PSF variants, predominantly the effect of several linking moieties between the phenyl rings and the groups substituted on the phenyl rings on gas permeation properties (Yampolskii 2012). Symmetric bulky substitutions (e.g., methyl groups) on the phenyl rings considerably increase gas permeability, while asymmetric substitution of these same groups decreases gas permeability (McHattie et al. 1991). For instance, substituting the isopropylidene connecting moiety with larger groups, like hexafluoro isopropylidene (-C(CF3)$_2$-), increases the permeability of polymers mainly due to the improved free volume (Aitken et al. 1992). However, these significant increases in CO_2 permeability can also influence the susceptibility of gas transport properties to plasticization by gases such as CO_2. A combination of hexafluoro isopropylidene groups and symmetric substitution led to the appearance of plasticization effects

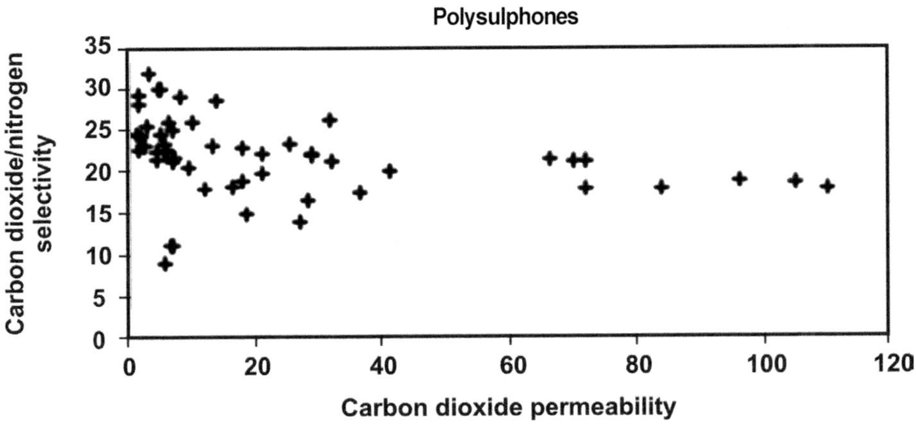

Figure 3.11. Carbon dioxide permeability vs. selectivity for polysulphones (Powell and Qiao 2006).

when CO_2 pressure surpassed approximately 15.2 bar (McHattie et al. 1992). Figure 3.11 shows the carbon dioxide permeability and carbon dioxide/nitrogen selectivity for a variety of polysulphones.

Polyimide

Polyimides combine an outstanding chemical and thermal stability with an extensive range of carbon dioxide permeabilities. The possessions of very high carbon dioxide permeability of polyimides are primarily by those integrating the group Dicarboxyphenyl Hexafluoropropane Dianhydride (6FDA). The publication made in 1996 was a lengthy review of the gas separation features of polyimides (Xu et al. 2002). The polycondensation of aromatic dianhydride and aromatic diamine monomers typically give rise to aromatic polyimides. The process of polyimide synthesis is typically through tetracarboxylic acid dianhydride which is added to a solution of diamine in a polar aprotic solvent at relatively low temperatures (15–75°C). The resulting poly (amic acid) is cyclodehydrated to the resultant polyimide by treatment with chemical dehydrating agents (i.e., chemical imidization), or by prolonged heating at high temperatures (i.e., bulk solid-state thermal imidization or solution imidization) (Kawakami et al. 1997). The formation of polyimides for separation of gas has been performed by a number of different diahydrides; these include Pyromellitic Dianhydride (PMDA), 6FDA and Benzophenone tetracarboxylic dianhydride (BTDA). Figure 3.13 exhibited the chemical structures of these dianhydrides. However, a number of researchers have produced novel dianhydrides in the development of novel polyimides and most polyimides are produced using either Biphenyltetracarboxylic Dihydride (BPDA), PMDA or 6FDA as the dianhydride (Al-Masri et al. 2000). The structures of these described starting materials are showed in Fig. 3.14. The reaction of these dianhydrides with a range of diamines will produce a number of polyimides. Very high permeabilities were observed using the polyimide, PI–TMMPA. This result is anticipated given the large steric bulky nature of the polymer as in Fig. 3.15. Much research is on course on polyimides integrating the group 6FDA as they tend to combine both high

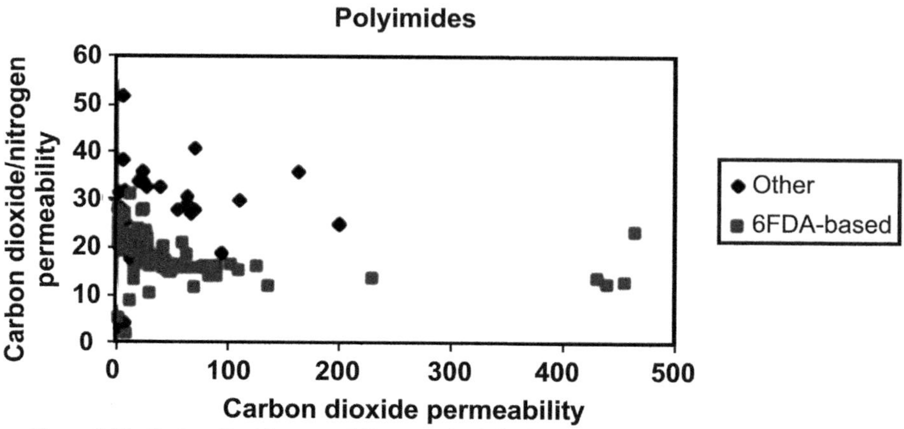

Figure 3.12. Carbon dioxide permeability vs. selectivity for polyimides (Powell and Qiao 2006).

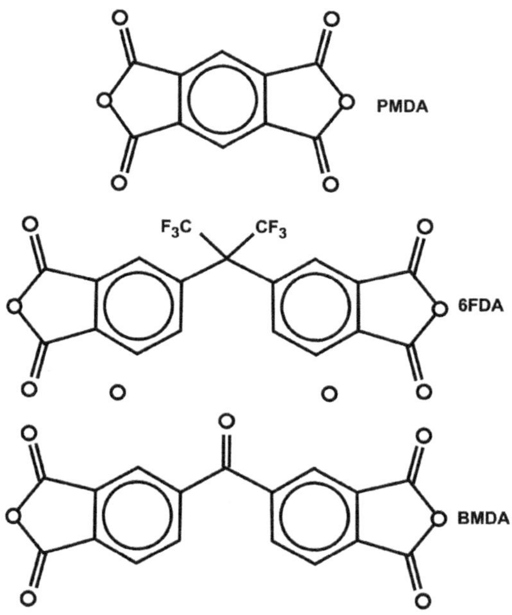

Figure 3.13. Common dianhydrides used in the synthesis of polyimides (Powell and Qiao 2006).

selectivities and high permeabilities. These behaviours were possible as a result of three reasons: (a) the toughness of the chain usually enhanced by the CF_3 groups substantially, allowing the membrane to efficiently separate molecules on the basis of steric bulk; (b) the large CF_3 groups that leads to an increase in the permeability is as a result of reduction in active chain packing; (c) the reduction in the formation of charge-transfer complexes that tend to reduce the likelihood of effective chain packing (Stern 1994).

Substantial research has been performed on commercial polyimide which is another target of polymer for membrane; one of polyimide specie is the Matrimid

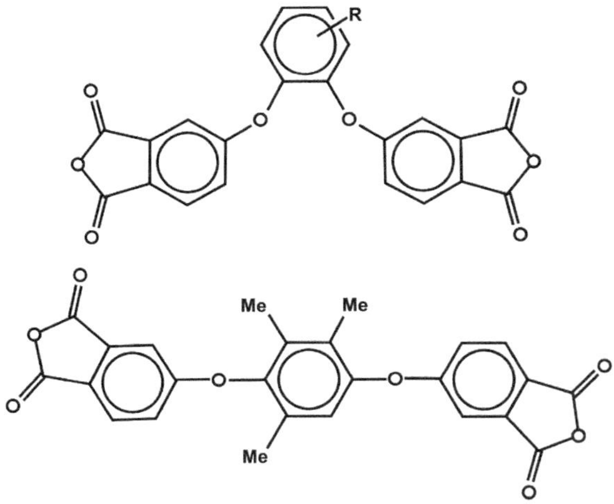

Figure 3.14. Catechol-based dianhydrides (Powell and Qiao 2006).

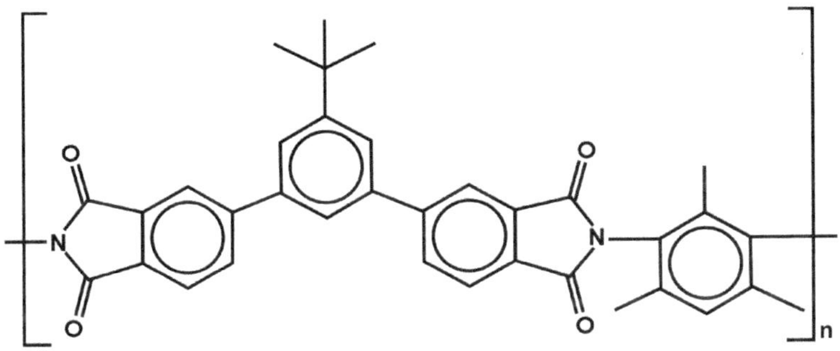

Figure 3.15. PI–TMMPA (Powell and Qiao 2006).

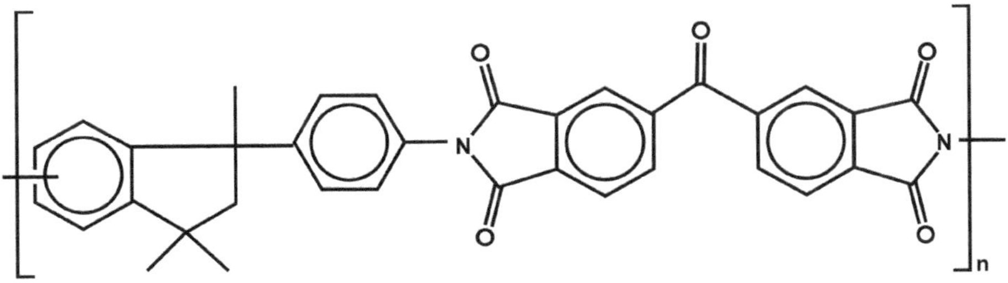

Figure 3.16. Matrimid 5218 (Powell and Qiao 2006).

5218 which is presented in Fig. 3.16. The possession of bromine by Matrimid has been functionalized to upsurge both the permeabilities of carbon dioxide and nitrogen, leading to a small decrease in the carbon dioxide/nitrogen selectivity (Guiver et al. 2002). Matrimid was initially developed for use in the microelectronics industry (Bos et al. 1998), but it is also used in gas separation membranes (Rowe et al. 2007) polysulfone and a polyimide, exhibited increasing refractive index and film thickness with increasing relative humidity. The effect of exposure to high water activity on dry glassy polymer film properties was studied. The specific refraction, as used in the Lorentz-Lorenz equation, was determined directly for these polymers, and its dependence on temperature and aging history was examined. Water vapor sorption in thick polymer films (l\u223c 100\u03bcm. Besides, its mechanical strength and high T_g make it better for more rigorous working environments, especially at high temperatures. Additionally, the good solubility of Matrimid® in common organic solvents permits it to be solution processed, which is a prerequisite for fabrication into a gas separation membrane (Farr et al. 1997). Figure 3.12 exhibits the carbon dioxide permeability and carbon dioxide/nitrogen selectivities of a variety of polyimides.

Polycarbonate

Studies were carried out on a number of synthesized polycarbonates and their carbon dioxide/nitrogen gas transport properties. Polycarbonates are types of polyesters of carbonic acid derived from phosgene or diphenyl carbonate that are tough engineering thermoplastics. A commercial polycarbonate commonly used is based on bisphenol A (4, 40-isopropylidene diphenol). Often, polycarbonate was synthesized through an interfacial phase-transfer catalyzed aqueous caustic process where alkali salts of bisphenol A (or its tetra-substituted variants) in aqueous solution are phosgenated in the presence of an inert solvent as in Fig. 17 (Thomas and Visakh 2011). Phosgene-free solution or melt processes have been established to synthesize polycarbonates through transesterification of diphenyl carbonate with bisphenol A. The melt transesterification process has been found to be commercially viable in recent years, mainly due to its strength to extirpate the want for toxic phosgene (Rogalsky et al. 2014). A common alteration to polycarbonates is the replacement of aromatic hydrogens with various functional groups as presented in Fig. 3.17. One of the bisphenol A polycarbonates of importance for gas separation applications is tetrabromo bisphenol A polycarbonate (TB-BisA-PC), where the four hydrogen atoms are proportionally substituted by Br atoms in the benzene ring (X = Br in Fig. 3.17) (Muruganandam et al. 1987). This material is assumed to be a vital material in gas separation membranes promoted by Innovative Gas Systems, formerly Generon (Puri 2007). A carbon dioxide permeability of fewer than 40 barriers and selectivities range from 15 to over 25 were tend to generate from polycarbonates polymers

Polycarbonate TMHFPC has an outstanding exception to this, this polycarbonate possesses a carbon dioxide/nitrogen selectivity of 15.0 and a carbon dioxide permeability of 111. Figure 3.18 exhibited the structure of this polymer. Figure 3.19 represents the carbon dioxide permeability and carbon dioxide/nitrogen selectivities for a number of polycarbonates.

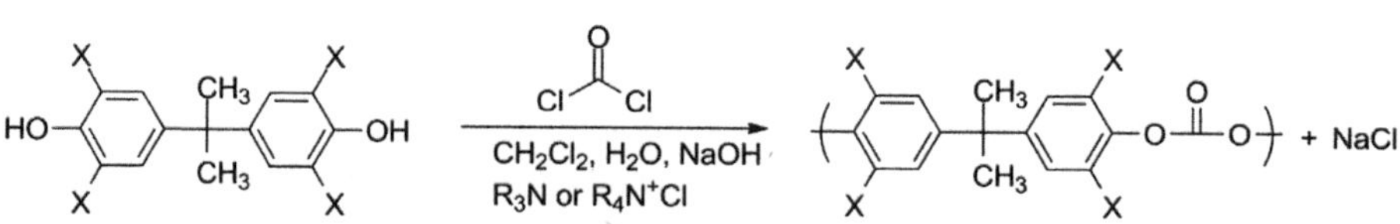

Figure 3.17. Interfacial polymerization of polycarbonates. Where X = CH_3 is tetra-methyl bisphenol A PC (TM-BisA-PC), X=Cl is tetrachlorobisphenol A PC (TC-BisA-PC), X = Br is tetrabromo bisphenol A PC (TB-BisA-PC) (Sanders et al. 2013).

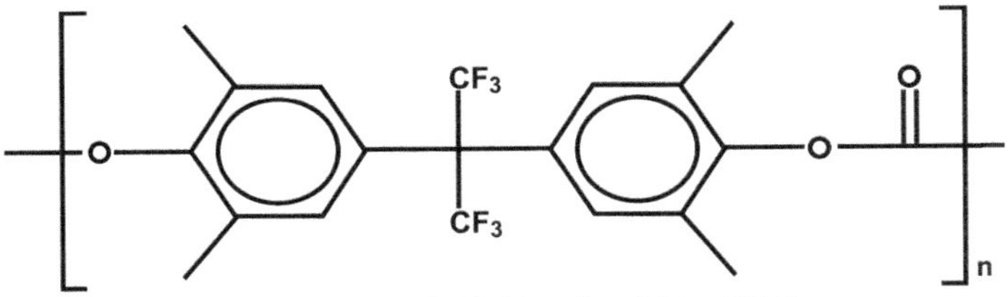

Figure 3.18. TMHFPC (Powell and Qiao 2006).

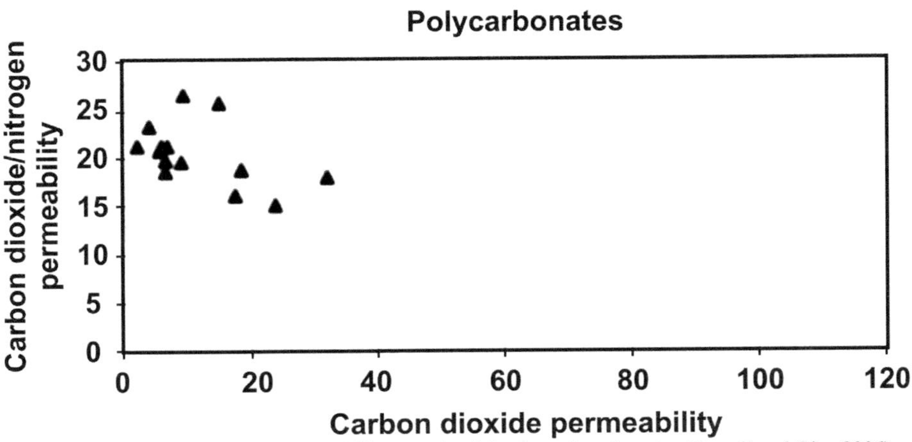

Figure 3.19. Carbon dioxide permeability vs. selectivity for polycarbonates (Powell and Qiao 2006).

Gas Separation Membrane Process for CO_2/N_2 Separation

There are certain criteria that have to be met for the polymers material used for CO_2/N_2 separation membranes: 1) permeation of the gas via the membrane; so a rational gas flux is attained during the separation, 2) the separation of carbon dioxide from other gases, 3) the tendency of the polymeric membrane to possess good mechanical and thermal properties so that the separation can be efficiently performed and rarely at high temperatures. These criteria are usually met by synthesis of a block polymer system as display in Fig. 3.20. The hard block and a soft block are often processes by the copolymer. Polymers with well-packed and more rigid structures are characterized as a hard block; thus a glassy section of the polymer chain will be formed by this feature (Powell and Qiao 2006). Contrariwise, a polymer with more flexible chains will easily produce a soft block and can form rubbery parts on the polymer chain. Meanwhile, the mechanical support needed by polymeric membrane formed by the use of these copolymers will be produced by the structural frame of the glassy polymer sections. The elevated temperature polymers such as polyimides are usually formed by the hard block and can help in provision of a superior thermal resistance. Conversely, the nature of the flexible chain structure of the rubbery sections allows the transportation of gas and often forms incessant micro-domains; thus provide a good permeability. Typically, the efficient separation without loss of permeability will be offered by the balance of the hard and soft block ratio. Glassy sections of the polymer are related to the lower free volumes while the higher free volumes are provided by the rubbery segments. The high free volume means efficient

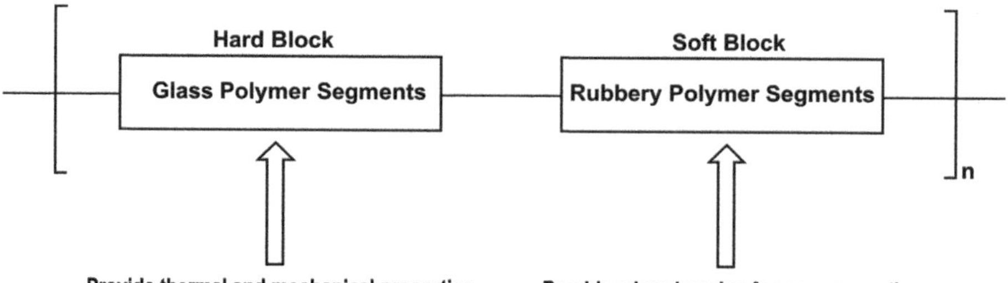

Figure 3.20. Artificial approach for polymer structure used for CO_2 separation membranes (Powell and Qiao 2006).

gas permeation. Introducing some bulky structures into the soft blocks will foster the high free volume in the polymer (Powell and Qiao 2006). The most often listed potential candidate for the application post-combustion capture is the membrane gas separation process. However, low concentration of CO_2 and pressure of the flue gas has been the main difficulty related to the restricted use of the membrane GS technology which requires the use of membranes with high selectivities to meet the requirement of the International Energy Agency, i.e., a CO_2 recovery of 80% with purity, of at least 80% (Bounaceur et al. 2006). Particularly, the requirements for the recovery are need to be equivalent to the international treaty (Kyoto Protocol, for instance). In contrast, the cost of CO_2 sequestration and the cost of the supplementary physical process downstream to the CO_2 capture system will adversely affect the final electricity cost in relation to the purity target. This leads to further research on single-stage and multiple-stage of membrane process.

Single-stage CO_2/N_2 Membrane Process

Figure 3.21 displays the operation of the flue gas track for an established pulverized coal combustion power plant. The position 1 has been generally recognized to be most efficient for the polymer membrane in comparison with a number of different kinds of membranes (e.g., sol–gel derived stainless-steel-based cermet membrane developed by IEF-1 Forschungszentrum Jülich, Germany; three different positions were projected for testing of PEBAX polymer membrane (working temperature lower than 70°C developed by GKSS, Germany) in the EnBW Energie Baden-Württemberg AG power plant, On the basis of the working conditions of coal-fired power plants, Position 1, analogous to amine stripping processes (Programme 2007) after the deNOx (SCR Selective Catalytic Reduction), dust removal (E-filter) and desulphurization (FGD, Flue-gas **desulphurization**) processes was chosen for the first test. There needs to be a resolution that the membrane can work in a tough environment before applying the cermet-type membrane at the other positions.

According to the purity requirements of CO_2 transport (> 95 mol% CO_2 purity) (Conturie 2006) and re-injection, a purification process should be included, i.e., CO_2 purity more than 99 vol% after capture (Strömberg et al. 2009).

Figure 3.22 displays the single-stage membrane process. Operating conditions such as pressure, temperature, CO_2 concentration and flow rate of the feed gas act on a certain membrane features such as selectivity, permeability and area, and then

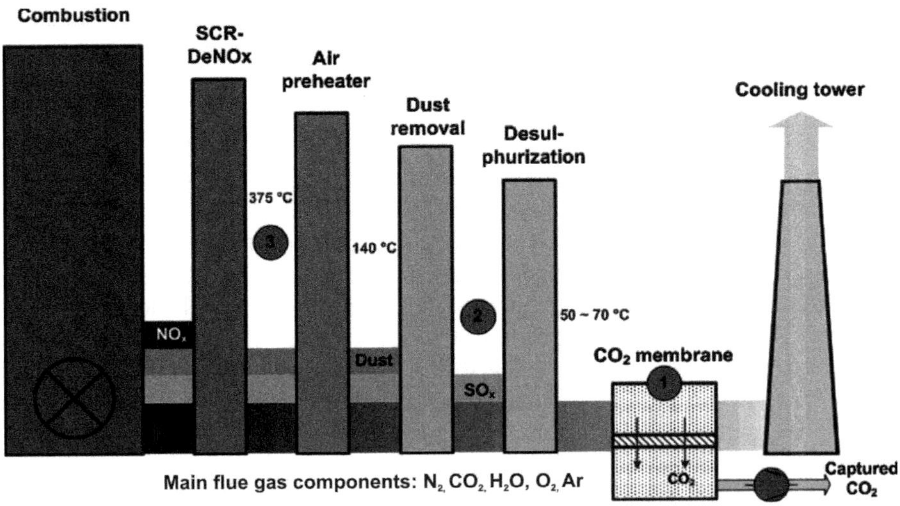

Figure 3.21. The position of CO_2 membrane in a post-combustion flue gas line. The 1, 2 and 3 numbers display the projected test positions for different membranes in the EnBW power plant (Zhao et al. 2008).

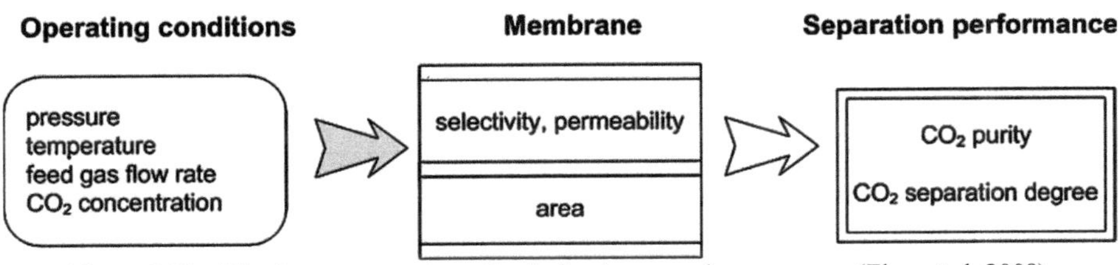

Figure 3.22. All relevant parameters of a single-stage membrane process (Zhao et al. 2008).

the performance of the membrane such as the CO_2 purity and the degree of CO_2 separation are important parameters needed. The literature also described the degree of CO_2 separation as the CO_2 recovery ratio (Sluijs et al. 1992).

Multi-stage Membrane Process

The ambient pressure in the feed flue gas and a relatively low CO_2 content in the post-combustion capture make it difficult for creating an efficient driving force while using CO_2/N_2 gas separation membranes. To actualize the separation target of –95 mol% CO_2 purity and suitable degree of CO_2 separation, multi-stage systems are the most feasible membranes in this aspect. In order to achieve a sufficient driving force, certain CO_2 partial pressure difference must be made artificially, owing to the restrictions of the operating conditions of post-combustion capture. The appropriate precautions can be taken: on the feed side by (1) increasing the pressure using a compressor; (2) recirculating an enriched CO_2 stream (increasing the CO_2 concentration); on the permeate side, (3) using a vacuum pump (decreasing the pressure); (4) using sweep gas (decreasing the CO_2 concentration) (Zhao et al. 2010).

The energy penalty of existing power plants will be caused by the consumed electrical energy by using a compressor and a vacuum pump. Part of the energy used by the compressor on the feed side can be recovered while coupled an expander on the retentate side; this is worthwhile to be noted. However, if the vacuum pump is applied on the permeate side, this will be impossible. The most energy-saving idea which is already on the radar of study by researchers is by using sweep gas on the permeate side (four-end membrane: feed, retentate + sweep gas, permeate) (Merkel et al. 2010). However, ways of finding a feasible source for a sweep gas and a way to separate the sweep gas and permeate so as to achieve the desired permeate product purity in the post-combustion process should be investigated in more depth. The strategy for multi-stage membrane process is demonstrated in Fig. 3.23. The most energy- and cost-effective technology for the applications of enhanced oil & gas recovery (EOR or EGR) in the last four decades is the use of high feed gas pressure (> 50 bar) membrane systems arranged in multi-stages (Kohl and Nielsen 1997). There are two forms of staged membranes; these are enricher and stripper. The enricher are connected in serial on the permeate side while the stripper is connected parallel on the permeate side. Figure 3.24 displays the various conditions of the feed gas and the way it was used to capture.

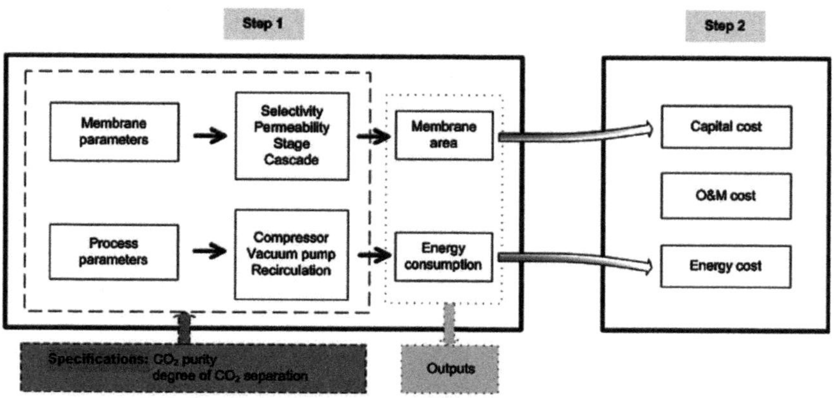

Figure 3.23. The strategy for multi-stage gas separation membrane systems (Zhao et al. 2010).

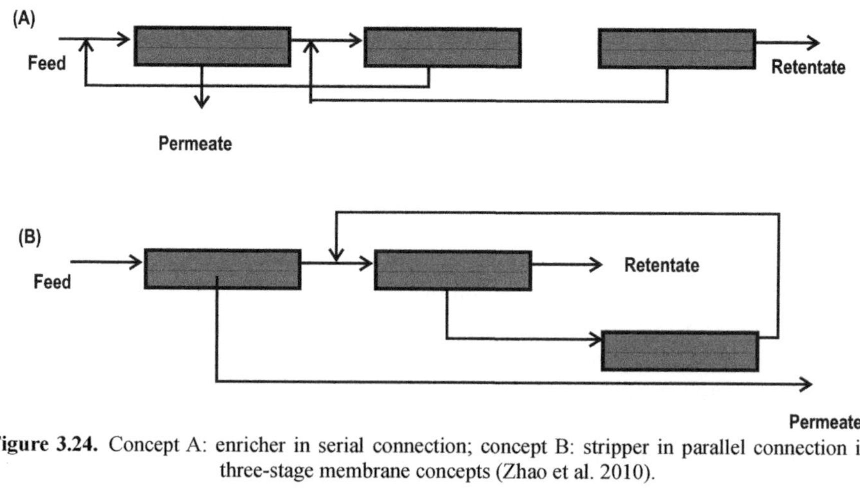

Figure 3.24. Concept A: enricher in serial connection; concept B: stripper in parallel connection in a three-stage membrane concepts (Zhao et al. 2010).

Separation of Minor Components from Carbon Dioxide

The separation of CO_2 from hydrogen (pre combustion systems), oxygen from nitrogen (in oxyfuel combustion system), CO_2 from flue gas (post-combustion system), and CO_2 from natural gas (natural gas processing) is a relatively new capture concept applying selective polymeric membranes. The composition of flue gases differs significantly subject to the power plant, fuel source and prior treatment. Flue gases can be any or the combination of any of the following compound; N_2, O_2, H_2O, CO_2, SO_2, NO*x* and HCl and tend to be oxidizing agent (Granite and Pennline 2002). Contrariwise, CO_2 can be separated from fuel gases before combustion. Fuel gases are commonly served as a reducing agent with CO, H_2, H_2O, CO_2 and H_2S (White et al. 2003). Moreover, substantial particulate matter is present at the unprocessed flue gas streams. Some industrial processes such as steel producing company, produces flue gases with a substantial amount of CO_2 concentration (Powell and Qiao 2006). The composition of flue gas varied from one source to the other. The capacity source of the CO_2 content can be more concentrated around 20–30% for steel and cement production plants, approximately 15% in coal power plants and very minimal in a gas turbine plant (around 4%). There are some exceptional cases where higher concentrations CO_2 is found such as in ammonia, syngas or biofuels (i.e., ethanol fermentation) plants (Liu et al. 2011). Furthermore, different compounds such as N_2, O_2, NOx, SOx can be found in typical coal combustion plants apart from CO_2 while compounds such as H_2, CO and N_2 will be found in the discharges of a blast furnace for steel production. Generally, the flue gas is wet and the condition is most often scale on saturated humidity. Nonetheless, majority of studies on carbon capture process were performed on dry CO_2/N_2 mixtures, usually approximately 15% CO_2 content, which manifest considerably to the largest part of CO_2 emissions type in the world i.e. coal power plants, irrespective of the broad ranges of flue gas compositions (Steeneveldt et al. 2006). Apart from minor components present, there are also major gases to be considered. SOx, NOx, and water are also found in flue gases during post-combustion process (Liu et al. 2011); while there is existence of H_2S, CO, NH_3, water and hydrocarbons in pre-combustion syngas process (Liu et al. 2011). Also, the presence of O_2 in combustion residual along with Argon is due to its presence in air. Similarly, heavy metals such as arsenic and mercury may present as well as chlorine and fluorine know as halogens, in their acidic form depending on the fuel (Ito et al. 2006). The existence of minor components may compete with CO_2 for separation during the membrane gas separation and thus may reduce permeability, in addition to damaging the membrane, fluctuating separation performance.

Minor Components of Gas in Separation of Carbon Dioxide

Sulphur oxide (SOx)

The trace amounts of SO_2 and SO_3 production is commonly found in the combustion of sulphur containing fuels, such as coal and the combination of this sulphide oxide is generally known as SOx (Ben-Gal et al. 2008). Releasing these sulphide oxide compounds to the atmosphere causes major havoc to the environment, leading to acid rain. Appropriate measure was put in place on a way of curbing sulphur emissions in the last four decades while research was on ground using membranes as an alternative means to scrubbing technologies (Ben-Gal et al. 2008). The evolution of this technology was deterred by the lack of significant industrial trials of membrane desulphurization. However, the antecedent of current CO_2 membrane capture technologies can be traced back to the much of the membrane research work performed on desulphurization in relation to the similar chemical properties observed between SO_2 and CO_2. The clearer provision of SO_2/CO_2 selectivity in the range of 5–40 has proven that many of the polymeric membranes have been recounted for their prospective CO_2 capture. The larger kinetic diameter SO_2 than CO_2 make it impossible for selectivity to be diffusion related (Scholes et al. 2009). Additionally, they have cognate relation with polymer because they are both are acidic gases in nature. Relatively, the greater loading in the Langmuir free volume sites and larger affinity constant can be related to the higher critical temperature which can easily be elucidated due to an improved permeability of SO_2 (Scholes et al. 2009). The permeation of SO_2 for different variety of polymeric membranes is plotted against selectivity over CO_2 in Fig. 3.25.

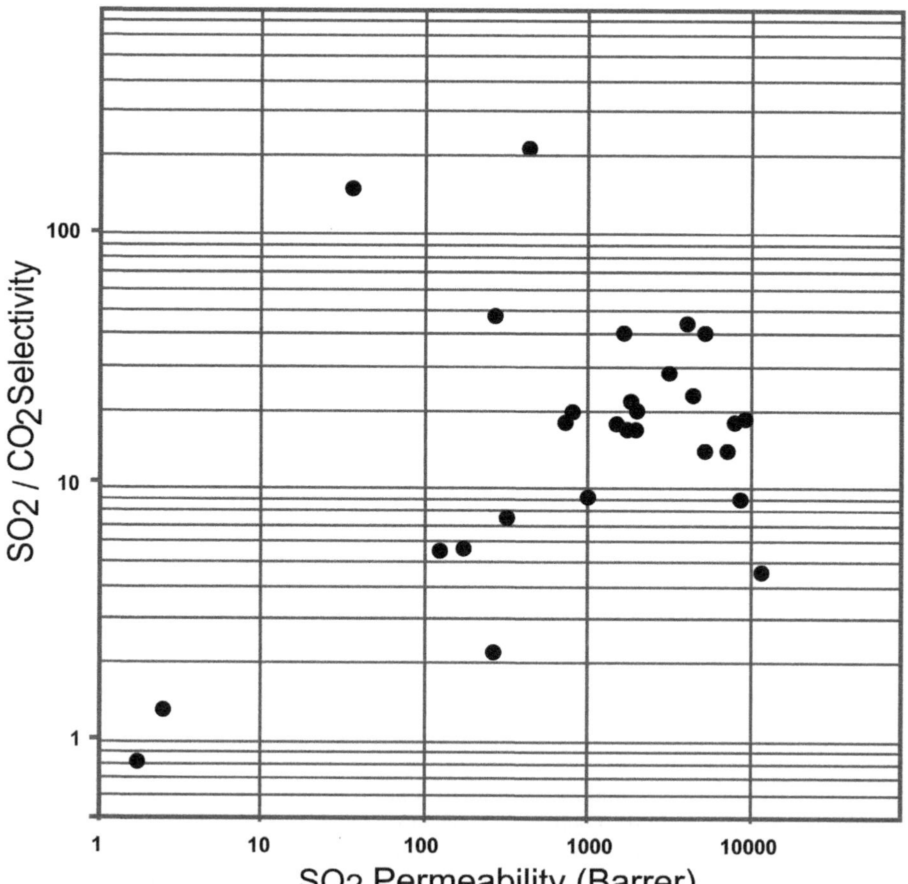

Figure 3.25. Polymeric membranes of SO_2 permeability and selectivity relative to CO_2 (Powell and Qiao 2006).

Meanwhile, the performance of SO_3 as a component of SOx in gas separation membranes has not been recounted in the literature to the best of the authors' knowledge. The reasons being that, the data on possible permeation via membranes is experimentally difficult to come by, and also, pure SO_3 is liquid at room temperature. However, comparing the condensable nature of SO_3 to SO_2 and CO_2, it can be rationally presumed that SO_3 will produce higher permeation via glassy polymeric membranes than either of the other two gases due to higher sorption (Scholes et al. 2009). It has to be recommended that, the future study need to address the deficiency of experimental information on SO_3 and should be included on the research of the general effects of SOx on polymeric membrane gas separation.

Nitric oxides (NOx)

The dominant component of NOx is nitrogen oxides (NO) and to a lesser extent is NO_2 which can be generated through the combustion at elevated temperature in air (Ben-Gal et al. 2008). The report available for the awareness into the effect of NO on polymeric membranes is not sufficient to the best of the authors' knowledge. NO with 3.17 will diffuse faster through membranes than CO_2 with 3.3 based on kinetic diameter. However, there is a suggestion that CO_2 will absorb more strongly in the membrane than NO, and subsequently dominate the solubility of the selective membranes; this is as a result of the greater difference in critical temperature. The permeability of 15.9 barrer and a NO_2/CO_2 selectivity of 1.6 (68) for poly tetrafluoroethylene was the only information available for NO_2 (Scholes et al. 2009). Expectedly, the stronger sorption NO_2 due to higher critical temperature in a glassy membrane compared to CO_2. Data on aging effects and plasticization of NOx are unknown. However, nitric acid will be produced by the presence of water in NOx and this acid will destroy polymeric membranes over a long period of time, similar to the report on SOx (Scholes et al. 2009). Hence, future research study should focus on NOx permeability and selectivity against CO_2 in polymeric membranes, in addition to the study on aging effect on the performance of NOx with presence of water.

Carbon monoxide (CO)

Generally, carbon monoxide is found prior and after combustion processes. It is a product of partial combustion in flue gas, especially when the oxygen supplied is limited. Notably, it is a foremost component of syngas, and the balance nature of the water-shift reaction allow substantial quantities to be found in fuel gasification (Steeneveldt 2007). The existence of some amount of hdrogen elevated temperatures make the performance of CO in polymeric membranes less found in the literature. The permeability of CO via a variety of polymeric membranes is exhibited in Fig. 3.26. The detailed of the quoted selectivity of CO/CO_2 which is based on permeabilities of CO_2 for the same membrane material can be found elsewhere (Powell and Qiao 2006), and the indication of syngas separation application is provided on H_2/CO selectivity. As mentioned earlier, the slower diffusion of CO via a glassy membrane is as a result of larger kinetic diameter with 3.76 compared to CO_2 with 3.3. Moreover, the critical temperature difference cause the solubility of CO to be lower (van Amerongen 1964).

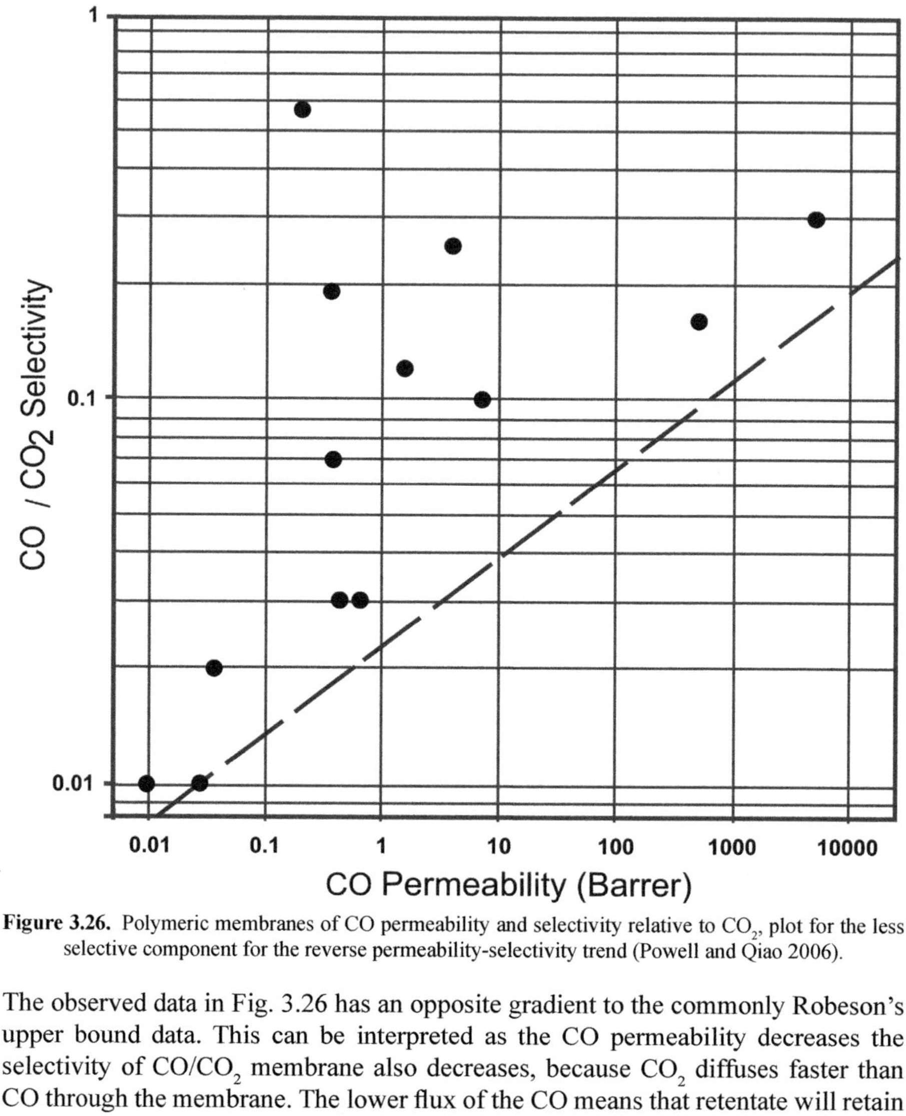

Figure 3.26. Polymeric membranes of CO permeability and selectivity relative to CO_2, plot for the less selective component for the reverse permeability-selectivity trend (Powell and Qiao 2006).

The observed data in Fig. 3.26 has an opposite gradient to the commonly Robeson's upper bound data. This can be interpreted as the CO permeability decreases the selectivity of CO/CO_2 membrane also decreases, because CO_2 diffuses faster than CO through the membrane. The lower flux of the CO means that retentate will retain the CO stream for pre-combustion gas separation via polymeric membranes (Scholes et al. 2009). No data available in the literature on effects of CO plasticization in polymeric membranes, but these are expected to be insignificant.

Hydrogen sulphide (H_2S)

Conversion of 'sour' gas to 'sweet' gas by Removal of H_2S is an important procedure in both the natural and bio-gas industries (Harasimowicz et al. 2007). H_2S is usually found in abundant quantities in syngas and also as a trace compound in the product of gasification to warrant removal as a result of its highly corrosive and toxic nature. The removal is conventionally accomplished via chemical scrubbing technology

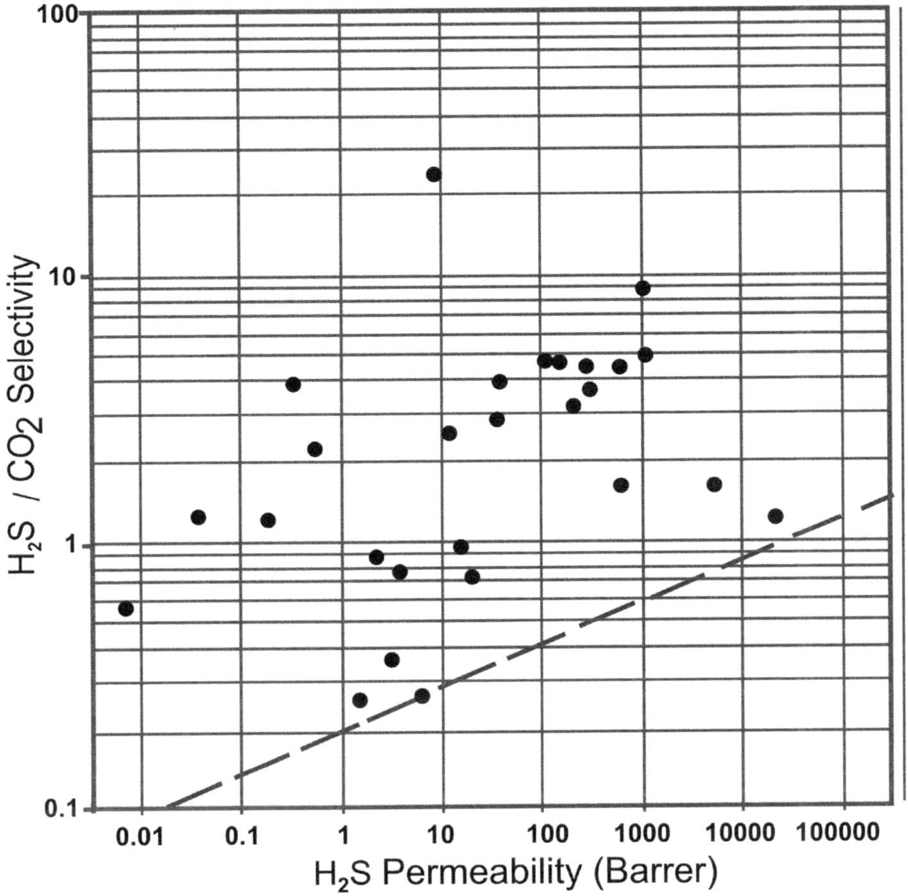

Figure 3.27. Polymeric membranes for H_2S permeability and selectivity relative to CO_2 (Powell and Qiao 2006).

(Orme et al 2004), and chemical conversion with lime (Hao et al. 2008). The available reports on H_2S in gas separation membranes were only for the treatment of natural gas, with a variety of materials demonstrating high permeability for H_2S and CO_2 with a large value of selectivity relative to methane (Orme et al. 2004). As a result of this, the variety of polymeric membranes is quite well recognized for the permeability and selectivity of H_2S and some of these data are presented in Fig. 3.27. Generally, the high condensability of hydrogen sulphide is as a result of an increased selectivity of H_2S over CO_2. Another area of interest is the separation of H_2S from H_2 with selectivities not provided here in relation to pre-combustion (Scholes et al. 2009). Protracted exposure of polymeric membranes to H_2S can lead to plasticization as for CO_2 and SO_2 (Pan 1986). A clear example of non-ideal behaviour for poly dimethylsiloxane (PDMS) rubbery polymeric membrane where the increase in pressure upsurges the permeability (Merket and Toy 2006). This behaviour shows that at low pressures, Henry's Law cannot be applied for H_2S, even when it is for N_2, H_2 and CO_2, given that this is a rubbery membrane. An area of interest for future research should be based on plasticization effect of H_2S on glassy polymers because no data is found reported on this (Scholes et al. 2009).

Ammonia (NH_3)

The removal of ammonia in pre-combustion is very significant due to its presence as a product of gasification. The existence of a variety of highly selective commercial membranes for separation of NH_3 makes its removal largely required and they are mostly considered for the Haber process (Tricoli and Cussler 1995). A variety of polymeric membranes can possibly be used for CO_2 capture and the permeability of NH_3 and selectivity over NN_3/CO_2 is presented in Fig. 3.28. Using the performance of polymeric membrane in justifying the greater permeability of NH_3 via polymeric membranes than CO_2, and thus in membrane capture, the permeate stream would become NH_3 rich. Membranes based on polyvinylammonium thiocyanate is a very good example of NH_3 that is identified to plasticize polymeric membranes (Bhown and Cussler 1991). The Haber process would benefit from the presence of NH_3 due to

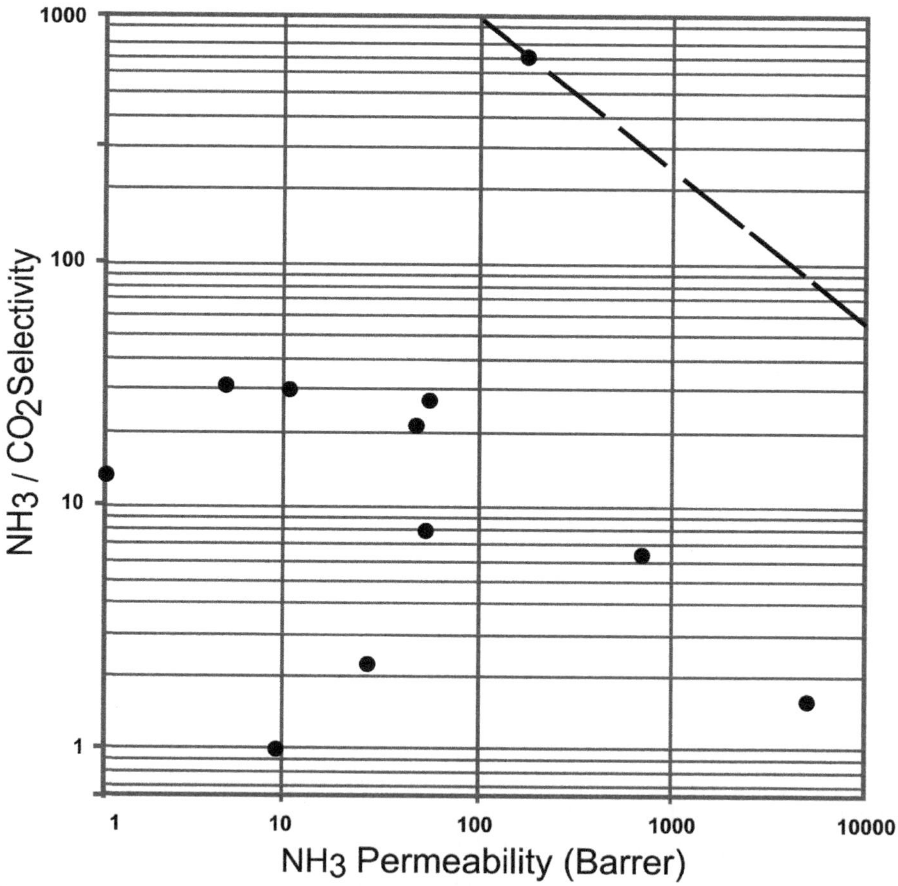

Figure 3.28. Polymeric membranes of NH_3 permeability and selectivity relative to CO_2 (Powell and Qiao 2006).

the fact that the presence of NH_3 would change the polymers from a glassy to rubbery state and this will increase both the permeability and selectivity of NH_3. Conversely, the plasticization effects of NH_3 are less possible to be pertinent in CO_2 capture due to the low partial pressure in pre-combustion process (Scholes et al. 2009).

Argon (Ar)

The introduction of Ar to the flue gas by the combustion oxyfuel is in a mol % concentrations (Liu et al. 2011). The selectivity of CO_2 separation is most often on a high side in all reviewed of the membranes. This is because the ability of the gas to be soluble within the polymeric matrix is restricted due to the inert nature of Ar, and thus diffusion is the driving mechanism. Ar will continue to be in the retentate stream during polymeric membrane gas separation because of the insignificant value of the kinetic diameter of Ar (Scholes et al. 2009).

Condensable components

The existence of water and hydrocarbons as a condensable component in the subcritical state of the feed gas is largely differing from the gas minor components because water exists in significant quantity. Therefore, separation performance can considerably change due to condensation within the membrane. However, there is difficulty in drawing comprehensive conclusions on the behaviour of condensable components due to the limited research undertaken on these. This is because of the wide difference between the fundamental relationship within the membrane's affinity for the various condensable components along with structure of the polymer matrix and the membranes that depends on the polymer and annealing history (Scholes et al. 2009).

Water (H_2O)

In general, CO_2 captures are inundated with water vapour in the pre- and post-combustion process. Hence, plasticization and aging affects, as well as viable water sorption in the membrane tend to have a significant stronger impact on membrane performance in comparsion to the minor gas components highlighted earlier. Usually, the water permeability via polymeric membranes is significantly higher than carbon dioxide in many cases as plotted in Fig. 3.29 and this is due to water having a smaller kinetic diameter of 2.65 compared with the kinetic diameter of CO_2 with 3.3 and as a result, water diffuses more rapidly and also possesses a higher critical temperature which leads to a greater sorption in the membrane (Scholes et al. 2009).

The membrane structure can permanently change by the influences of water-induced plasticization, and the dry membrane would not allow the initial performance to return to its stage. This can obviously be seen in a variety of polymide based membranes (Pye et al. 1976) while the influence on the performance would be related but the CO_2 is not precisely reflected. The competitive sorption of water usually decreases the permeability of the gas. Applying the process of dry feed gas in drying the membranes, original performance conditions are not usually attained with

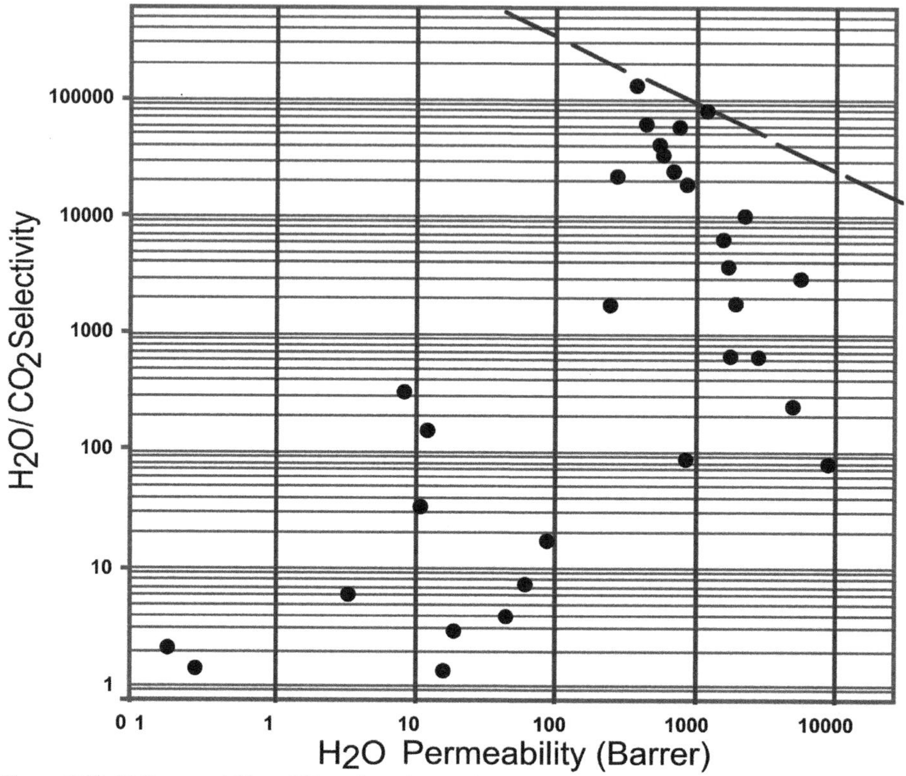

Figure 3.29. H_2O permeability within polymeric membranes and selectivity relative to CO_2 (Powell and Qiao 2006).

permanent permeability losses of up to 20% as observed in 2, 5-diaminobenzoic acid polyimide membrane, the H_2 is observed to decrease to around 60% (Pye et al. 1976). Remarkably, the flow regime across the membrane surface may be dictated by the water presence in the feed gas. Typically, the conditions of membrane gas separation are under counter-current flow. However, Matsumiya et al. 1999, hypothesized that the presence of water in a co-current flow configuration will enhance the separation of the membrane (Matsumiya et al. 1999). The reason being that the entire active area in a co-current flow configuration would allow the water vapour to be diluted by the permeate stream, whereas narrow region only accommodate the water vapour in the counter-current flow configuration. Thus, the co-current flow type engenders an extensive membrane area with an enormous CO_2 driving force than the counter-current system (Francisco et al 2007).

Hydrocarbons

Typically, trace amount of several aromatic and paraffinic hydrocarbons can be found in natural gas processing streams. A prior combustion process as a result of partial reformation of the fuel (Steeneveldt 2007) can also be found. The performance of polymeric membranes can completely be altered by the occurrence of trace amounts

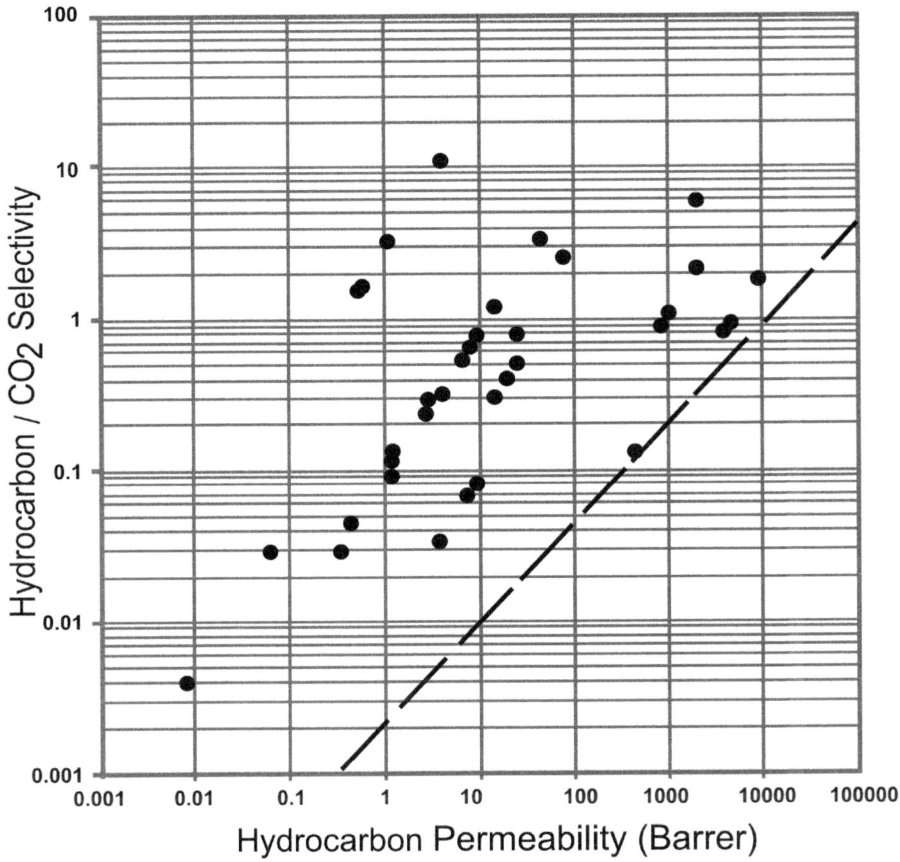

Figure 3.30. Polymeric membranes of aromatic and paraffins (Hydrocarbon) permeability with selectivity relative to CO_2 (Powell and Qiao 2006).

of C6 and higher condensable hydrocarbons. The permeability versus selectivity plot of various hydrocarbons via polymeric membranes compared with carbon dioxide is presented in Fig. 3.30. The coverage of a number of hydrocarbons shows that the permeabilities of hydrocarbon are smaller in number than that of CO_2 for most polymeric membrane and is as a result of kinetic diameter of hydrocarbon with 5.85, which favours CO_2 with Fig. 3.3. The hydrocarbon solubility in PDMS-based membranes is much higher compared with CO_2 with higher permeability and selectivities greater than 1 because of it non-polar rubbery nature (Scholes et al. 2009).

Conclusion

The gas transport properties of various types of polymeric membranes through carbon dioxide and nitrogen have been evaluated. The usefulness as materials for effective gas separation membranes performance of different classes of polymers has also been evaluated. Combining different types of polymers, such as polyimides, polycarbonates, and polysulphones, possess good performance with substantial scope for variation in structure. Constructing polymers which consist of both hard and soft sections are the tactical way of synthesizing high performance gas separation membranes. Membrane selectivity decide the CO_2 purity while permeability decide the degree of CO_2 separation in a single-stage membrane, and has a strong influence on the electricity used to drive the compression machines (energy consumption) in relation with the energy cost and total membrane area in relation with the capital cost for a multi-stage membrane system. Meanwhile, consumption of energy is a significant problem for gas-separation membrane processes. Theoretically, due to the optimization and integration of CO_2 capture and liquefaction as well, multi-stage membrane system consumes more energy than a single-stage process. It can be projected that the need for energy consumption will make membrane gas separation become more competitive with the conventional gas separation counterpart.

However, the effect of minor components and the performance on membrane gas separation for application in pre- and post-combustion CO_2 capture have been thoroughly examined in this book. Particularly, various gases such as SOx, NOx, CO, H_2S, NH_3, Ar, water and hydrocarbons have been evaluated for polymeric membrane performance. Typically, the permeability of SOx, NOx, H_2S, NH_3 and water is most often greater through glassy polymeric membranes than CO_2 and thus will enrich the permeate stream. This permeability enhancement can be linked to the higher critical point of the gas species. The lower permeability of CO, Ar and hydrocarbons makes them remain in the retentate stream compared with CO_2 with higher permeability.

References

Abedini, Reza and Amir Nezhadmoghadam. 2010. Application Of Membrane In Gas Separation Processes: Its Suitability And Mechanisms. Journal of Petroleum and Coal 52(2): 69–80.

Aguilar-Vega, M. and Paul. D.R. 1993. Gas transport properties of polycarbonates and polysulfones with aromatic substitutions on the bisphenol connector group. J. Polym. Sci. B: Polymer Physics 31(11): 1599–1610.

Aitken, C.L., William J. Koros and Donald R. Paul. 1992. Gas Transport Properties of Biphenol Polysulfones. Macromolecules 25: 3651–58.

Al-Masri, Majdi, Detlev Fritsch and Hans R. Kricheldorf. 2000. New Polyimides for Gas Separation. 2. Polyimides Derived from Substituted Catechol Bis (Etherphthalic Anhydride) S. Macromolecules 33(19): 7127–35.

Alentiev, A. Yu., Yu. P. Yampolskii, V.P. Shantarovich, S.M. Nemser and N.A. Platé. 1997. High transport parameters and free volume of perfluorodioxidole copolymers. J. Membr. Sci. 126: 123–32.

Anderson, Mark R., Benjamin R. Mattes, Howard Reiss and Richard B. Kaner. 1991. Conjugated Polymer Films for Gas Separations. Science 252(5011): 1412–1415.

van Amerongen, G.J. 1964. Diffusion in elastomers. Rubber Chem. Technol. 37(5): 1065–1152.

Azeez, Asif Abdul, Kyong Yop Rhee, Soo Jin Park and David Hui. 2013. Epoxy clay nanocomposites - processing, properties and applications: a review. Compos. Part B-ENG 45(1): 308–20.

Baker, R.L., Mwamachi, D.M., Audho, J.O., Aduda, E.O. and Thorpe, W. 1998. Resistance of galla and small east african goats in the sub-humid tropics to gastrointestinal nematode infections and the peri-parturient rise in faecal egg counts. Vet. Parasitol. 79(1): 53–64.

Baker, R.W. 2004. Membrane Technology and Applications. J Wiley.

Baker, Richard, W. 2002. Future directions of membrane gas separation technology. Ind. Eng. Chem. Res. 41: 1393–1411.

Barbari, T.A., Koros, W.J. and Paul, D.R. 1988. Gas transport in polymers based on bisphenol-A. J. Polym. Sci. B.: Polymer Physics 26(4): 709–27.

Ben-Gal, Irad, Roni Katz and Yossi Bukchin. 2008. Robust eco-design: A new application for air quality engineering. IIE Trans. 40(10): 907–18.

Berichte, Glastechnische. 1993. United States Patent 191 Date of Patent 4–5.

Bernardo, P., Drioli, E. and Golemme, G. 2009. Membrane gas separation: A review/state of the art. Ind. Eng. Chem. Res. 48(10): 4638–4663.

Bhown, Abhoyjit and E.L. Cussler. 1991. Mechanism for selective ammonia transport through Poly(Vinylammonium Thiocyanate) membranes. J. Am. Chem. Soc. 113(3): 742–49.

Bos, A., Pünt, I.G.M., Wessling, M. and Strathmann, H. 1998. Plasticization-resistant glassy polyimide membranes for CO_2/CO_4 separations. Sep. Purif. Technol. 14(1-3): 27–39.

Bounaceur, Roda, Nancy Lape, Denis Roizard, C??cile Vallieres, and Eric Favre. 2006. Membrane processes for post-combustion carbon dioxide capture: a parametric study. Energy 31(14): 2220–34.

Chen, Chien Chiang, Wulin Qiu, Stephen J. Miller and William J. Koros. 2011. Plasticization-resistant hollow fiber membranes for CO_2/CH_4 separation based on a thermally crosslinkable polyimide. J. Membr. Sci. 382(1-2): 212–21.

Coker, D.T. and B.D. Freeman. 1998. Modeling multicomponent gas separation using hollow-fiber membrane contactors. AIChE Journal 44(6): 1289–1302.

Collins, Michael C. and Fred Ramirez, W. 1979. Transport through polymeric membranes. J. Phys. Chem. 83(17): 2294–2301.

Conturie, M. 2006. Reduction of carbon dioxide emissions by capture and re-injection, Renewable Energy Sources and Environment.

Doshi, David Dortmundt and Kishore. 1999. Recent Developments in CO_2 Removal Membrane Technology. Uop Llp 1–32.

Falbo, F., Tasselli, F., Brunetti, A., Drioli, E. and Barbieri, G. 2012. Polyimide Hollow Fiber Membranes for CO_2 Separation from Wet Gas Mixtures. 31(04): 1023–34.

Farr, I.V., T.E. Glass, Q. Ji and J.E. McGrath. 1997. Synthesis and characterization of diaminophenylindane based polyimides via ester-acid solution imidization. High Performance Polymers 9(3): 345–52.

Favre, Eric. 2011. Membrane processes and postcombustion carbon dioxide capture: challenges and prospects. J. Chem. Eng. 171(3): 782–93.

Fick, Adolph. 1995. On liquid diffusion. J. Membr. Sci. 100: 33–38.

Francisco, Gil, J., Amit Chakma and Xianshe Feng. 2007. Membranes comprising of alkanolamines incorporated into Poly(Vinyl Alcohol) matrix for CO_2/N_2 separation. J. Membr. Sci. 303(1-2): 54–63.

Freeman, Benny, D. and Ingo Pinnau. 1999. Polymer Membranes for Gas and Vapor Separation: ACS Publications.

Ghalei, Behnam and Mohamad Ali Semsarzadeh. 2007. A novel nano structured blend membrane for gas separation. Macromolecular Symposia 249-250: 330–35.

Graham, T. 1866. On the absorption and dialytic separation of gases by colloid septa. Philosophical Transactions of the Royal Society of London 156(0): 399–439.

Granite, E.J. and H.W. Pennline. 2002. Photochemical removal of mercury from flue gas. Ind. Eng. Chem. Res. 41(22): 5470–5476.

Guiver, Michael, D., Gilles P. Robertson, Ying Dai, François Bilodeau, Yong Soo Kang, Kwi Jong Lee, Jae Young Jho and Jongok Won. 2002. Structural Characterization and Gas-Transport Properties of Brominated Matrimid Polyimide. J. Polym. Sci. A 40(23): 4193–4204.

Guo, R. and McGrath, J.E. 2012. 5.17—Aromatic Polyethers, Polyetherketones, Polysulfides, and Polysulfones. pp. 377–430. *In*: Matyjaszewski, K. and Möller, M. (eds.). Polymer Science: A Comprehensive Reference. Amsterdam: Elsevier.

Hamad, F., G. Chowdhury and T. Matsuura. 2002. Effect of metal cations on the gas separation performance of sulfonated Poly(Phenylene Oxide) membranes. Desalin. 145(1-3): 365–70.

Hao, J., Rice, P.A. and Stern, S.A. 2008. Upgrading low-quality natural gas with H2S- and CO_2-selective polymer membranes. Part II. Process design, economics, and sensitivity study of membrane stages with recycle streams. J. Membr. Sci. 320(1-2): 108–22.

Harasimowicz, M., Orluk, P., Zakrzewska-Trznadel, G. and Chmielewski, A.G. 2007. Application of polyimide membranes for biogas purification and enrichment. J Hazard Mater. 144(3): 698–702.

Henis, Jay M.S. and Mary K. Tripodi. 1981. Composite hollow fiber membranes for gas separation: the resistance model approach. J. Membr. Sci. 8(3): 233–46.

Hu, Chien Chieh, Kueir Rarn Lee, Ruoh Chyu Ruaan, Jean, Y.C. and Juin Yih Lai. 2006. Gas separation properties in cyclic olefin copolymer membrane studied by positron annihilation, sorption, and gas permeation. J. Membr. Sci. 274(1-2): 192–99.

Huang, Shu Hsien, Chien Chieh Hu, Kueir Rarn Lee, Der Jang Liaw and Juin Yih Lai. 2006. Gas separation properties of aromatic Poly(Amide-Imide) membranes. Eur. Polym. J. 42(1): 140–48.

Iarikov, Dmitri, D. and Ted Oyama, S. 2011. Review of CO_2/CH_4 Separation Membranes. 1st ed. Elsevier B.V.

Ismail, Ahman Fauzi and Lai Ping Yean. 2003. Effects of phase inversion and rheological factors on formation of defect-free and ultrathin-skinned asymmetric polysulfone membranes for gas separation. Sep Purif Technol. 33: 442–51.

Ito, Shigeo, Takahisa Yokoyama, and Kazuo Asakura. 2006. Emissions of mercury and other trace elements from coal-fired power plants in Japan. Sci. Total Environ. 368(1): 397–402.

Iupac. 1996. International, union of pure glossary of basic terms in polymer. Pure Appl. Chem. 68(12): 2287–2311.

Javaid, Asad. 2005. Membranes for solubility-based gas separation applications. J. Chem. Eng. 112(1–3): 219–26.

Kawakami, Hiroyoshi, Masato Mikawa and Shoji Nagaoka. 1997. Formation of surface skin layer of asymmetric polyimide membranes and their gas transport properties. J. Chem. Eng. 137(1-2): 241–50.

Kesting, R.E. 1985. Phase inversion membranes. ACS Symposium Series ACS Symposium Series Vol 269(1967): 131–64.

Kim, Il-Won, Kwi Jong Lee, Jae Young Jho, Hyun Chae Park, Jongok Won, Yong Soo Kang, Michael D. Guiver, Gilles P. Robertson and Ying Dai. 2001. Correlation between structure and gas transport properties of silyl-modified polysulfones and poly (Phenyl Sulfone) S. Macromolecules 34(9): 2908–2913.

Kohl, A.L. and Nielsen, R. 1997. Gas Purification. Elsevier Science.

Lin, H. and Freeman, B.D. 2004. Gas solubility, diffusivity and permeability in Poly(Ethylene Oxide). J. Membr. Sci. 239(1): 105–17.

Lin, Wen-Hui, Rohit H Vora and Tai-Shung Chung. 2000. Gas transport properties of 6fda-durene/1, 4-phenylenediamine (Ppda) copolyimides. J. Polym. Sci. B 38(21): 2703–2713.

Liu, Yamin, Qing Ye, Mei Shen, Jingjin Shi, Jie Chen, Hua Pan and Yao Shi. 2011. Carbon dioxide capture by functionalized solid amine sorbents with simulated flue gas conditions. Environ. Sci. Technol. 45(13): 5710–5716.

Loeb, S. 1962. Sea Water Demineralization by Means of a Semipermeable Membrane : Progress Report Jan. 1 - June 30, 1961. Los Angeles: [s.n.].

Matsumiya, N., Inoue, N., Mano, H. and Haraya, K. 1999. Effect of water vapor on CO_2 separation performance of membrane separator. Kagaku Kogaku Ronbunshu 25(3): 367–73.

McGrail, P.T. 1996. Polyaromatics. Polymer International 41(2): 103–21.

McHattie, J.S., Koros, W.J. and Paul, D.R. 1991. Gas transport properties of polysulphones: 1. role of symmetry of methyl group placement on bisphenol rings. Polymer 32(5): 840–50.

McHattie, J.S., Koros, W.J. and Paul, D.R. 1992. Gas Transport properties of polysulphones: 3. Comparison of tetramethyl-substituted bisphenols. Polymer 33(8): 1701–11.

Merkel, Tim, C., Haiqing, Lin, Xiaotong, Wei and Richard, Baker. 2010. Power plant post-combustion carbon dioxide capture: an opportunity for membranes. J. Membr. Sci. 359(1-2): 126–39.

Merket, T.C. and Toy, L.G. 2006. Comparison of hydrogen sulfide transport properties in fluorinated and nonfluorinated polymers. Macromolecules 39(22): 7591–7600.

Mulder, M. 1996. Basic principles of membrane technology. Zeitschrift Für Physikalische Chemie 564.

Muruganandam, N., Koros, W.J. and Paul, D.R. 1987. Gas sorption and transport in substituted polycarbonates. J. Polym. Sci. B: Polymer Physics 25(9): 1999–2026.

Orme, Christopher, J., John, R. Klaehn and Frederick, F. Stewart. 2004. Gas Permeability and Ideal Selectivity of Poly[Bis-(Phenoxy)Phosphazene], Poly[Bis-(4-Tert-Butylphenoxy)Phosphazene], and Poly[Bis-(3,5-Di-Tert- Butylphenoxy)1.2(Chloro)0.8phosphazene]. J. Membr. Sci. 238(1-2): 47–55.

Pan, C.Y. 1986. Gas Separation by high-flux, asymmetric hollow-fiber membrane. AIChE Journal 32(12): 2020–27.

Pandey, P. and Chauhan, R.S. 2001. Membranes for gas separation. Prog. Polym. Sci. 26(6): 853–93.

Paul, D.R. 1995a. Gas transport properties based polysulfones of adamantane. Polymer 36(16): 3165–72.

Paul, D.R. 1995b. S Transport Properties of Polyarylate D on 9 , 9-Bis (4-Hydroxyphenyl) Anth. 36: 2745–51.

Petersen, Joachim, Masaji Matsuda and Kenji Haraya. 1997. Capillary carbon molecular sieve membranes derived from kapton for high temperature gas separation. J. Membr. Sci. 131(1-2): 85–94.

Powell, Clem, E. and Greg, G. Qiao. 2006. Polymeric CO_2/N_2 gas separation membranes for the capture of carbon dioxide from power plant flue gases. J. Membr. Sci. 279(1-2): 1–49.

Programme, IEA Greenhouse Gas R&D. 2007. CO_2 Capture Ready. Energy (May).

Puri, Pushpinder, S. n.d. Commercial Applications of Membranes in Gas Separations 1: 215–44.

Pye, D.G., Hoehn, H.H. and Panar, M. 1976. Measurement of gas permeability of polymers. ii. apparatus for determination of permeabilities of mixed gases and vapors. J. Appl. Polym. Sci. 20(2): 287–301.

Steeneveldt, R., Berger, B. and Torp, T.A. 2007. CO_2 capture and storage projects. Energy (x): 1–52.

Robeson, L.M. 2012. 8.13 - Polymer membranes. pp. 325–47. *In*: Matyjaszewski, K. and Möller, M. (eds.). Polymer Science: A Comprehensive Reference. Amsterdam: Elsevier.

Robeson, Lioyd, M. 2007. Polymer blends: a comprehensive review. Carl Hanser GmbH 10–23.

Rogalsky, Sergiy, P., Olena V. Moshynets, Lyudmila G. Lyoshina and Oksana P. Tarasyuk. 2014. Antimicrobial polycarbonates for biomedical applications. EPMA Journal 5(Suppl 1): A133.

Rowe, B.W., B.D. Freeman and D.R. Paul. 2007. Effect of sorbed water and temperature on the optical properties and density of thin glassy polymer films on a silicon substrate. Macromolecules 40(8): 2806–13.

Sanders, David F., Zachary P. Smith, Ruilan Guo, Lloyd M. Robeson, James E. McGrath, Donald R. Paul and Benny D. Freeman. 2013. Energy-efficient polymeric gas separation membranes for a sustainable future: a review. Polym. J. 54(18): 4729–4761.

Scholes, Colin A., Sandra E. Kentish and Geoff W. Stevens. 2009. Effects of minor components in carbon dioxide capture using polymeric gas separation membranes. Sep. Purif. Rev. 38(1): 1–44.

Shamsabadi, Ahmad Arabi, Ali Kargari, Masoud Bahrami Babaheidari and Saeed Laki. 2013. Separation of hydrogen from methane by asymmetric pei membranes. nd. Eng. Chem. Res. 19(5): 1680–88.

Sluijs, J.P., Van Der, C.A. Hendriks and K. Blok. 1992. Feasibility of polymer membranes for carbon dioxide recovery from flue gases. Energy Convers. Manag. 33(5–8): 429–36.

Steeneveldt, R., Berger, B. and Torp, T.A. 2006. CO_2 capture and storage: closing the knowing–doing gap. Chemical Engineering Research and Design 84(9): 739–763.

Stern, S.A., Y. Liu and W.A. Feld. 1993. Structure/permeability relationships of polyimides with branched or extended diamine moieties. J. Polym. Sci. B: Polymer Physics 31(8): 939–51.

Stern, S. Alexander. 1994. Polymers for gas separations: the next decade. J. Membr. Sci. 94(1): 1–65.

Strömberg, Lars, Göran Lindgren, Jürgen Jacoby, Rainer Giering, Marie Anheden, Uwe Burchhardt, Hubertus Altmann, Frank Kluger and Georg-Nikolaus Stamatelopoulos. 2009. Update on vattenfall's 30 MWth oxyfuel pilot plant in schwarze pumpe. Energy Procedia 1(1): 581–589.

Thomas, Sabu and P.M. Visakh. 2011. Polyethers and Polyesters.

Tricoli, Vincenzo and E.L. Cussler. 1995. Ammonia selective hollow fibers. J. Membr. Sci. 104(1-2): 19–26.

White, Curt M., Brian R. Strazisar, Evan J. Granite, James S. Hoffman and Henry W. Pennline. 2003. Separation and Capture of CO_2 from Large Stationary Sources and Sequestration in Geological Formations—Coalbeds and Deep Saline Aquifers. J. Air Waste Manag Assoc. 53(6): 645–715.

Wijmans, J.G. and R.W. Baker. 1995. The solution-diffusion model: a review. J. Membr. Sci. 107(1-2): 1–21.

Xu, Z. 2002. Novel Poly(Arylene Ether) as membranes for gas separation. J. Membr. Sci. 205(1-2): 23–31.

Xu, Zhi-kang, Li Xiao, Jian-li Wang and Jürgen Springer. 2002. Gas separation properties of PMDA/ODA polyimide membranes filling with polymeric nanoparticles. J. Membr. Sci. 202: 27–34.

Yampolskii, Yuri. 2011. Membrane Gas Separation - Google Books.

Yampolskii, Yuri. 2012. Polymeric gas separation membranes. Macromolecules (ASC Publications) 45(8): 3298–3311.

Yuri Yampolskii and Benny Freeman. 2010. Membrane Gas Separation.

Zhao, Li, Ernst Riensche, Ludger Blum and Detlef Stolten. 2010. Multi-stage gas separation membrane processes used in post-combustion capture : energetic and economic analyses. J. Membr. Sci. 359(1-2): 160–72.

Zhao, Li, Ernst Riensche, Reinhard Menzer, Ludger Blum and Detlef Stolten. 2008. A parametric study of CO_2/N_2 gas separation membrane processes for post-combustion capture. J. Membr. Sci. 325(1): 284–94.

Zimmerman, Catherine, M. and William J. Koros. 1999. Entropic selectivity analysis of a series of polypyrrolones for gas separation membranes. Macromolecules 32(10): 3341–46.

4

Recent Progress on Asymmetric Membranes Developed for Natural Gas Purification

Gongping Liu,[1,*] *Zhi Xu*[2] and *Wanqin Jin*[1]

INTRODUCTION

Natural gas is a naturally formed hydrocarbon gas mixture consisting primarily of methane, but commonly including a certain amount of carbon dioxide, hydrogen sulfide, higher alkanes and other inert gases(Baker and Lokhandwala 2008). The composition of gas delivered to commercial pipeline must be controlled to avoid undesired corrosion. For instance, raw gas usually requires treatment to reduce the CO_2 concentration to less than 2% and H_2S concentration below 4 ppm according to the U.S. natural gas pipeline specifications. Until now, processing of natural gas has been the largest industrial gas separation application. Conventional separation technologies such as amine sorption are very energy-consuming and non-environmental friendly. Ninety percent energy will be saved if applying membrane technology for removing CO_2 from the industrial streams (Leung et al. 2014). The key of this membrane process is developing high-performance asymmetric gas separation membranes—highly permeable and selective for CO_2 molecules. Polymers, inorganic materials, mixed-matrix materials and carbon molecular sieves have been processed into asymmetric membranes via different techniques. This chapter will discuss the latest progresses in design and preparation of asymmetric membranes for natural gas purification, particularly focusing on the CO_2/CH_4 separation.

[1] Nanjing Tech University, Department of Chemical Engineering, 5 Xinmofan Road, Nanjing 210009, China.

[2] University of Oxford, Oxford Institute of Biomedical Engineering, OX3 7DQ, Oxford, UK.
Emails: gpliu@njtech.edu.cn; wqjin@njtech.edu.cn

* Corresponding author: sleepxuzhi@gmail.com

General Structures and Fabrication Approaches of Asymmetric Membranes

For practical application, the dense separation layer of a membrane must be as thin as possible (typically less than 0.5 μm and often less than 0.1 μm[1]), while strong enough to withstand transmembrane pressure (Koros and Zhang 2017). Such requirement is ideally achieved with asymmetric membranes consisting of a thin dense layer and a porous support layer. As shown in Fig. 4.1, the thin separation layer provides selective permeation, and the porous support layer offers mechanical strength meanwhile minimizes the transport resistance. It has two main types: flat sheet and hollow fiber. Since hollow fiber membrane could achieve packing density as high as 4000 m^2/m^3, hollow fiber modules allow larger membrane areas to be packaged into membrane modules compared with spiral-wound modules (Fane et al. 2011).

The asymmetric membranes can be prepared either by precipitation or coating approach, which are also called anisotropic Loeb-Sourirajan membranes or composite membranes, respectively (Baker and Lokhandwala 2008). In the conventional Loeb-Sourirajan membrane, the relatively dense surface layer and the porous support layer are formed at the same time from the same material. This limits the types of material that can be used to make the membrane, and may also increase the membrane cost if the membrane materials are expensive. On the other hand, the composite membrane can use different materials to form the microporous support layer and the thin separation layer. The two layers are able to be optimized separately, and the materials best suited for each function can be used. Composite membranes make high-cost membrane materials to be used cost-effectively in the separation layer.

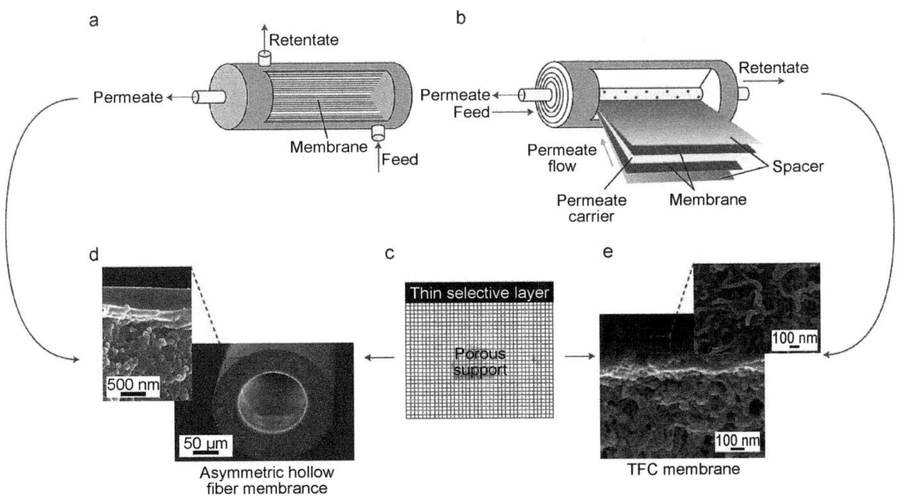

Figure 4.1. Asymmetric membrane module and structures. (a) schematic illustration of hollow fiber format; (b) schematic illustration of spiral wound format; (c) simplified structure of asymmetric or composite membranes, both consisting of a thin selective layer supported by a porous substrate; (d) SEM images of an asymmetric hollow fiber membrane formed by Matrimid polyimide; (e) SEM images of a thin-film composite (TFC) membrane that comprises a polyamide selective layer interfacially polymerized atop a porous polysulfone substrate. Reproduced with permission (Koros and Zhang 2017), Copyright 2017 Nature Publishing Group.

In the content of defect-free asymmetric membranes, it always pursues thinner separation layer to maximize the gas permeance. Baker et al. (Baker and Low 2014) pointed out that the thickness of selective layer should be 0.1–1.0 μm in order to minimize the membrane area required for industrial application. Besides of the classic Loeb-Sourirajan method, dry-jet/wet-quench spinning and interfacial polymerization are two effective approaches to process polymers into high-performance asymmetric membranes. A typical schematic for spinning single-layer hollow fibers is shown in Fig. 4.2a. Polymer solutions (dopes) and bore fluid are co-extruded through a spinneret into the air gap during the "dry-jet" process, and then into an quench bath during the "wet quench" process where polymer solution solidifies due to phase separation prior to being collected on a take-up drum. When the dope is extruded through the air gap, evaporation of the volatile solvents and non-solvents causes the formation of dense separation layer. After that, the fiber enters into the quench bath, the non-solvent diffuses into the nascent fiber causing the phase separation of the dope to form a porous support layer. The interfacial polymerization approach is adopted from the development of Thin-Film Composite (TFC) membranes for water purification (e.g., reverse osmosis process for water desalination), where interfacial polymerization reaction occurs at the solvent interface between one monomer dissolved in aqueous solution and another monomer dissolved in organic solution (Fig. 4.2b).

Inorganic composite membrane is usually formed by growing the molecular sieve layer (e.g., zeolite) on a porous support. Since the nucleation and subsequent crystal growth is difficult to control, secondary growth method is well established to fabricate thin and defect-free molecular sieve layers. As shown in Fig. 4.2c, seed crystallites of the desired zeolite are firstly synthesized and then introduced via electrostatic force, covalent bonding or capillary force during dip-coating or

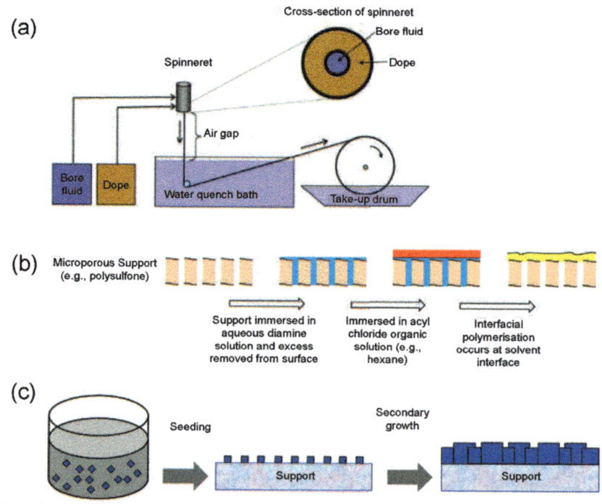

Figure 4.2. Schematic of some typical approaches for fabricating asymmetric membranes for natural gas purification. (a) dry-jet/wet-quench spinning, Reproduced with permission (Ma and Koros, 2013b), Copyright 2013 Elsevier; (b) interfacial polymerization Reproduced with permission (Davey et al., 2016), Copyright 2016 MDPI AG; the blue represents the aqueous diamine solution, red represents the acyl chloride organic solution, and yellow refers to the separation layer formed by the interfacial polymerization between the diamine in the water solution and acyl chloride in the organic (hexane) solution; (c) secondary growth for fabricating inorganic membranes.

spin-coating to the support surface. These seeded crystallites can grow *in situ* to a continuous thin separation layer.

Polymeric Membranes

Gas separation membranes in current commercial use are polymeric membranes that separate gas mixtures based on the solution-diffusion mechanism. Hundreds of polymers have been reported for CO_2/CH_4 separation, most of them are glassy polymers, some new materials such as polymers of intrinsic microporosity (PIMs) and Thermally Rearranged (TR) polymers are also included (Baker and Low 2014). Nevertheless, only a few of them have been fabricated as asymmetric membranes showing promising performance in practical CO_2/CH_4 separation. Cellulose Acetate (CA) and polyimide membranes have been mainly used for CO_2/CH_4 separation commercially (Han and Lee 2011), the typical chemical structures are shown in Fig. 4.3a.

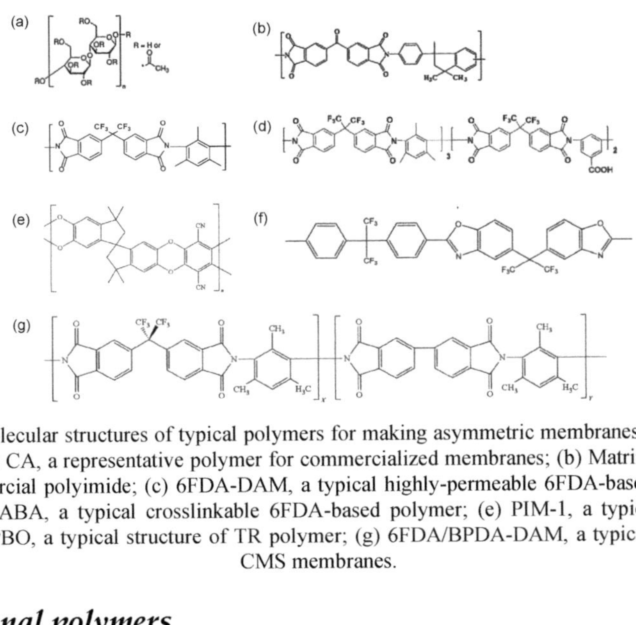

Figure 4.3. Molecular structures of typical polymers for making asymmetric membranes for natural gas purification. (a) CA, a representative polymer for commercialized membranes; (b) Matrimid®, a widely-studied commercial polyimide; (c) 6FDA-DAM, a typical highly-permeable 6FDA-based polymer; (d) 6FDA-DAM:DABA, a typical crosslinkable 6FDA-based polymer; (e) PIM-1, a typical structure of PIMs; (f) TR-PBO, a typical structure of TR polymer; (g) 6FDA/BPDA-DAM, a typical precursor for CMS membranes.

Conventional polymers

CA was initially developed for reverse osmosis process by UOP (Universal Oil Products) and now has dominated the industrial CO_2/CH_4 separation (Houde et al. 1996). CA-based materials have a pure-gas CO_2 permeability of 1.8–6.5 barrer (1 barrer = 10^{-10} cc(STP) cm·cm^{-2}·s^{-1}·cmHg^{-1}) and a CO_2/CH_4 selectivity of 32–35 at 1 bar and 35°C (Baker and Low 2014). The relatively low permeability must require a very thin separation layer to achieve a good permeance. CO_2 permeance of 2.5 GPU (1 GPU = 10^{-6} cc(STP) cm^{-2}·s^{-1}·cmHg^{-1}) was reported for a CA asymmetric membrane with CO_2/CH_4 selectivity of 20 for the 50/50 mixed-gas feed at 8 bar and 35°C (Visser et al. 2007). Less than 70 nm thick and defect-free dense layer is required to raise the CO_2 permeance of CA membrane over 100 GPU, which is considered to be very challenging for the current membrane fabrication technology.

Alternatively, polyimides offer good intrinsic CO_2/CH_4 separation properties, thermal, chemical and mechanical stability and are processable using current commercial fabrication processes. In particular, fluorinated polyimides containing 2,2'-bis (3,4'dicarboxyphenyl) hexafluoropropane dianhydride (6FDA) have been identified as the promising materials since they exhibited much higher permeability compared with the non-fluoropolyimides (e.g., Matrimid, as shown in Fig. 4.3b) (Hillock and Koros 2007, Staudt-Bickel and Koros 1999, Wind et al. 2004). It can be attributed to the bulky—$C(CF_3)_2$—group which effectively hinders intra-segmental mobility and disrupts inter-chain packing and stiffens backbones (Qiu et al. 2013). One of the most permeable 6FDA-polyimides, 6FDA-DAM (2,4,6-trimethyl-1,3-diaminobenzene) as shown in Fig. 4.3c, could obtain CO_2 permeability of more than 1000 barrer with mixed-gas CO_2/CH_4 selectivity of ~ 20 below 4 bar. Recently, defect-free asymmetric 6FDA-DAM hollow fiber membranes have been developed (Xu et al. 2014b) .

Crosslinkable polymers and asymmetric membranes

One of the challenges for polymeric membranes is plasticization, especially in practical natural gas purification that usually involves aggressive feed conditions (high pressure, high CO_2 content). Plasticization causes swelling of the polymer and increases chain segmental motion, eventually causing the polymer to lose its size and shape discriminating ability, leading the loss of selectivity and hence separation efficiency. Several approaches have been used to address the plasticization of polymers. Among them, covalent crosslinking is efficient to increase plasticization resistance by suppressing swelling and segmental chain mobility in the polymer, thereby preserving the membrane selectivity (Hillock and Koros 2007). Polyimides can be crosslinked by thermal, UV, or chemical treatments. Diol crosslinking of polyimides containing carboxylic acid groups, developed by Koros group (Hillock and Koros 2007, Omole et al. 2010, Omole et al. 2008), is most studied.

6FDA-DAM:DABA (diaminobenzoic acid), as shown in Fig. 4.3d, is a typical 6FDA-based polyimide, where the carboxyl groups provide a functional platform for further chemical treatments including thermal crosslinking and esterification. Decarboxylation-induced thermal crosslinking occurs at elevated temperatures (330–400°C) for 6FDA-DAM:DABA (Qiu et al. 2011, Qiu et al. 2013). The plasticization resistance of the crosslinked 6FDA-DAM:DABA dense film was demonstrated up to 48 bar for pure CO_2 gas or 69 bar for 50/50 CO_2/CH_4 mixed-gas feed (Qiu et al. 2011). However, such high temperatures might cause the densification of transition layers and porous substructures of asymmetric hollow fibers, resulting in loss of the permeance. To reduce the crosslinking temperature, ester crosslinking of 6FDA-DAM:DABA membranes was proposed using a diol to form ester bonds among polyimide chains (Hillock and Koros 2007). A typical schematic is shown in Fig. 4.4. The 6FDA-DAM:DABA(3:2) first undergoes monoesterification with 1,3-propanediol to form a crosslinkable polyimide that is referred to as "PDMC", meaning **p**ropane**d**iol **m**onoesterified **c**ross-linkable polyimide (Omole et al.

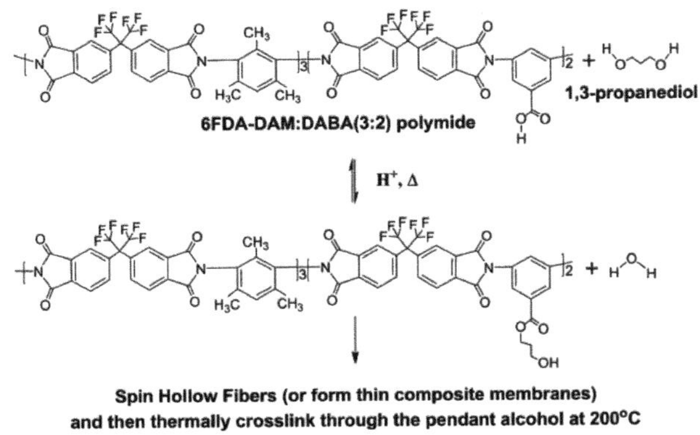

Figure 4.4. Schematic of crosslinking polyimide hollow fiber membranes for natural gas separations Reproduced with permission (Ma and Koros 2013b), Copyright 2013 Elsevier.

2008). The modified polyimide can be formed into an asymmetric membrane (most commonly in hollow fiber form) and then undergoes transesterification and crosslinking by simply heating at 200°C for 2 hours. By optimizing the molecular weight of the polymer, dope composition and dry-jet/wet-quench spinning variables, Ma et al. 2013b developed thin-skinned defect-free ester-crosslinkable polyimide (PDMC) asymmetric hollow fiber membranes. The crosslinked PDMC hollow fiber showed high separation productivity and efficiency with CO_2 plasticization resistance with 50/50 CO_2/CH_4 feed at 55 bar and 35°C: CO_2 permeance is above 100 GPU with CO_2/CH_4 selectivity of over 25.

Despite of the high performance, collapse of the thin nanoporous transition layer that lies between the dense skin layer and underlying porous support still occurred in the PDMC asymmetric hollow fiber membranes, thereby causing an increase in effective skin thickness during crosslinking (Figs. 5a–d). As a result, the CO_2 permeance can be reduced by 50% (Fig. 5g). To overcome this obstacle, Liu et al. 2016b synthesized new rigid polyimides by incorporating bulky CF_3 groups into PDMC backbone and translated the polymers into asymmetric hollow fiber membranes. Typically, the 6FDA-DAM:DABA(3:2) structure was modified by 5 mol% 5-(trifluoromethyl) benzene-1,3-diamine and monoesterified by 1,3-propanediol to form a crosslinkable polyimide that is referred to as PDMC-CF_3. It was demonstrated that the enhanced rotational barrier provided by properly positioned CF_3 side groups prohibited fiber transition layer to collapse during crosslinking (Fig. 5e–f) and preserved the high CO_2 permeance (Fig. 5g), thereby greatly improving CO_2/CH_4 separation performance compared to conventional materials for aggressive natural gas feeds (Fig. 5h).

To further reduce the membrane cost, dual-layer hollow fiber membranes can be developed to reduce the amount of polyimide required, while still achieving high separation performance. Generally, dual-layer hollow fiber membranes comprise two layers made of different materials, where the sheath layer offers selective dense skin layer while the low-cost supporting core layer is porous to provide the mechanical strength. Dual-layer hollow fibers are spun by coextruding two spinning dopes (sheath dope and core dope) and the bore fluid from a composite spinneret with three annular channels (sheath dope channel, core dope channel, and bore fluid

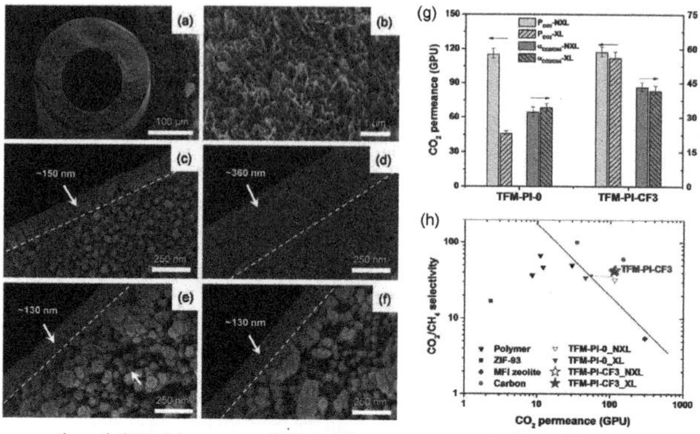

Figure 4.5. Cross-sectional SEM images of PDMC asymmetric hollow fiber membranes: (a) overview; (b) porous substrate layer; dense skin layer upon transition layer of PDMC fiber (c) before and (d) after cross-linking; dense skin layer upon transition layer of PDMC-CF_3 fiber (e) before and (f) after crosslinking; Gas separation performance of hollow fiber membranes: (g) comparison between PDMC (TFM-PI-0) and PDMC-CF_3 (TFM-PI-CF_3) hollow fiber membrane before and after cross-linking, non-cross-linked (NXL), crosslinked (XL), feed conditions: 50:50 CO_2/CH_4 mixtures, 14 bar, 35°C; (c) comparison of CO_2 permeance and CO_2/CH_4 selectivity reported for hollow fiber membranes under mixed-gas permeation. The green eye-guiding line is only used to indicate the general "trade-off" between permeance and selectivity in hollow fiber membranes. Reproduced with permission (Liu et al. 2016b), Copyright 2016 John Wiley and Sons.

channel). Ma and Koros 2013a fabricated a defect-free dual-layer PDMC/Torlon composite hollow fiber membranes showing a CO_2 permeance of 40 GPU with a CO_2/CH_4 selectivity of 39 with 50/50 CO_2/CH_4 feed at 6.9 bar and 35°C. The core layer polyamide-imide polymer, Torlon, showed excellent adhesion with PDMC spinning and ester-crosslinking, and maintained the open substructure of the core layer due to its superior thermal stability.

Other high-performing polymers

Facilitated transport membranes contain reactive carriers that could interact with CO_2 molecules through reversible reactions were studied in the past decades. Towards non-interacting gases like CH_4 having negligible binding interactions with carriers, CO_2 transport is greatly enhanced via the facilitated transport mechanism. Various types of basic groups, such as $-NH_2$, F^-, PO_4^{3-}, CO_3^{2-}, $-COO^-$ have been employed as CO_2 carriers (Wang et al. 2016a). Although some asymmetric facilitated transport membranes with both high CO_2 permeance and CO_2/CH_4 selectivity were developed, the loss of the carrier and low-pressure operation (CO_2 pressure is often below 1 bar) limit this type of membranes applied for natural gas purification process. As the feed gas pressure is high, the CO_2 permeance is severely reduced since most of the carriers are saturated and thus lost the facilitated transport function.

Some new polymers have been synthesized to improve the free volumes of conventional glassy polymers (~ 10 – 20%). Two types among these high free volume materials are relatively stable and promising for development of asymmetric membranes (Baker and Low 2014). One is thermally rearranged (TR) polymers proposed by Lee and Freeman (Park et al. 2007). They are aromatic polymers with heterocyclic rings, such as polybenzox-azoles (PBO) or polybenzothiazoles (PBT) formed via thermal rearrangement (usually at 300–450°C) of polyimide with ortho-functional group prepared from poly(amic acid). Generally, precursor of the TR-polymer is soluble and can be processed into asymmetric membranes (e.g., spun into hollow fibers) via non-solvent induced phase separation process. Kim et al. 2012 produced highly permeable TR-PBO (Fig. 4.3f) hollow fiber membranes by thermal imidization of the spun hydroxyl poly(amic acid) and then thermal rearrangement of the hydroxyl polyimide. The synthetic scheme of TR-PBO is shown in Fig. 4.3. The TR-PBO hollow fibers thermally treated at 450 °C showed a CO_2 permeance of 1938 GPU and CO_2/CH_4 selectivity of 14 with pure-gas feed at 1–6 bar and 25°C. The other polymers possessing large free volume is polymers of intrinsic microporosity (PIMs) synthesized by Budd and McKeown 2010. The structure of first PIM family polymers, PIM-1, is shown in Fig. 4.3e. These polymers have rigid fused-ring ladder structure blocks connected by sites of contortion to produce kinks in the polymer chain. Recently, the translation of PIMs dense films into defect-free asymmetric membranes was realized by Lively and co-workers (Jue et al. 2017). The lack of suitable nonvolatile solvents of PIM-1 was overcome by using a dual-bath spinning via a triple orifice spinneret. PIM-1 fibers with skin thicknesses from 3–6 μm were fabricated with a CO_2 permeance of 360 GPU and CO_2/CH_4 selectivity of 23 with pure-gas feed at 6.9 bar and 35°C. Separation performance of these asymmetric membranes made of new polymers under mixed-gas, high-pressure or impurities such as water and hydrocarbons in the feed are unknown that need further investigations.

Inorganic Membranes

Inorganic membranes are particularly useful for CO_2 separation under harsh conditions such as high temperature and pressure where polymer-based membranes often suffer challenges, although the scalable fabrication of inorganic membrane remains challenging. Two types of inorganic materials can be used: microporous zeolites and MOFs (Metal-Organic Frameworks), as well as nonporous dual-phase composite containing mixed-conducting oxide ceramic and molten carbonate phases. The dual-phase membranes have absolute selectivity for CO_2, while low permeability, and they must be operated at very high temperature (> 400°C) (Chung et al., 2005), which currently are unattractive for practical natural gas purification.

Zeolite membranes

Zeolite membranes with sub-nanometer pores are promising candidates for CO_2/CH_4 separation owing to their excellent chemical resistance to plasticization and superior selectivity compared to polymeric membranes. Separation CO_2 from CH_4 in zeolites is performed by two functions: (1) favorable adsorption of CO_2 over CH_4 on most zeolites, and/or (2) molecular sieving properties due to the molecular kinetic diameters of CO_2 (3.3 Å) over CH_4 (3.8 Å). Generally, using MFI (description of the Zeolite Framework Type ZSM-FIve)-type zeolite membranes with pore size of 5.5 Å to separate CO_2/CH_4 is mainly based on the CO_2 preferential adsorption, in which low CO_2/CH_4 selectivities of 2.3–8.2 were observed under room temperature

(Zhang et al. 2013). As for FAU (Faujasite)- and T-type zeolite membranes, the CO_2 adsorption capacities are close to saturation at 1 bar, causing reduced CO_2 permeance under high pressure (Cui et al. 2004). In contrast, SAPO-34 or DDR type zeolites having pores of 3.8 Å and 3.6 Å enable very high selectivity (~ 50–500) for CO_2/CH_4 in the zeolite membranes, which should be owing to the combination of differences in diffusivity and competitive adsorption of CO_2 over CH_4 (Zhang et al. 2013).

The SAPO-34 structure is a Si-substituted aluminophosphates (SAPOs) with the composition $Si_xAl_yP_zO_2$, where $x = 0.01–0.98$, $y = 0.01–0.60$, and $z = 0.01–0.52$. Falconer's group has contributed great efforts in developing SAPO-34 membranes. In 2000s, they reported the synthesis of continuous SAPO-34 membranes on porous stainless steel and α-Al_2O_3 (aluminum oxide) tubes via *in situ* crystallization method (Li et al. 2004). Typical SEM images of the membrane surface and cross-section are shown in Fig. 4.6a and 6b, respectively. These membranes exhibited CO_2 permeance of 480 GPU with CO_2/CH_4 selectivity of 67 for 50/50 CO_2/CH_4 mixture at 2.2 bar and 24°C in the feed. The performance was stable in the presence of H_2O, N_2, C_2H_4, C_3H_8, and n-C_4H_{10} impurities that are often present in natural gas stream (Li et al. 2005). Various Si/Al gel ratios of 0.15, 0.2 and 0.3 were studied to optimize the synthesis conditions. It was found that Si/Al ratio of 0.15 remarkably increased the CO_2/CH_4 selectivity to 170 while slightly reduced the CO_2 permeance to 420 GPU (50/50 CO_2/CH_4, 1.4 bar and 22°C) (Li et al. 2006). As the feed pressure increased to 70 bar, the selectivity dropped to 100 and permeance decreased to ~ 50 GPU. This might be due to the highly reduced CO_2 sorption (or even saturated) at high pressure. Overall, the separation performance kept stable for a week at 57 bar and 22°C with 75/25 CO_2/CH_4 as the feed.

Several approaches have been proposed to further advance the permeance and/or selectivity of SAPO-34 membranes. Li et al. 2006 employed secondary growth method to synthesize SAPO-34 membranes using the same Si/Al ratio (0.15) for the *in situ* growth approach as discussed above. The resulting membranes had lower CO_2/CH_4 separation selectivity (115) but higher CO_2 permeance (1200 GPU). By using combinations of structure directing agents (SDAs) for both seed and membrane preparation of SAPO-34, higher CO_2 permeance (up to 1080 GPU) and CO_2/CH_4 selectivity (up to 227) were achieved (Fig. 5c–d) (Carreon et al. 2008). The highly improved performance could be attributed to the thin and defect-free SAPO-34 separation layer formed on the substrate. They also scaled-up the tubular SAPO-34 membranes from 5 cm to 25 cm length (Li et al. 2010). Aluminum sources, gel compositions, and synthesis procedures were systematically studied to reduce the membrane cost and optimize the separation performance and reproducibility. The 25-cm SAPO-34 membranes exhibited CO_2 permeances of ~ 150 GPU and CO_2/CH_4 selectivity greater than 200 for 50/50 CO_2/CH_4 mixtures at 1.4 bar and 22°C. Very recently, Gu and co-workers (Chen et al. 2017) synthesized SAPO-34 membranes on α-Al2O3 four-channel hollow fibers, showing excellent CO_2/CH_4 separation performance: CO_2 permeance of 3540 GPU and CO_2/CH_4 selectivity of 160 for 50/50 CO_2/CH_4 mixtures at 2 bar and 25°C.

Aluminophosphates (AlPOs) are classes of crystalline materials built of equimolar $AlO4^-$ and $PO4^+$ tetrahedral units. The AEI-type AlPO-18 has a three-dimensional pore system possessing 8-membered intersecting channels with a

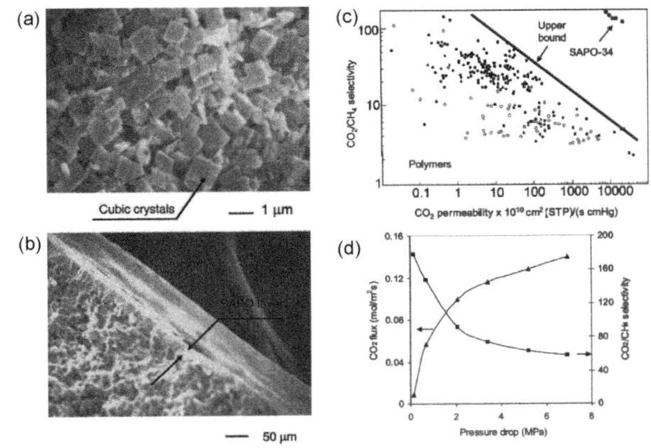

Figure 4.6. SAPO-34 membrane supported on α-Al_2O_3 substrate: (a) SEM top and (b) SEM cross sectional view. Reproduced with permission (Li et al. 2004), Copyright 2004 Elsevier; (c) CO_2/CH_4 separation selectivity versus CO_2 permeability for polymeric membranes and SAPO-34 membranes; (d) CO_2 flux and selectivity for a CO_2/CH_4 mixture (50:50) at 22°C as a function of pressure drop for SAPO-34 membrane. Reproduced with permission (Carreon et al. 2008), Copyright 2008 John Wiley and Sons.

diameter of 0.38 nm, which is an excellent molecular sieve for CO_2/CH_4. Zhou and co-workers (Wu et al. 2014) synthesized AlPO-18 membranes on the outside surface of microporous tubular α-alumina supports (average pore size: 1.3 μm) in a single hydrothermal synthesis step. The AlPO-18 membranes had an average CO_2 permeance of 540 GPU and CO_2/CH_4 selectivity of 101 for 50/50 CO_2/CH_4 mixtures at 1 bar and 25°C. Different from the 'round' shaped pores of zeolites such as SAPO-34 or AlPO-18 (window size of 3.8 Å × 3.8 Å), ERI-type AlPO-17 and SAPO-17 have 'ellipsoid' shaped pores with window size of 3.6 Å × 5.1 Å nm to produce an effective pore size of 3.6 Å. These slimmer pore windows may provide better shape selectivity for CO_2/CH_4 pair. Nevertheless, the obtained CO_2/CH_4 selectivity of 53 for SAPO-17 membrane was lower than that reported for SAPO-34 or AlPO-18 membranes, suggesting some non-selective defects might exist in the SAPO-17 membrane (Zhong et al. 2016).

Compared with SAPO-34, DDR-type zeolite has smaller pore size (3.6 × 4.4 Å), offering higher molecular sieving selectivity. Also, it is essentially comprised of silicon and oxygen atoms that is expected to be affected less by water adsorption (Zhang et al. 2013). Tomita et al. (Himeno et al. 2007) fabricated DDR membranes on porous alumina tube by hydrothermal process. The membranes showed 200 GPU for CO_2 permeance and 220 for CO_2/CH_4 selectivity for 50/50 CO_2/CH_4 mixtures at 5 bar and 24°C. The permeance and selectivity were reduced by half when increasing the temperature from 24 to 100°C. Gu and co-workers (Wang et al. 2017) found that as-prepared DDR membranes were susceptible to crack during high temperature activation (> 600°C) for the template removal. To overcome this challenge, zone environment was proposed to lower the activation temperature to 200°C, which prevented the formation of cracks. SEM images of the as-prepared membranes are shown in Fig. 4.7a–d. The high-quality DDR zeolite membranes on α-Al_2O_3 four-channel hollow fibers showed CO_2 permeance of 105 GPU with CO_2/CH_4 selectivity

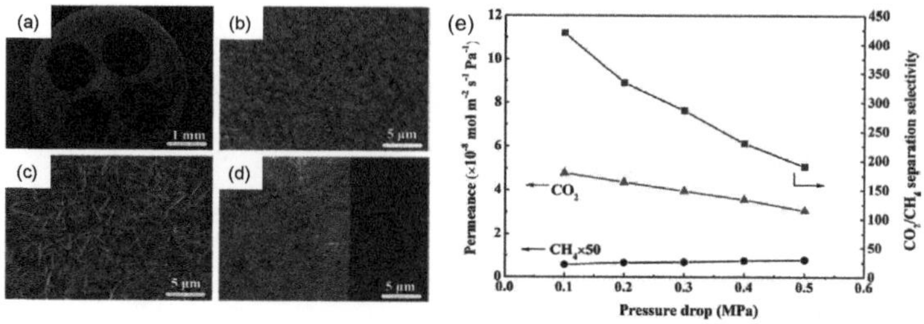

Figure 4.7. SEM images for (a) cross-section and (b) seeded surface of four-channel hollow fiber support; (c) surface and (d) cross-section of DDR zeolite membranes; (e) separation results of DDR zeolite membrane as a function of transmembrane pressure drop for equimolar CO_2/CH_4 mixture at 25°C, 10^{-6} mol m^{-2} s^{-1} Pa^{-1} = 3000 GPU. Reproduced with permission (Wang et al. 2017), Copyright 2017 Elsevier.

of 500 for 50/50 CO_2/CH_4 mixtures at 2 bar and 25°C, as displayed in Fig. 4.7e. Similar permeance and selectivity drops were observed by increasing the feed pressure from 1 to 5 bar.

MOF membranes

MOF membranes have emerged as an additional type of crystalline microporous membranes, which combined the highly accessible porosity, wide range of pore sizes, and unique surface chemistry properties, making them ideal candidates for gas separation. Several MOFs have been prepared in membrane form, and have been applied for CO_2/CH_4 gas mixtures separation (Venna and Carreon 2015), including MOF-5, MIL-53, HKUST-1, ZIF-8, ZIF-7, ZIF-22, ZIF-69, ZIF-95 and ZIF-90. Unfortunately, none of them displayed attractive separation performance for natural gas purification. The CO_2/CH_4 selectivity of current MOF membranes is below 8, which is even much lower than that of the-state-of-the-art polymeric membranes. This has been related to the relatively large window size, as well as flexible pore structure induced by the organic linkers, which may limit a potential molecular sieving effect. MOFs with smaller window size and more rigid framework may improve the CO_2/CH_4 separation performance. Moreover, main challenges for developing high-quality MOF membranes for gas separation include poor intergrowth at the membrane-support interface, limited stability in moisture, and insufficient control over "non-porous pathways" (Venna and Carreon 2015).

Venna and Carreon 2015 summarized the performance of (a) zeolite membranes and (b) MOF membranes for CO_2/CH_4 separation in Fig. 4.8. It is clearly seen that zeolite membranes exhibit much higher CO_2/CH_4 selectivity than MOF membranes, which should be associated to the molecular sieving effect of zeolite pores. Zeolite DD3R shows the highest CO_2/CH_4 separation selectivity but low CO_2 permeance, because the 2-dimensional pore structure with pore sizes of 3.6 × 4.4 Å offers excellent molecular sieving and activated transport. Another zeolite presented on the performance limit line is SAPO-34, showing the best overall separation performance for CO_2/CH_4 mixtures. Higher CO_2 permeance is often achieved by preparing thinner membranes.

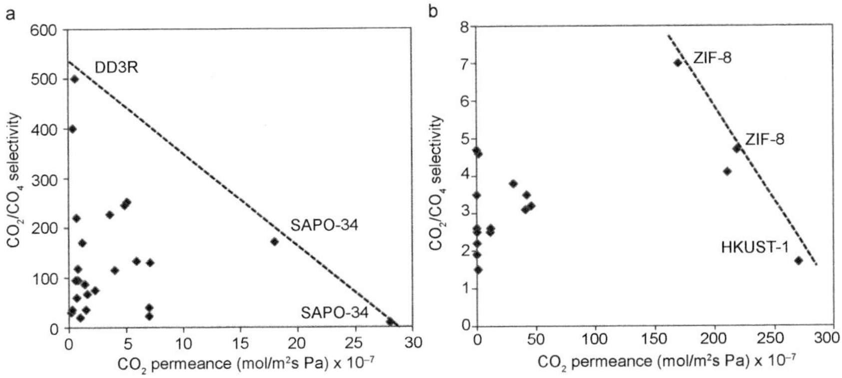

Figure 4.8. CO_2/CH_4 separation selectivity versus CO_2 permeance for (a) zeolite and (b) MOF membranes. Reproduced with permission (Venna and Carreon 2015), Copyright 2015 Elsevier.

Mixed-matrix Membranes

Polymeric membranes generally suffer the permeability-selectivity trade-off, such as the well-known 1991 and 2008 Robeson upper-bounds (Robeson 1991, 2008) for CO_2/CH_4 separation. A practically promising route to enhance the transport properties of polymers is the development of mixed-matrix membranes combining the processibility of polymers with the excellent transport properties of high-performing molecular sieves such as zeolite and MOFs described above (Koros and Zhang 2017). Mixed-matrix membranes were started in the late 1980s and have received increasing attention especially with the advancement of MOFs since 2010. Most mixed-matrix membrane studies focused on the materials design and dense film fabrication at small scale. Only a few efforts have been made to advance mixed-matrix membranes into asymmetric membranes, which is eventually demanded by the practical application.

Zeolite mixed-matrix membranes

Initial studies in dense films identified many material issues for optimizing the selection of molecular sieves and polymers for mixed-matrix membranes. Two critical challenges include aggregation of molecular sieves and defective sieve-polymer interfaces. Considerably more issues must be addressed when transitioning dense mixed-matrix films to asymmetric mixed-matrix membranes with minimized skin defects and attractive selectivity (Husain and Koros 2007). Ideally, asymmetric mixed-matrix membrane should show economically attractive selectivity and permeance that are simultaneously enhanced over the pure polymeric membrane. Meanwhile, the membrane should be easily and inexpensively processed. As shown in Fig. 4.9, Koros and co-workers (Zhang et al. 2014) defined desirable characteristics for developing scalable mixed-matrix hollow fiber membrane.

At the beginning, fabrication of mixed-matrix membranes was based on zeolites that required complex surface modifications to adhere with glassy polymer matrices. Husain and Koros 2007 compared two separate techniques to modify an

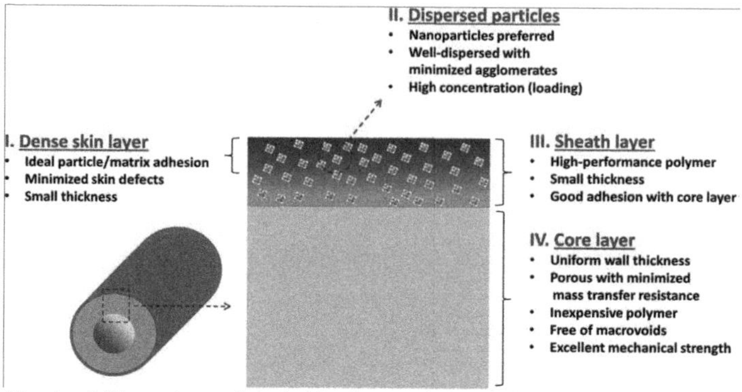

Figure 4.9. Structural illustration and desirable characteristics of dual-layer mixed-matrix hollow fiber membranes. Reproduced with permission (Zhang et al. 2014), Copyright 2014 John Wiley and Sons.

aluminosilicate zeolite, HSSZ-13, and incorporated these modified zeolites into an polyetherimide (Ultem) 1000 matrix to fabricate asymmetric mixed-matrix hollow fiber membranes. They found that increasing zeolite-polymer compatibility via silane modification and subsequent polymer "sizing" did not increase the permeation selectivity. It is due to the observed sieve-in-a-cage defects in the fibers, which might be generated by nucleation of non-solvent and/or polymer lean phase around the zeolite during the phase separation process. To restrict such nucleation, the hydrophobicity of the zeolite surface was increased by capping surface hydroxyls with hydrophobic organic molecules via a Grignard treatment. The asymmetric mixed-matrix hollow fibers incorporating 10.3 vol% (with respect to polymer) of Grignard treated zeolites, after post-treatment, showed selectivity improvement of 17% for pure gas and 25% for mixed-gas CO_2/CH_4 separation over pure polymer results.

MOF mixed-matrix membranes

Despite the progress that has been made, the moderate performance enhancements of zeolite-based mixed-matrix membranes did not justify the great efforts needed to scale-up the materials into asymmetric membranes. Compared with purely inorganic zeolites, MOF with organic linkers offers a more compatible interface with organic polymers, showing great potential in remarkable performance improvement, as well as scalable fabrication of asymmetric membranes. Vankelecom and co-workers (Basu et al. 2011) prepared asymmetric Matrimid mixed-matrix membranes containing three different MOFs [$Cu_3(BTC)_2$], ZIF-8 or MIL-53(Al) in flat sheet form via phase inversion process. An optimized priming approach of the fillers prior to membrane preparation resulted in a uniform dispersion of the MOFs in polymer matrix. Nevertheless, some non-selective voids were still present, which was overcome by coating a highly permeable silicone layer on top of the asymmetric membranes. The CO_2 permeance is nearly doubled while the CO_2/CH_4 selectivity was increased by 11-33% in the Matrimid-based asymmetric membranes by incorporating MOF particles with loading up to 30 wt%. The moderate improvement in selectivity might be due to the relatively large pore size of these MOFs that could not offer high selectivity for

CO_2/CH_4. Recently, Koros group (Zhang et al. 2014) successfully fabricated dual-layer ZIF-8/6FDA-DAM mixed-matrix hollow fiber membranes with ZIF-8 loading up to 30 wt % using the dry-jet/wet-quench spinning technique. The resulting mixed-matrix hollow fibers showed significantly enhanced C_3H_6/C_3H_8 selectivity that was consistent with mixed-matrix dense films. Although post-treatment is required at high MOF loading (30 wt%), this work represents a notable step in advancing the concept of mixed-matrix membranes beyond academic research.

2D-material mixed-matrix membranes

Two-dimensional materials, including graphene-based materials and clays such as montmorillonite (MT), having oxygen-containing groups and inter-layer spaces, can be utilized as CO_2 selective-transport channels. Alignment of the interlayer channels to enable high-speed molecular transport pathways is continuously pursued. Qiao et al. 2016 reported highly permeable aligned MT/PSf mixed-matrix membranes for CO_2 separation. As shown in Fig. 4.10, the MT framework is composed of a two-layer silica tetrahedron and a single-layer alumina octahedron. Aligned MT (AMT) layers with thickness of ~ 100 nm were attached onto a porous polysulfone (PSf) substrate (pore size of ~ 45 nm) using polyvinylamineacid as a linker and an alignment agent for MT. MT was exchanged by Na ions to form Na-exchanged MT, and the surface-hydroxylated PSf substrate was grafted with polyvinylamineacid. AMT/PSf

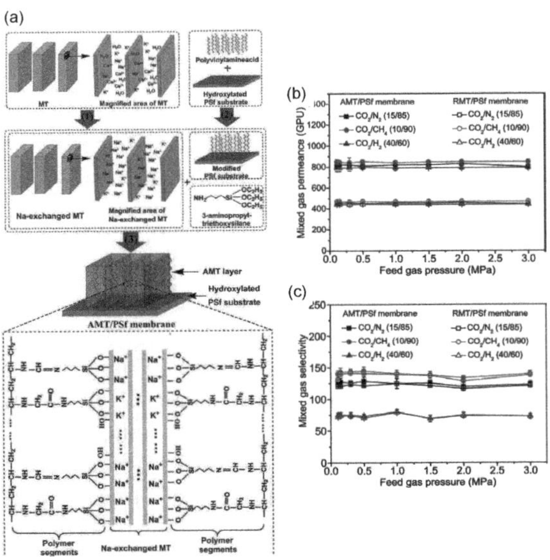

Figure 4.10. (a) formation of the AMT/PSf membrane: (1) Na ion exchange to form Na-exchanged MT; (2)surface modification by grafting polyvinylamineacid onto the hydroxylated PSf substrate to form the modified PSf substrate; (3) immobilization of Na-exchanged MT onto the modified PSf substrate through uncoiled polyvinylamineacid to obtain the AMT/PSf membrane; (4) one segment of an AMT layer; (b) CO_2 permeance and (c) selectivity of the AMT/PSf (filled symbols) and RMT/PSf (open symbols) membranes tested with gas mixtures. AMT/PSf has a much higher gas permeance than RMT/PSf, but a similar gas pair selectivity. Reproduced with permission (Qiao et al. 2016), Copyright 2016 John Wiley and Sons.

mixed-matrix membranes were obtained by intercalating Na-exchanged MT with polyvinylamineacid using 3-aminopropyltriethoxysilane as a coupling agent. It was found that the AMT interlayer spaces are aligned along the stretching orientation of the polyvinylamineacid chains, which served as high-speed CO_2 transport channels. High CO_2 permeance of ~ 800 GPU was achieved combined with a high mixed-gas CO_2/CH_4 selectivity of ~ 140 at 1–30 bar and 50°C, and the performance is stable over 600 h and independent of the water content in the feed (Fig. 4.10 b–c).

Carbon Molecular Sieve Membranes

Molecular sieving materials, such as inorganic zeolites, MOFs and Carbon Molecular Sieves (CMS), have shown great potential to exceed the polymer performance limitation (e.g., Robeson upper-bound). Among them, CMS includes open micropores connected by rigid slit-like ultramicropores to provide high sorption and diffusion coefficients, along with the entropically enabled diffusion selectivity (Koros and Zhang 2017). In contrast to well-defined pore dimensions of crystalline zeolites and MOFs, CMS has distributions of micropores and ultramicropores, resulting in an amorphous structure. Nevertheless, reproducible permeability and selectivity can be obtained by using controlled precursors and pyrolysis conditions (Steel and Koros 2005). The ability to form CMS materials from easily processable polymeric precursors makes them an especially promising membrane candidate (Vu et al. 2002). In the past decade, significant efforts have been devoted to the development of CMS membranes.

Formation of CMS membranes

Wenz and Koros 2017 proposed the structure of CMS materials as follows: CMS are formed by the high temperature pyrolysis of polymer precursors such as polyimides that yield nongraphitized products with desirable molecular scale morphologies. The resulting CMS materials are envisioned to consist of highly disordered sp^2-hybridized graphene-like carbon sheets, with very little long-range order, so they are viewed as being essentially amorphous. In the short-range order, the sp^2-hyridized carbon sheets are thought to align into laminar structures, yielding a relatively ordered structure depicted in Fig.4.11a. Imperfect packing between the laminar sp^2-hybrized carbon sheets results in the molecular sieving structure of CMS membranes, which can be described by "slit-like" pores and represented via a bimodal distribution, shown in Fig. 4.11b. The edges of adjacent carbon laminates are believed to form the slit-like ultramicropores (< 7 Å), while larger micropore (7–20 Å) "galleries" are formed by packing imperfections between the planes of adjacent carbon sheets. The micropores provide sorption sites for penetrant gases, yielding highly productive materials, while the ultramicropores are responsible for molecular size selectivity. Moreover, the relatively long jump lengths between the equilibrium micropore sorption sites generate high diffusion coefficients. The combination of characteristic pores yields a microstructure allowing CMS to surpass the upper-bound limit for conventional polymers.

Generally, CMS membranes are formed by pyrolyzing polymer precursors under vacuum or inert gas. The initial step of pyrolysis at temperature between

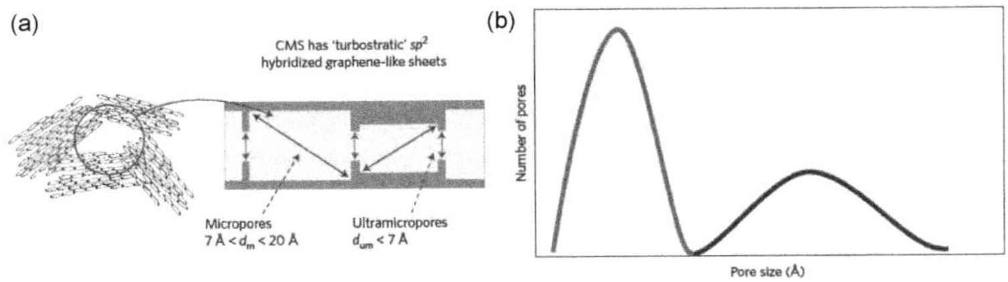

Figure 4.11. (a) Representation of CMS materials with diffusion jump length set by (b) micropore distribution and molecular size discrimination set by ultramicropore distribution. Reproduced with permission (Koros and Zhang 2017), Copyright 2017 Nature Publishing Group.

300 and 500°C involves bond cleavage, hydrogen transfer and formation of free radicals, followed by intramolecular coupling and intermolecular cross-linking of the generated radicals. At ~ 500°C, dehydrogenation starts and aliphatic carbon is converted to aromatic carbon, resulting in higher C/H ratio. Separate conjugated systems become interconnected to form disordered carbon lamellae. Meanwhile, small molecules such as H_2O, CH_4, CO_2, CO and H_2 are released and thus microporosity within the rigid macromolecular system is generated. The carbonization of polymer is complicated and dependent on various precursors and pyrolysis conditions. The pore size distribution and the transport properties of resulting CMS membranes are controlled by several key parameters including polymer precursor, pre-treatment, heating protocol and pyrolysis atmosphere.

Tuning structures of CMS membranes

CMS membrane is derived from easily processed polymeric membranes that can be formed into the practically desirable asymmetric membranes, mostly in hollow fiber form. However, gas permeance has been much lower than expected when translating from dense film to hollow fiber. The low permeance in asymmetric CMS hollow fiber membranes is attributed to the densification of microporous sub-structure morphology, with an increase in effective selective layer thickness during pyrolysis of polymer precursor. To address this great challenge, Koros's group (Bhuwania et al. 2014) developed a pre-pyrolysis treatment to restrict the morphology collapse in asymmetric CMS hollow fiber membranes, which was referred as V-treatment. The idea is using a sol-gel crosslinking reaction between vinyltrimethoxysilane and moisture to introduce vinyl crosslinked silica on precursor fiber pore walls, prohibiting the collapse of porous support during pyrolysis (Fig. 4.12). The V-treatment was successfully applied for two widely studied polyimide precursors: Matrimid and 6FDA:BPDA-DAM (BPDA: 3,3'-4,4'-biphenyl tetracarboxylic acid), as shown in Figs. 4.13a and 4.13 b. Significant reductions up to 5–6 folds in apparent membrane skin thickness were observed compared to the CMS fiber from untreated precursors. As a result, CO_2 permeance was increased by 3–4 times for both pure and mixed-gas CO_2/CH_4 feeds in V-treated Matrimid and 6FDA:BPDA-DAM CMS hollow fiber membranes. Ideally, the V-treatment not involving any chemical reaction with precursor should be a general approach to restrict morphology collapse of all asymmetric CMS polymer precursors.

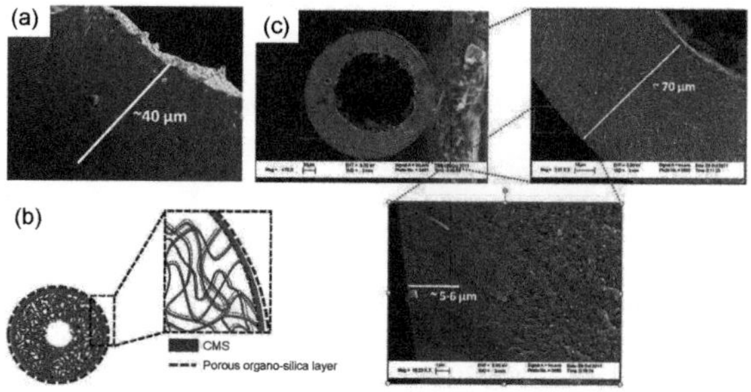

Figure 4.12. (a) SEM images of CMS hollow fiber membrane wall from untreated Matrimid precursors pyrolyzed at 550°C; (b) schematic representation of silicon distribution on the asymmetric CMS V-treated structure; (c) SEM images of CMS hollow fiber from V-treated Matrimid precursor showing improved substructure morphology. Reproduced with permission (Bhuwania et al. 2014), Copyright 2014 Elsevier.

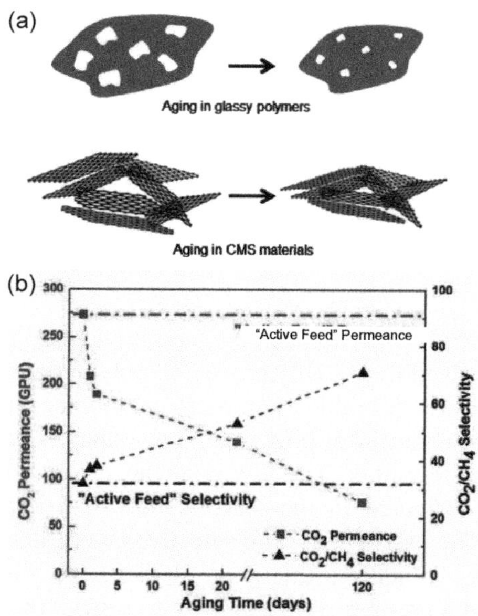

Figure 4.13. (a) schematic cartoon representation of envisioned physical ageing in glassy polymers and CMS materials (Xu et al. 2014a); (b) Ageing of 6FDA/BPDA-DAM 550 °C CMS fibers under vacuum. Pure gas permeation at 3.45 bar and 35°C. Horizontal dash-dot lines indicate the approximate permeance (red, upper) and selectivity (black, lower) values for a CMS fiber under "active feed" conditions. Reproduced with permission (Wenz and Koros 2017), Copyright 2017 John Wiley and Sons.

It has been noticed that rapid changes of transport properties at the early stage after the CMS membrane fabrication. This phenomenon was observed in both dense film and asymmetric hollow fiber membranes (Xu et al. 2014a). As shown in Fig. 4.13b, initial CO_2/CH_4 separation performance of 6FDA:BPDA-DAM-derived CMS hollow fiber membranes is a CO_2 permeance of 273 GPU and an ideal selectivity of 32. After exposure of the membrane to active vacuum for 120 days, the CO_2 permeance dropped to 76 GPU while the CO_2/CH_4 selectivity increased to 71 (Wenz and Koros 2017). Physical ageing appears to be the primary cause, as demonstrated by recent study of Koros and co-workers. The CMS pores are believed to age analogously to the "unrelaxed free volume" in glassy polymers (Fig. 4.13a) (Vu et al. 2002). Over time, these pores tend to shrink so as to achieve thermodynamically more stable states. Sorption measurements indicated decreases of Langmuir sorption capacity over time, supporting the hypothesis of physical aging of CMS materials. Despite of this, such ageing has been shown to be readily suppressed under active feeds in practical applications.

Conclusions and Perspectives

In conclusion, this chapter discussed the state-of-the-arts asymmetric membranes for natural gas purification with emphasis on CO_2/CH_4 separation. Polymeric membranes still dominated the gas separation membrane market owing to the distinct advantages in low-cost and large-scale fabrication of asymmetric membranes either in flat sheet or hollow fiber form. Inorganic zeolite membranes with rigid and well-defined subnanosized pores having molecular sieving property, generally exhibited 5–10 folds higher separation performance than conventional polymeric membranes. Mixed-matrix membranes integrating the polymers and molecular sieves via a facile and economic approach, showed great potential in scalable improving the selectivity and/or permeance of current polymeric membranes. CMS membranes offer simpler economical processing options than pure zeolites or MOFs to achieve attractive selectivity and permeability properties beyond polymers. Despite the importance, removing H_2S from natural gas using asymmetric membranes has rarely been studied, which would require more robust membranes due to the strong corrosivity of H_2S.

The key desirable properties for successful membranes include high permeance and selectivity, good processability and stability, as well as low cost. Future research and development effort would ideally provide more durable membranes with improved performance and can be easily scaled-up to fulfil these membranes applied for natural gas purification. Specifically, it is still worthwhile to improve the permeance meanwhile maintain sufficient plasticization-resistance of polymeric membranes. Current fabricating methods would be modified to realize the translation of new-generation polymers into asymmetric membranes. More studies need to be performed to improve the scalable fabrication while reduce the cost of inorganic membranes. Although the advancement of mixed-matrix membranes is mainly determined by the high-performing molecular sieves, matching the compatibility and transport properties between the matrix and fillers is still crucial to highly enhance the separation performance of polymer. Fully translating the ultra-high permeability and selectivity of dense film into highly permeable and selective asymmetric membranes would strongly push CMS membranes into real-world application. Finally, it is always pursued for making a defect-free membrane as thin as possible to maximize the productivity of membrane process, which riles on both membrane materials and fabrication techniques. The emerging two-dimensional materials with atomic thickness, such as graphene oxide, that can be simply stacked on porous supports to develop membranes with ultra-thin selective layers (< 10 nm) (Liu et al. 2016a, Wang et al. 2016b), might provide a promising platform.

Acknowledgements

This work was financially supported by the National Natural Science Foundation of China (Grant Nos. 21776125, 21490585, 21476107), National Key Basic Research Program (2017YFB0602500), the Innovative Research Team Program by the Ministry of Education of China (Grant No. IRT17R54). Dr. Xu is sponsored by China Regenerative Medicine International (CRMI) through a technology centre grant.

References

Baker, R.W. and Lokhandwala, K. 2008. Natural Gas Processing with Membranes: An Overview. Ind. Eng. Chem. Res. 47: 2109–2121.

Baker, R.W. and Low, B.T. 2014. Gas Separation Membrane Materials: A Perspective. Macromolecules 47: 6999–7013.

Basu, S., Cano-Odena, A. and Vankelecom, I.F.J. 2011. MOF-containing mixed-matrix membranes for CO2/CH4 and CO2/N2 binary gas mixture separations. Sep. Purif. Technol. 81: 31–40.

Bhuwania, N., Labreche, Y., Achoundong, C.S.K., Baltazar, J., Burgess, Karwa, S.K. et al. 2014. Engineering substructure morphology of asymmetric carbon molecular sieve hollow fiber membranes. Carbon 76: 417–434.

Budd, P.M. and McKeown, N.B. 2010. Highly permeable polymers for gas separation membranes. Polym. Chem. 1: 63–68.

Carreon, M.A., Li, S., Falconer, J.L. and Noble, R.D. 2008. SAPO-34 Seeds and Membranes Prepared Using Multiple Structure Directing Agents. Adv. Mater. 20: 729–732.

Chen, Y., Zhang, Y., Zhang, C., Jiang, J. and Gu, X. 2017. Fabrication of high-flux SAPO-34 membrane on α-Al2O3 four-channel hollow fibers for CO_2 capture from CH4. J CO_2 Util. 18: 30–40.

Chung, S.J., Park, J.H., Li, D., Ida, J.I., Kumakiri, I. and Lin, J.Y.S. 2005. Dual-Phase Metal–Carbonate Membrane for High-Temperature Carbon Dioxide Separation. Ind. Eng. Chem. Res. 44: 7999–8006.

Cui, Y., Kita, H., Okamoto, K.-i., 2004. Preparation and gas separation performance of zeolite T membrane. J. Mater. Chem. 14: 924-932.

Davey, C., Leak, D., Patterson, D., 2016. Hybrid and Mixed Matrix Membranes for Separations from Fermentations. Membranes 6, 17.

Fane, A.G., Wang, R. and Jia, Y. 2011. Membrane technology: past, present and future. pp. 1–45. *In*: Wang, L.K., Chen, J.P., Hung, Y.-T. and Shammas, N.K. (eds.). Membrane and Desalination Technologies. Humana Press, Totowa, NJ.

Han, S.H. and Lee, Y.M. 2011. Chapter 4 Recent High Performance Polymer Membranes for CO_2 Separation, Membrane Engineering for the Treatment of Gases: Volume 1: Gas-separation Problems with Membranes. The Royal Society of Chemistry, pp. 84–124.

Hillock, A.M.W. and Koros, W.J. 2007. Cross-Linkable Polyimide Membrane for Natural Gas Purification and Carbon Dioxide Plasticization Reduction. Macromolecules 40: 583–587.

Himeno, S., Tomita, T., Suzuki, K., Nakayama, K., Yajima, K. and Yoshida, S. 2007. Synthesis and Permeation Properties of a DDR-Type Zeolite Membrane for Separation of CO_2/CH_4 Gaseous Mixtures. Ind. Eng. Chem. Res. 46: 6989–6997.

Houde, A.Y., Krishnakumar, B., Charati, S.G. and Stern, S.A. 1996. Permeability of dense (homogeneous) cellulose acetate membranes to methane, carbon dioxide, and their mixtures at elevated pressures. J. Appl. Polym. Sci. 62: 2181–2192.

Husain, S. and Koros, W.J. 2007. Mixed matrix hollow fiber membranes made with modified HSSZ-13 zeolite in polyetherimide polymer matrix for gas separation. J. Membr. Sci. 288: 195–207.

Jue, M.L., Breedveld, V. and Lively, R.P. 2017. Defect-free PIM-1 hollow fiber membranes. J. Membr. Sci. 530: 33–41.

Kim, S., Han, S.H. and Lee, Y.M. 2012. Thermally rearranged (TR) polybenzoxazole hollow fiber membranes for CO2 capture. J. Membr. Sci. 403: 169–178.

Koros, W.J. and Zhang, C. 2017. Materials for next-generation molecularly selective synthetic membranes. Nat. Mater. 16: 289–297.

Leung, D.Y.C., Caramanna, G. and Maroto-Valer, M.M. 2014. An overview of current status of carbon dioxide capture and storage technologies. Renw. Sust. Energ. Rev. 39: 426–443.

Li, S., Alvarado, G., Noble, R.D. and Falconer, J.L. 2005. Effects of impurities on CO_2/CH_4 separations through SAPO-34 membranes. J. Membr. Sci. 251: 59–66.

Li, S., Carreon, M.A., Zhang, Y., Funke, H.H., Noble, R.D. and Falconer, J.L. 2010. Scale-up of SAPO-34 membranes for CO_2/CH_4 separation. J. Membr. Sci. 352: 7–13.

Li, S., Falconer, J.L. and Noble, R.D. 2004. SAPO-34 membranes for CO_2/CH_4 separation. J. Membr. Sci. 241: 121–135.

Li, S., Falconer, J.L. and Noble, R.D. 2006. Improved SAPO-34 Membranes for CO_2/CH_4 Separations. Adv. Mater. 18: 2601–2603.

Liu, G., Jin, W. and Xu, N. 2016a. Two-Dimensional-Material Membranes: A New Family of High-Performance Separation Membranes. Angew. Chem. Int. Ed. 55: 13384–13397.

Liu, G., Li, N., Miller, S.J., Kim, D., Yi, S., Labreche, Y. et al. 2016b. Molecularly Designed Stabilized Asymmetric Hollow Fiber Membranes for Aggressive Natural Gas Separation. Angew. Chem. Int. Ed. 55: 13754–13758.

Ma, C. and Koros, W.J. 2013a. Ester-Cross-linkable Composite Hollow Fiber Membranes for CO_2 Removal from Natural Gas. Ind. Eng. Chem. Res. 52: 10495–10505.

Ma, C. and Koros, W.J. 2013b. High-performance ester-crosslinked hollow fiber membranes for natural gas separations. J. Membr. Sci. 428: 251–259.

Omole, I.C., Adams, R.T., Miller, S.J. and Koros, W.J. 2010. Effects of CO_2 on a High Performance Hollow-Fiber Membrane for Natural Gas Purification. Ind. Eng. Chem. Res. 49: 4887–4896.

Omole, I.C., Miller, S.J. and Koros, W.J. 2008. Increased molecular weight of a cross-linkable polyimide for spinning plasticization resistant hollow fiber membranes. Macromolecules 41: 6367–6375.

Park, H.B., Jung, C.H., Lee, Y.M., Hill, A.J., Pas, S.J., Mudie, S.T. et al. 2007. Polymers with cavities tuned for fast selective transport of small molecules and ions. Science 318: 254–258.

Qiao, Z., Zhao, S., Wang, J., Wang, S., Wang, Z. and Guiver, M.D. 2016. A Highly Permeable Aligned Montmorillonite Mixed-Matrix Membrane for CO2 Separation. Angew. Chem. Int. Ed. 55: 9321–9325.

Qiu, W., Chen, C.-C., Xu, L., Cui, L., Paul, D.R. and Koros, W.J. 2011. Sub-Tg Cross-Linking of a Polyimide Membrane for Enhanced CO_2 Plasticization Resistance for Natural Gas Separation. Macromolecules 44: 6046–6056.

Qiu, W., Xu, L., Chen, C.-C., Paul, D.R. and Koros, W.J. 2013. Gas separation performance of 6FDA-based polyimides with different chemical structures. Polymer 54: 6226–6235.

Robeson, L.M. 1991. Correlation of separation factor versus permeability for polymeric membranes. J. Membr. Sci. 62: 165–185.

Robeson, L.M. 2008. The upper bound revisited. J. Membr. Sci. 320: 390–400.

Staudt-Bickel, C. and J. Koros, W. 1999. Improvement of CO_2/CH_4 separation characteristics of polyimides by chemical crosslinking. J. Membr. Sci. 155: 145–154.

Steel, K.M. and Koros, W.J. 2005. An investigation of the effects of pyrolysis parameters on gas separation properties of carbon materials. Carbon 43: 1843–1856.

Venna, S.R. and Carreon, M.A. 2015. Metal organic framework membranes for carbon dioxide separation. Chem. Eng. Sci. 124: 3–19.

Visser, T., Masetto, N. and Wessling, M. 2007. Materials dependence of mixed gas plasticization behavior in asymmetric membranes. J. Membr. Sci. 306: 16–28.

Vu, D.Q., Koros, W.J. and Miller, S.J. 2002. High Pressure CO_2/CH_4 Separation using carbon molecular sieve hollow fiber membranes. Ind. Eng. Chem. Res. 41: 367–380.

Wang, L., Zhang, C., Gao, X., Peng, L., Jiang, J. and Gu, X. 2017. Preparation of defect-free DDR zeolite membranes by eliminating template with ozone at low temperature. J. Membr. Sci. 539: 152–160.

Wang, S., Li, X., Wu, H., Tian, Z., Xin, Q., He, G. et al. 2016a. Advances in high permeability polymer-based membrane materials for CO2 separations. Energy Environ. Sci. 9: 1863–1890.

Wang, S., Wu, Y., Zhang, N., He, G., Xin, Q., Wu, X. et al. 2016b. A highly permeable graphene oxide membrane with fast and selective transport nanochannels for efficient carbon capture. Energy Environ. Sci. 9: 3107–3112.

Wenz, G.B. and Koros, W.J. 2017. Tuning carbon molecular sieves for natural gas separations: A diamine molecular approach. AIChE J. 63: 751–760.

Wind, J.D., Paul, D.R. and Koros, W.J. 2004. Natural gas permeation in polyimide membranes. J. Membr. Sci. 228: 227–236.

Wu, T., Wang, B., Lu, Z., Zhou, R. and Chen, X. 2014. Alumina-supported AlPO-18 membranes for CO_2/CH_4 separation. J. Membr. Sci. 471: 338–346.

Xu, L., Rungta, M., Hessler, J., Qiu, W., Brayden, M., Martinez, M. et al. 2014a. Physical aging in carbon molecular sieve membranes. Carbon 80: 155–166.

Xu, L., Zhang, C., Rungta, M., Qiu, W., Liu, J. and Koros, W.J. 2014b. Formation of defect-free 6FDA-DAM asymmetric hollow fiber membranes for gas separations. J. Membr. Sci. 459: 223–232.

Zhang, C., Zhang, K., Xu, L., Labreche, Y., Kraftschik, B. and Koros, W.J. 2014. Highly scalable ZIF-based mixed-matrix hollow fiber membranes for advanced hydrocarbon separations. AIChE J. 60: 2625–2635.

Zhang, Y., Sunarso, J., Liu, S. and Wang, R. 2013. Current status and development of membranes for CO_2/CH_4 separation: A review. Int. J. Greenh. Gas Con. 12: 84–107.

Zhong, S., Bu, N., Zhou, R., Jin, W., Yu, M. and Li, S. 2016. Aluminophosphate-17 and silicoaluminophosphate-17 membranes for CO2 separations. J. Membr. Sci. 520: 507–514.

5

Membranes for Post-combustion CO_2 Capture

How Far to Commercialization

*Xuezhong He**

INTRODUCTION

The International Energy Outlook 2011 (IEO 2011) reference case reported that world energy-related carbon dioxide (CO_2) emissions would increase from 30.2 billion metric tons in 2008 to 35.2 billion metric tons in 2020 and 43.2 billion metric tons in 2035. Control of anthropogenic emissions of greenhouse gases (GHG), especially CO_2, is one of the most challenging environmental issues related to global climate change. Three different options could be employed to reduce CO_2 emissions, i.e., (1) improve energy efficiency in industrial processes, (2) switch to use less carbon-intensive and renewable energy processes, and (3) CO_2 capture and storage (CCS). The global energy consumption will grow by 53% from 2008 to 2035 as reported by IEO 2011 highlight, and the renewable energy sources are not sufficient to meet the increased energy demands in the world due to the limitation of technology development and cost of the renewable energy (Hoffert et al. 2002). Thus, CCS is considered as the most promising way to combat global warming by reducing CO_2 emission to the atmosphere. The main application of CCS is likely to be at large CO_2 point sources: fossil fuel power plants and industrial plants (particularly the manufacture of iron, steel, refinery, cement and chemicals, and natural gas/biogas plants). Among them, fossil fuel power plants are responsible for the largest CO_2

Norwegian University of Science and Technology, Department of Chemical Engineering, Sem Sælands Vei 4, Trondheim, Sør-Trøndelag, Norway, 7034.
Email xuezhong.he@ntnu.no

emissions of 78%, followed by cement factory (7%), refinery (6%), and iron/steel plant 4.8% (2005), and post-combustion power plants being the main contributor which need to be firstly tackled. Moreover, CO_2 capture from the exhaust gases in process industries, such as cement factory, refinery, iron and steel production plants should also receive attention due to its large CO_2 amount.

Different technologies such as chemical absorption (e.g., monoethanolamine (MEA), methyldiethanolamine (MDEA)), physical absorption (e.g., Selexol, Rectisol), physical adsorption (e.g., molecular sieves, metal organic frameworks), cryogenic distillation and membrane separation have the potential to be used for CO_2 capture from flue gas in power plant and off-gas from industry (Brunetti et al. 2010, D'Alessandro et al. 2010, He et al. 2013c, Samanta et al. 2011). Chemical absorption is one of the mature technology for post-combustion CO_2 capture. Some large scale CO_2 capture plants have been built up around the world to demonstrate the process feasibility of CO_2 capture from flue gas as summarized in the Table 5.1 (2017b), and these contributions can promote the amine-based CO_2 capture system to be commercialized in the near future.

However, conventional amine absorption technology still faces the challenges of energy intensive and solvent degradation and emissions, which causes a large incremental cost and a high environmental impact. Some emerging separation technologies based on the novel absorbents of ionic liquids (high CO_2 solubility) and solid sorbents of microporous materials (solid adsorbents) such as zeolite, metal organic frameworks (MOFs) and metal oxides (chemical looping) have been recently investigated for CO_2 capture. These 3rd generation materials showed a good potential and cost reduction benefit, but most of them are in the early research phase. For more detail and in-depth understanding of absorption and adsorption materials for CO_2 capture, the reader can refer to the previous reviews (Brunetti et al. 2010, D'Alessandro et al. 2010, He and Hägg 2012a, He et al. 2013c, Samanta et al. 2011, Yu et al. 2012, Zhang et al. 2012).

Table 5.1. Examples of worldwide large scale post-combustion CO_2 capture plants using chemical absorption (> 0.1 Mt CO_2/yr) (adapted from literature (2017b)).

Project*	Location	CO_2 sources	Capacity (Mt CO_2/yr)	Status
Technology Centre Mongstad (TCM)	Mongstad, Norway	Refinery, and Natural gas power plant (Statoil)	0.1	Operative
Boundary Dam integrated CCS plant	Saskatchewan, Canada	Coal-fired power plant (Saskpower)	1.0	Operative
Petra Nova Carbon Capture	Texas, USA	Coal-fired power plant (WA Parish Generating Station)	1.4	Operative
ROAD	Rotterdam, The Netherland	Coal and biomass fired power plant (Uniper Benelux)	1.1	Advanced development
Sinopec Shengli Power Plant CCS	Shangdong, China	Coal-fired power plant (Sinopec)	1.0	Advanced development

* Projects at the early stage development are not included.

Membrane separation has already been considered as an alternative and competitive technology for selected gas separation processes such as air separation and natural gas sweetening during the last two or three decades. Strong interest was put on CO_2 capture using gas separation membranes in the last decade, examples can be found in the literature (Bredesen et al. 2004, Carapellucci and Milazzo 2003, Deng et al. 2009, Hagg and Lindbrathen 2005, Huang et al. 2008, Lin and Freeman 2005, Reijerkerk 2010, Sandru et al. 2010, Yang et al. 2008). However, there are still some challenges related to the membranes for post-combustion CO_2 capture, e.g., the limitation of membrane separation performance (the trade-off of permeance and selectivity existed in most polymeric membranes), membrane stability and lifetime. Thus, high performance membranes with relatively low production cost should be developed. Moreover, the membranes should also possess long-term stability by being exposed to the acid gases of SO_2, NO_x and water as well as some other impurities. In 2006, a large EU project (NanoGLOWA) launched to develop a high performance Fixed-Site-Carrier (FSC) membrane (developed by membrane research group at NTNU) for CO_2 capture from flue gas in power plants. A small pilot-scale membrane system was tested at Sines coal-fired power plant in Portugal over 8 months in 2011 (Sandru et al. 2013). Later on, a pilot hollow fiber FSC membrane system (membrane area 20 m^2) was built up and tested at Norcem cement factory where the CO_2 feed concentration is ca. 17 mol.% (wet-base) since 2015 at Brevik, Norway (Hagg et al. 2016). Moreover, MTR (Membrane Technology & Research, Inc.) built up a pilot-scale membrane system for post-combustion carbon capture in a 1 MW coal-fired power plant using the high permeable Polaris™ membranes. The PolyActive™ membranes from Helmholtz-Zentrum Geesthacht was also tested for CO_2 capture from real flue gas in the pilot scale (Pohlmann et al. 2016). Those efforts significantly promote to bring gas separation membrane technology into the commercial CO_2 capture application in the near future. In this work, we are aiming at providing a review on the recent progress of membrane materials and process development related to post-combustion CO_2 capture from power plants and industries was conducted.

Membrane Materials for CO_2 Separation

Various membranes such as common polymer membranes, Microporous Organic Polymers (MOPs), Fixed-Site-Carrier (FSC) membranes, Mixed Matrix Membranes (MMMs) can be used for CO_2 separation (Ramasubramanian and Ho 2011). Different membrane materials possess various separation properties, thermal and chemical stability, mechanical strength as well as production cost, and presents their own suitable applications. Most polymeric membranes based on solution-diffusion transport mechanism suffer the trade-off of gas permeability and selectivity, and the relatively low stability or short lifetime when exposed to acid gases (e.g., SO_2 and NO_x). Seeking novel high performance membrane materials is crucial to bring membrane technology to future commercial application in CO_2 capture. The FSC membranes presented quite good membrane performance, the engineering challenge to maintain high water vapor content in the gas stream should be addressed. MMMs are important research field, but the compatibility between the polymer and filler, and scaling-up are the main challenges in this regard. Choosing a suitable membrane material for a specific application mainly depends on membrane material properties, feed gas composition/impurities, process operating conditions as well the separation requirements. Recently, membrane separation performance has been significantly improved owning to the great effort that has been taken in the membrane community to develop novel materials. Even though most membrane materials are still in the fundamental research and will take quite long time to bring into any commercial application or may not be successful in the end, there are some membranes that are quite promising for CO_2 separation form flue gas due to the high performance and the good stability. Among them, the FSC membranes developed by Norwegian University of Science and Technology (NTNU) (He et al. 2017b), the PolyActive™ membranes from Helmholtz-Zentrum Geesthacht (Pohlmann et al. 2016), and the Polaris membrane developed by Membrane Technology and Research, Inc. (MTR) are the frontiers in post-combustion CO_2 capture using membrane technology.

Common polymer membranes

Gas permeability and selectivity are the two key parameters to characterize separation performance of a dense polymer membrane, and should be as high as possible to achieve separation requirements with lower costs. However, there is a trade-off between permeability and selectivity in common polymer membranes as reported by . Gas permeability is mainly dependent on a thermodynamic factor (solubility (*S*) of penetrants into a membrane) and a kinetic factor (diffusivity (*D*) of the gas species transport through a membrane) (Baker 2004). Many researches have used the more polar nature of CO_2 molecule to increase its solubility and therefore, its permeability since the size difference between CO_2 and N_2 molecules are quite small to get sufficient high diffusivity selectivity (Ramasubramanian and Ho 2011). The Poly Ethylene Oxide (PEO) based block copolymer materials (e.g., Pebax®) are widely used to develop membranes for CO_2 capture. (Car et al. 2008, Scholes et al. 2016, Yave et al. 2010). Highly ordered block segments was developed by University of Twente (Reijerkerk et al. 2010), they reported a high CO_2 permeability of 530 Barrer in the PEBAX®1657/PDMS–PEG blend membranes. The Polaris™ membrane which is a thin-film composite (TFC) structure based on Pebax polyether-polyamide copolymers (Figueroa et al. 2008) has a CO_2 permeance of 1000 GPU (1000 GPU = 2.76 m^3(STP)/($m^2 \cdot h \cdot$bar)) with a CO_2/N_2 selectivity of 40–50 (Merkel et al. 2010), and the membrane performance has been recently improved and reach CO_2 permeance 2200 GPU and a CO_2/N_2 selectivity of 50 (Merkel et al. 2013). The Polyactive® (poly(butylene terephthalate)) has relatively low CO2 permeability of 100–200 Barrer (Metz et al. 2004), but the very thin nanocomposite membrane has been reported to achieve CO_2 permeance 2000 GPU with CO_2/N_2 selectivity 60 (W et al. 2010). Those investigation results indicated that extremely thin defect-free films of less than 50 nm thickness could be produced by dip-coating and scaled up successfully in future (Ramasubramanian and Ho 2011).

Microporous organic polymers

Strong interests have been put on development of Microporous Organic Polymers (MOPs) due to its large surface area which can be comparable to typical microporous inorganic materials such as zeolites and carbons. The representative MOPs include thermally rearranged (TR) polymers (Han et al. 2012, Kim and Lee 2012, Park et al. 2010, Park et al. 2007) and polymers of intrinsic microporosity (PIMs) (Ahn et al. 2010, Budd et al. 2004, Budd et al. 2008, Budd et al. 2005, Du et al. 2011, McKeown 2012, Thomas et al. 2009, Yong et al. 2013). Polyimide-based TR polymers with an average pore size 0.4–0.9 nm and a narrow pore size distribution was firstly prepared by Park et al. 2007, which presented a molecular sieving transport mechanism in gas permeation through a membrane. The flexible structures provide feasibility and easiness of module construction. Moreover, TR polymer membranes were also found to exhibit an excellent gas separation performance, especially for CO_2 related separation processes (e.g., CO_2/CH_4 separation without any significant plasticization effects) (Park et al. 2010) and also for high temperature H_2/CO_2 separation in pre-combustion process (Han et al. 2012). However, most efforts are still focused on preparation of lab-scale films of the TR polymer membranes, only a little literature reported the fabrication of hollow fiber TR membranes (Kim et al. 2012, Woo et al. 2015). Kim et al. 2012 prepared the lab-scale TR-PBO hollow fiber membranes with a CO_2 permeance of 1938 GPU and a CO_2/N_2 selectivity about 13 (Kim et al. 2012), and the selectivity obviously needs to be further improved to reach the industry attractive region.

Polymers of Intrinsic Microporosity (PIMs) attracted great interests due to their relatively slow physical ageing, high gas permeability, as well as high selectivity compared to poly(1-trimethylsilyl-1-propyne) (PTMSP) (Du et al. 2011). PIMs showed a high surface area (600–900 m^2/g) as reported by Budd et al. (Budd et al. 2004) and a high fractional free volume (22–24% (McKeown et al. 2005)) which is comparable to PTMSP (32–34.3% (Morisato et al. 1996, Pope et al. 1994)). Du et al. reported that PIMs functionalized with CO_2-philic pendant tetrazole groups (TZPIMs) can further improve CO_2 permeance by increasing CO_2 solubility due to a strong interaction between CO_2 and N-containing organic heterocyclic groups (Du et al. 2011). Their results indicated that CO_2/N_2 separation performance of TZPIMs can surpass the Robeson upper bound. Moreover, a systematic review on the preparation, characterization and application of PIMs has been conducted by McKeown (McKeown 2012). They pointed out that composite membrane consisting of PIMs and other polymers showed a promising strategy for tailoring membrane properties to improve gas separation performance.

Fixed-site-carrier membranes

Fixed-Site-Carrier (FSC) membranes for gas separation, especially for CO_2 removal from flue gas have attracted more attention due to its high CO_2 performance and high CO_2/N_2 selectivity. The carriers ($-NH_2$) are chemically bonded onto polymer matrix via covalent bonding, hence exhibiting a much higher stability compared to traditional Supported Liquid Membrane (SLM) and Emulsion Liquid Membrane

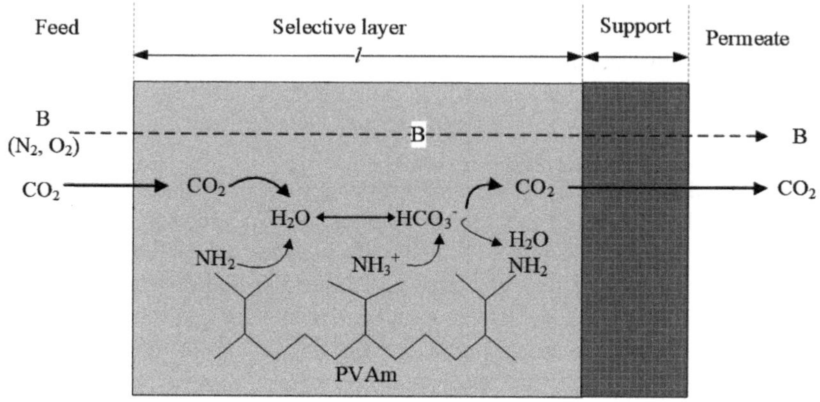

Figure 5.1. Gas transport through a PVAm-based FSC membrane (He 2011).

(ELM). CO_2 can react with amino functional groups when water is available and transport through membranes by a combination of Solution-Diffusion (S-D) and Facilitated Transport (FT) mechanisms, while the other non-reactive gas species such as N_2, O_2 can only transport via solution-diffusion as documented by Kim et al. 2004. A schematic diagram for the gas permeation through a FSC membrane is shown in Fig. 5.1 (He 2011). The gas flux of reactive component A (such as CO_2) will be the sum of both solution-diffusion and carrier-mediated diffusion (i.e., facilitated transport), which can be expressed as follows (Kim et al. 2004, Paul and Jampol'skij 1994)

$$J_A = \frac{D_A}{l}\left(c_{A,0} - c_{A,l}\right) + \frac{D_{AC}}{l}\left(c_{AC,0} - c_{AC,l}\right) \tag{1}$$

Where D_A and D_{AC} are diffusion coefficient of Fickian diffusion and carrier mediated (complex) diffusion, respectively, and l is membranes thickness of a selective layer. CO_2 will transport through the membranes based on the solution-diffusion and facilitated transport mechanism, and presents a high CO_2 permeance as illustrated in Fig. 5.1. CO_2 will react with amino neutral functional group when water is present, and diffuse through the matrix in the complex, and finally release CO_2 in the permeate side. While N_2 and O_2 can only transport based on solution-diffusion mechanism. Thus, high selectivities of CO_2 over other gas species are expected. It is also worth noting that feed pressure is crucial to get high flux by enhancing the contribution from both S-D and FT. However, after carrier saturation, further increasing feed CO_2 partial pressure will not enhance the FT. Even though CO_2 flux will continue increase due to S-D contribution, the trade-off between energy consumption and reduced membrane area (flux increase) should be identified to determine the optimal operating condition (He et al. 2017c). Thus, a moderate feed pressure (e.g., 2.5–3 bar) was recommended as the optimal operation condition of FSC membranes (He and Hägg 2014).

Table 5.2 shows some representative facilitated transport membranes that have been reported in the literature. Facilitated transport mechanism (Kreuzer and

Table 5.2. Representative fixed-site-carrier membranes for gas separation.

Material	Support	Membrane module	Gas separation	Reference
Poly(amidoamine)/ Poly(vinyl alcohol)	-	Flat sheet	CO_2/H_2	(Duan et al. 2012)
Polyallylamine (PAAm) / poly (vinyl alcohol) (PVA) blend	Polysulfone	Flat sheet	$CO_2/H_2/N_2/$ CO	(Zou and Ho 2006)
PVAm & PVAm/PVA blend	Polysulfone, polyphenylene oxide (PPO)	Flat sheet, Hollow fiber	CO_2/N_2, CO_2/CH_4	(Deng 2009, He et al. 2013b, Kim et al. 2004, Kim et al. 2013, Sandru, Uddin and Hägg 2012a, Uddin and Hägg 2012b)
PVA	-	Flat sheet	$CO_2/H_2/N_2$	(Huang et al. 2008)
CNT-reinforced PVAm/PVA blend	Polysulfone	Flat sheet	CO_2/CH_4	(He et al. 2014a, He and Hägg 2012b, He et al. 2014b)
High temperature ionic liquids	Nylon	Flat sheet	CO_2/H_2	(Myers et al. 2008)
Room temperature ionic liquids (RTILs)	PVDF	Flat sheet	CO_2/air, SO_2/air, $CO_2/N_2/H_2$	(Luis et al. 2009, Neves et al. 2009)

Hoofd 2011). Among them, the NTNU patented polyvinyl amine (PVAm)-based FSC membrane showed a very high CO_2 permeance (up to 5 m^3 (STP)/(m^2·h·bar)) and CO_2/N_2 (> 500) selectivity under humidified conditions (Kim et al. 2013). This membrane is extremely promising for CO_2 capture from flue gas in post combustion process where water vapor is usually involved in flue gas stream (He et al. 2015, He and Hägg 2014, Kim et al. 2013). The use of Ionic Liquids (ILs) as carrier to transport CO_2 in membrane separation processes is one of the fast growing research interests in the last years, and Supported Ionic Liquid Membranes (SILMs) are preferred to be used for CO_2 separation due to their high selectivity and permeability as well as relatively good mechanical stability compared to conventional liquid membranes (Cserjési et al. 2010, Hanioka et al. 2008, Kim et al. 2011, Neves et al. 2010, Uchytil et al. 2011, Zhao et al. 2012). However, most of ionic liquids are still not commercially available.

Mixed matrix membranes

MMMs comprise rigid permeable or impermeable particles, such as zeolites, carbon molecular sieves, silica and Carbon Nanotubes (CNTs), and disperse in a continuous polymeric phase to present interesting materials for improving separation performance of common polymer membranes (Aroon et al. 2010). Two types of inorganic fillers can be added into polymer matrix such as microporous fillers (e.g., carbon molecular sieves) and nonporous nanoparticles (e.g., SiO_2, TiO_2). MMMs with microporous fillers could improve selectivity based on molecular sieving or surface flow transport mechanism, and it might also be able to get an increased permeability if the preferred

Table 5.3. Latest polymers and inorganic fillers used for CO_2 selective MMMs.

Polymer matrix	Inorganic filler	Source
Poly(vinyl acetate) (PVAc)	Zeolite 4A, TiO_2	(Adams et al. 2011, Ahmad and Hägg 2013, Ahmad and Hågg 2013)
Polymers of intrinsic microporosity (PIMs)	ZIF-8, CNTs	(Bushell et al. 2013, Khan et al. 2013)
poly (ethylene oxide) (PEO)	Graphene oxide	(Quan et al. 2017)
Pebax	1 D multi-walled carbon nanotube/graphene oxide nanoribbon	(Pan et al. 2017)
PIMs	metal–organic framework (MOFs)	(Ghalei et al. 2017)

solid phase has a higher diffusion coefficient. While MMMs made from the adding of nonporous nanoparticles could improve gas permeability due to the increase of free volume. Chung et al. 2007 reported that the properties for both polymer materials and inorganic fillers could affect the morphology and separation performance of MMMs. The rigid structure glassy polymers with high selectivity are more suitable for polymer matrix compared to rubbery polymers. However, the adhesion between glassy polymer phase and inorganic filler phase is a challenging issue for preparation of MMMs. Moreover, the thermal and chemical stabilities of MMMs are mainly dependent on physical property of a polymer matrix, which may suffer from the acid gases of SO_2 or NOx that are usually involved in flue gas. MMMs normally present an enhanced mechanical strength compared to pure polymer membranes, and a reduced cost compared to pure inorganic membranes. However, the main challenge for preparation of MMMs is to choose proper materials for both polymeric and inorganic phases to get a high gas separation performance and good compatibility. Examples for selection of polymer and inorganic filler for making CO_2 selective MMMs are reviewed by He et al. 2013, and here only shows the latest materials in Table 5.3.

Membranes for CO_2 Capture in Power Plant

Gas membrane separation technology is an energy efficient and environmentally friendly process which has already been commercially used for many years for selected gas purification processes such as air separation and natural gas sweetening (He and Hägg 2012a, Yang et al. 2008), and is judged to be an alternative and competitive next generation CO_2 capture technology. Much effort is being put into the development of high performance membranes for this potential application, selected examples are given in the following references (Bredesen et al. 2004, Hagg and Lindbrathen 2005, He and Hägg 2011, He et al. 2009, He et al. 2013c, Kim et al. 2012, Kim et al. 2013, Labreche et al. 2014, Reijerkerk et al. 2011, Sandru et al. 2010, Tong and Ho 2016). However, there are some challenges related to the limited application of a membrane system in post combustion CO_2 capture. Flue gas in coal-fired power plant usually has a low feed pressure (a little over atmosphere pressure) and contains ca. 12–14 vol.% CO_2, and has an even lower CO_2 content (e.g., 4 vol.%) from natural gas fired power plants. Thus, the driving force for CO_2 transport through a standard membrane system (i.e., a polymeric membrane based on solution-diffusion mechanism) will be very low without using feed compression and/or vacuum suction. Moreover, the chemical stability by exposure to the impurities such as SO_2 and NO_x which usually exist in flue gas, may also be challenging related to membrane durability and lifetime. Therefore, a highly CO_2 permeable, selective and chemically stable membrane at low cost is required for a membrane system to compete with other CO_2 capture technologies, typically the benchmark amine absorption. The main focus on the development of membrane materials for post-combustion CO_2 capture is to produce high performance membranes (high CO_2 permeance and relatively good selectivity over other gas molecules) with long lifetime at a low cost.

The flat-sheet FSC membranes developed by NTNU has been tested in EDP's power plant in Sines (Portugal) and E.ON's plant in Scholven (Germany) in 2011, and the membranes showed a stable performance over 6 months (Hägg et al. 2012). Later on, the hollow fiber FSC membranes were tested at Sintef CO_2 laboratory at Tiller (Norway) with a flue gas produced from a propane burner. He et al. reported that single stage membrane system (8.4 m^2) can achieve > 60% permeate CO_2 purity at a feed and permeate pressure of 2 bar and 0.2 bar, respectively (He et al. 2017a), and the system also showed quite a fast response when changing feed CO_2 composition. The reported pilot FSC membrane system provided great flexibility on testing the influence of process operating parameters, especially the operating temperature, but the challenges related to the module and process design should be further investigated. In December 2016, Air Products and Chemicals, Inc. licensed the technology and will bring the FSC membranes for post-combustion carbon capture to a higher Technology Readiness Level (TRL), and the commercialization (2017a).

From the membrane material point of view, low pressure ratio (i.e., low feed pressure and low vacuum) is preferred to achieve high CO_2 permeance for the FSC membranes. However, from the engineering point of view, a relatively high pressure ratio (increasing driving force) will give higher CO_2 flux, and reduce the required membrane area. It is however important to balance this against the operating conditions where the facilitated transport can be of advantage. Thus, the trade-off between capital expenditure (CAPEX) related to the required membrane area and the operation expenditure (OPEX) related to power consumption of driving equipment) needs to be well balanced. He et al. (He et al. 2015, He and Hägg 2014, He et al. 2013a) and (Hussain and Hägg 2010) conducted the process feasibility analysis by HYSYS integrated with an in-house membrane program (ChemBrane, developed by Grainger (Grainger 2007)) to investigate the influence of process parameters on energy demand and flue gas processing cost using the CO_2-selective FSC membranes. Their results showed that membrane process using a high performance FSC membrane was feasible for CO_2 capture, even with a low CO_2 concentration (~ 10 %) in the feed flue gas, compared to the amine absorption in terms of the energy requirement, and it was possible to achieve > 90 % CO_2 recovery and a purity above 95 % CO_2 in permeate stream.

The PolyActive™ membranes developed by Helmholtz-Zentrum Geesthacht was tested for CO_2 capture from real flue gas in the pilot scale with a Membrane area of 12.5 m^2 (Pohlmann et al. 2016). The membrane system also shows stable performance over 740 hours continuously. They reported that membrane processes using PolyActive™ membranes seem to be well suited for post combustion CO_2 separation, and a CO_2 purity 68.2 mol.% in the permeate and a recovery of 42.7% can be achieved at the tested condition in single stage process.

The Polaris® membranes developed by Membrane Technology & Research, Inc. (MTR) has been demonstrated on the pilot scale for CO_2 capture from a Natural Gas Combined Cycle (NGCC) power plant (Casillas et al. 2015). A 20 TPD skid was tested to validate the advanced modules (multi-tube and plate-and-frame) designed for low pressure drop and small footprint, and the system showed quite stable performance over ca. 1000 hours (Merkel 2016). MTR patented their process by feeding high CO_2 content air stream (air as sweep gas in the permeate side of the 2nd stage membrane unit) into the boiler to increase the CO_2 concentration in the flue gas (Baker et al. 2009), which can potentially reduce the energy consumption by avoiding the vacuum pump. However, how the CO_2 contained air influences the boiler operation should be further tested.

Merkel et al. (Merkel et al. 2010) also pointed out that it is more important to reduce the cost of CO_2 capture from flue gas by improving membrane permeance instead of increasing selectivity (if selectivity > 30). They reported that membrane with a CO_2/N_2 selectivity above 50 and a 4000 GPU permeance could offer a capture cost below 15 US$/tonne CO_2, which is lower than US Department of Energy's (DOE) target goal of 20 US$/tonne CO_2. Even though the required high performance membrane has not yet been achieved, their researches emphasized quantitatively the need to improve the present membranes to realize a purely membrane-based process for CO_2 capture. Therefore, this environmentally friendly technique with further improved membrane performance could promote the membrane systems as a promising candidate for CO_2 capture from flue gas in post-combustion process if the above mentioned challenges can be well addressed.

Membranes for CO_2 Capture from Process Industry

CO_2 capture from the cement factory

The cement factory is pursuing solutions for carbon capture from high CO_2 content flue gas (ca. 17 vol.% wet base) as it represents 7% of global anthropogenic CO_2 emissions. Application of CO_2 capture from cement kilns would have great potential to reduce CO_2 emission from this industry, but will naturally influence cement production cost. Thus, the European cement industry (through Heidelberg Cement) is taking big interest in low-cost CCS technologies.

The cement production releases greenhouse gas emissions both directly and indirectly: the heating of limestone (calcination) directly releases ~ 50% of all CO_2 emission in cement production; the burning of fossil fuels to heat the kiln indirectly results in CO_2 emissions. Employment of CCS is considered as one of the most important techniques to achieve the Norcem Zero CO_2 Emission Vision 2030—a test site for carbon capture technologies is placed in Brevik, Norway, and funding of this project is mainly provided by the CLIMIT program through the Research Council of Norway. The project in Brevik was launched 2013 to test the process feasibility with four different technologies (amine absorption, membranes, solid adsorbent, and chemical looping). This is the first pilot-scale membrane system tested in a cement factory (He and Hägg 2015), and the PVAm based flat-sheet FSC membranes (developed by NTNU) was chosen for CO_2 capture from a 17 vol.% (wet base) CO_2 flue gas.

Many challenges related to the process and module design were revealed, and it was difficult to achieve a stable and high performance membrane system, but a CO_2 purity up to 72% was achieved for short periods when all process parameters were well controlled in the single stage FSC membrane system (Hägg et al. 2015). However, the membrane efficiency in the plat-and-frame module was quite low, and the designed system suffered water condensation/corrosion issues. Thus, the hollow fiber FSC membrane module with a pilot membrane area (ca. 20 m^2) was constructed by the joining force from Air Products and Chemicals, Inc. (MemCCC project funded by GASSNOVA) in 2016. In that project, the FSC membrane system was evaluated to be at "Technology readiness level" (TRL) level 5 (according to the EU-definition). Figure 5.2 shows a 3D simulation snapshot of the layout of the containerized pilot membrane system at Norcem, Brevik (Hagg et al. 2016). The system was tested over 6 months at different conditions, and the stable performance was found even at a high NO_x and SO_2 loading (average 100 ppm and 5 ppm, respectively) flue gas. They reported that stable permeate CO_2 purity of 65% over the accumulated 24 days was achieved (Hagg et al. 2016). The techno-economic feasibility analysis was also reported to achieve 80% CO_2 recovery and > 90% CO_2 purity. However, the designed two stage membrane system was difficult to achieve high CO_2 purity (> 95%) requirement for enhanced oil/gas recovery (EOR/EGR) (especially the O_2 limitation). The potential solutions are to introduce a third stage membrane unit

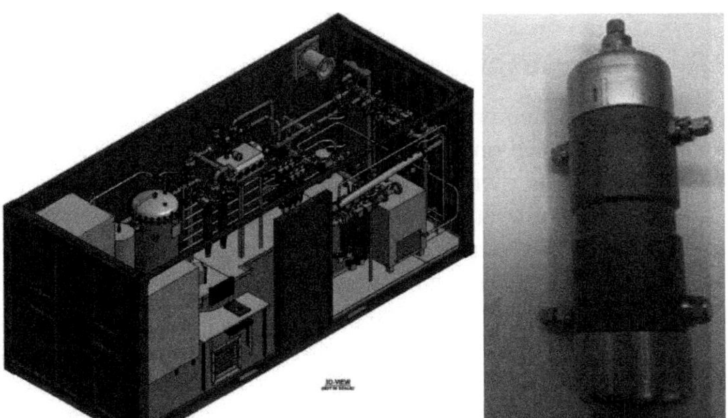

Figure 5.2. (A) 3D simulation of the arrangement inside the Pilot plant in MemCCC; (B) a 4.2m^2 hollow fiber FSC membrane module (Hagg et al. 2016).

or a low-temperature liquefaction unit. The CEMCAP project under EU H2020 looked into the membrane-assisted CO_2 liquefaction process for CO_2 capture in cement factory, which might provide an energy efficient solution. However, such investigation is still in the early conceptual design phase.

CO_2 capture from other industries

Iron and steel production industries are quite large energy consuming manufacturing sectors around the world, and CO_2 emissions from these manufacturing sectors represent about 10% of total global CO_2 emissions (2005). CO_2 capture in power plants has received a lot of attention as described in Section 3, but very little attention has been put on CO_2 capture from iron and steel production plants so far (CO_2 concentration of flue gas in industries is usually higher than that in power plants). There are only a few studies reported on CO_2 capture from iron and steel production industries (Farla et al. 1995, Gielen 2003, Lie et al. 2007, Wang et al. 2009). Membrane technology as an energy efficient process has been widely used for different gas separation processes, and can potentially compete with chemical absorption in terms of energy requirement, and could be favorable to be used in steelmaking industries as reported by Favre et al. 2007. In September 2004, one of the largest European projects, Ultra Low CO_2 Steelmaking (ULCOS) was launched to develop a new steel production technology that could drastically cut CO_2 emissions to 50% by 2030 (base year 2004), and membrane system was chosen for a candidate for CO_2 capture from Nitrogen Free Blast Furnace (NFBF) exhaust gases ($N_2/CO_2/CO/H_2$: 10%/36%/47%/7%) (Lie et al. 2007). The experimental testing and process simulation for CO_2 capture from blast furnace effluents were conducted by a PVAm based Fixed-Site-Carrier (FSC) membrane, and their results indicated that FSC membranes could be a potential candidate for CO_2 capture from flue gas in the steelmaking industry. However, no membrane materials has been tested for CO_2 capture from real exhaust gas in steel/iron industry so far.

Conclusions and Future Perspectives

The PVAm based FSC membrane, the PolyActive™ membrane and the Polaris™ membrane have been tested on the pilot scale post-combustion CO_2 capture. Each type of membrane has its own challenges on engineering design of module and process (especially the pre-treatment) for commercialization. Further development is required to reach higher TRL level, and two/multi-stage membrane system should be tested to achieve the separation requirement related to CO_2 purity and CO_2 recovery. The main challenges in commercializing membrane technology in CO_2 capture include the design of high performance membrane module to effectively use the membrane are, and the process design and optimization to achieve the balance of energy consumption and cost. Nevertheless, gas separation membrane systems could be an environmentally friendly technology for post combustion CO_2 capture in the near future by the joint force from membrane science and engineering.

Acknowledgements

The author wants to acknowledge the Research Council of Norway for the funding of this work in the membranes for CO_2 removal project (#267615).

References

Adams, R.T., Lee, J.S., Bae, T.-H., Ward, J.K., Johnson, J.R., Jones, C.W. et al. 2011. CO_2–CH_4 permeation in high zeolite 4A loading mixed matrix membranes. J. Membr. Sci. 367: 197–203.

Ahmad, J. and Hägg, M.-B. 2013. Preparation and characterization of polyvinyl acetate/zeolite 4A mixed matrix membrane for gas separation. J. Membr. Sci. 427: 73–84.

Ahmad, J. and Hågg, M.B. 2013. Polyvinyl acetate/titanium dioxide nanocomposite membranes for gas separation. J. Membr. Sci.

Ahn, J., Chung, W.-J., Pinnau, I., Song, J., Du, N., Robertson, G.P. et al. 2010. Gas transport behavior of mixed-matrix membranes composed of silica nanoparticles in a polymer of intrinsic microporosity (PIM-1). J. Membr. Sci. 346: 280–287.

Aroon, M.A., Ismail, A.F., Matsuura, T. and Montazer-Rahmati, M.M. 2010. Performance studies of mixed matrix membranes for gas separation: A review. Sep. Purif. Technol. 75: 229–242.

Baker, R. 2004. Membrane Technology and Applications, 2nd ed. McGraw-Hill.

Baker, R.W., Wijmans, J.G., Merkel, T.C., Lin, H., Daniels, R. and Thompson, S. 2009. Gas separation process using membranes with permeate sweep to remove CO_2 from combustion gases Membrane Technology & Research, Inc, US.

Bredesen, R., Jordal, K. and Bolland, O. 2004. High-temperature membranes in power generation with CO_2 capture. Chem. Eng. Process 43: 1129–1158.

Brunetti, A., Scura, F., Barbieri, G. and Drioli, E. 2010. Membrane technologies for CO_2 separation. J. Membr. Sci. 359: 115–125.

Budd, P.M., Elabas, E.S., Ghanem, B.S., Makhseed, S., McKeown, N.B., Msayib, K.J. et al. 2004. Solution-processed, organophilic membrane derived from a polymer of intrinsic microporosity. Adv. Mater. 16: 456–459.

Budd, P.M., McKeown, N.B., Ghanem, B.S., Msayib, K.J., Fritsch, D., Starannikova, L. et al. 2008. Gas permeation parameters and other physicochemical properties of a polymer of intrinsic microporosity: Polybenzodioxane PIM-1. J. Membr. Sci. 325: 851–860.

Budd, P.M., Msayib, K.J., Tattershall, C.E., Ghanem, B.S., Reynolds, K.J., McKeown, N.B. et al. 2005. Gas separation membranes from polymers of intrinsic microporosity. J. Membr. Sci. 251: 263–269.

Bushell, A.F., Attfield, M.P., Mason, C.R., Budd, P.M., Yampolskii, Y., Starannikova, L. et al. 2013. Gas permeation parameters of mixed matrix membranes based on the polymer of intrinsic microporosity PIM-1 and the zeolitic imidazolate framework ZIF-8. J. Membr. Sci. 427: 48–62.

Car, A., Stropnik, C., Yave, W. and Peinemann, K.-V. 2008. PEG modified poly(amide-b-Tethylene oxide) membranes for CO_2 separation. J. Membr. Sci. 307: 88–95.

Carapellucci, R. and Milazzo, A. 2003. Membrane systems for CO_2 capture and their integration with gas turbine plants proceedings of the institution of mechanical engineers, part A: Journal of Power and Energy 217: 505–517.

Casillas, C., Chan, K., Fulton, D., Kaschemekat, J., Kniep, J., Ly, J. et al. 2015. Pilot Testing of a Membrane System for Post-Combustion CO_2 Capture, NETL CO_2 Capture Technology Meeting Pittsburgh.

Chung, T.-S., Jiang, L.Y., Li, Y. and Kulprathipanja, S. 2007. Mixed matrix membranes (MMMs) comprising organic polymers with dispersed inorganic fillers for gas separation. Prog. Polym. Sci. 32: 483–507.

Cserjési, P., Nemestóthy, N. and Bélafi-Bakó, K. 2010. Gas separation properties of supported liquid membranes prepared with unconventional ionic liquids. J. Membr. Sci. 349: 6–11.

D'Alessandro, D.M., Smit, B. and Long, J.R. 2010. Carbon Dioxide Capture: Prospects for New Materials. Angewandte Chemie International Edition 49: 6058–6082.

Deng, L. 2009. Development of novel PVAm/PVA blend FSC membrane for CO_2 capture. Norwegian University of Science and Technology, Trondheim.

Deng, L., Kim, T.-J. and Hägg, M.-B. 2009. Facilitated transport of CO_2 in novel PVAm/PVA blend membrane. J. Membr Sci. 340: 154–163.

Du, N., Park, H.B., Robertson, G.P., Dal-Cin, M.M., Visser, T., Scoles, L. et al. 2011. Polymer nanosieve membranes for CO_2-capture applications. Nat. Mater. 10: 372–375.

Duan, S., Taniguchi, I., Kai, T. and Kazama, S. 2012. Poly(amidoamine) dendrimer/poly(vinyl alcohol) hybrid membranes for CO2 capture. J. Membr. Sci. 423-424: 107–112.

Farla, J.C., Hendriks, C.A. and Blok, K. 1995. Carbon dioxide recovery from industrial processes. Climatic Change 24: 439–461.

Favre, E. 2007. Carbon dioxide recovery from post-combustion processes: Can gas permeation membranes compete with absorption? J. Membr. Sci. 294: 50–59.

Figueroa, J.D., Fout, T., Plasynski, S., McIlvried, H. and Srivastava, R.D. 2008. Advances in CO_2 capture technology—The U.S. Department of Energy's Carbon Sequestration Program. IJGGC 2: 9–20.

Ghalei, B., Sakurai, K., Kinoshita, Y., Wakimoto, K., Isfahani, Ali P., Song, Q. et al. 2017. Enhanced selectivity in mixed matrix membranes for CO_2 capture through efficient dispersion of amine-functionalized MOF nanoparticles. Nature Energy 2: 17086.

Gielen, D. 2003. CO_2 removal in the iron and steel industry. Energy Convers. Manage. 44: 1027–1037.

Grainger, D. 2007. Development of carbon membranes for hydrogen recovery, Department of Chemical Engineering. Norwegian University of Science and technology, Trondheim.

Hägg, M.-B., He, X., Sarfaraz, V., Sandru, M. and Kim, T.-J. 2015. CO_2 Capture using a Membrane Pilot Process at Cement factory, in Brevik Norway- lessons learnt, The 8th Trondheim CCS Conference (TCCS8), Trondheim.

Hagg, M.-B., Lindbråthen, A., He, X., Nodeland, S. and Cantero, T. 2016. Pilot demonstration – Reporting on CO2 capture from a cement plant using a hollow fiber membrane process GHGT-13, Lausanne.

Hagg, M.B. and Lindbrathen, A. 2005. CO_2 Capture from natural gas fired power plants by using membrane technology. Ind. Eng. Chem. Res. 44: 7668–7675.

Hägg, M.B., Sandru, M., Kim, T.J., Capala, W. and Huijbers, M. 2012. Report on pilot scale testing and further development of a facilitated transport membrane for CO_2 capture from power plants Euromembrane 2012, London, UK.

Han, S.H., Kwon, H.J., Kim, K.Y., Seong, J.G., Park, C.H., Kim, S. et al. 2012. Tuning microcavities in thermally rearranged polymer membranes for CO_2 capture. Phys. Chem. Chem. Phys.

Hanioka, S., Maruyama, T., Sotani, T., Teramoto, M., Matsuyama, H., Nakashima, K. et al. 2008. CO_2 separation facilitated by task-specific ionic liquids using a supported liquid membrane. J. Membr. Sci. 314: 1–4.

He, X. 2011. Development of Hollow Fiber Carbon Membranes for CO_2 Separation, Department of Chemical Engineering. Norwegian University of Science and Technology, Trondheim.

He, X., Yu, Q. and Hägg, M.-B. 2013. CO_2 Capture. *In*: Hoek, E.M.V. and Tarabara, V.V. (eds.). Encyclopedia of Membrane Science and Technology. John Wiley & Sons, Inc.

He, X., Fu, C. and Hägg, M.-B. 2015. Membrane system design and process feasibility analysis for CO2 capture from flue gas with a fixed-site-carrier membrane. Chem. Eng. J. 268: 1–9.

He, X. and Hägg, M.-B. 2011. Hollow fiber carbon membranes: Investigations for CO_2 capture. J. Membr. Sci. 378: 1–9.

He, X. and Hägg, M.-B. 2012a. Membranes for Environmentally Friendly Energy Processes. Membranes 2: 706–726.

He, X. and Hägg, M.-B. 2014. Energy efficient process for CO_2 Capture from flue gas with novel fixed-site-carrier membranes. Energy Procedia 63: 174–185.

He, X. and Hägg, M.-B. 2015. Can Energy Efficient Membrane Process be An Alternative for CO_2 Capture? 8th Trondheim Conference on CO2 Capture, Transport and Storage (TCCS-8), Trondheim.

He, X., Hägg, M.-B. and Kim, T.-J. 2014a. Hybrid FSC membrane for CO_2 removal from natural gas: Experimental, process simulation, and economic feasibility analysis. AIChE J. 60: 4174–4184.

He, X. and Hägg, M.B. 2012b. Hybrid Fixed-site-carrier Membranes for CO_2/CH_4 Separation, Euromembrane 2012, London, UK.

He, X., Kim, T.-J. and Hägg, M.-B. 2014b. Hybrid fixed-site-carrier membranes for CO_2 removal from high pressure natural gas: Membrane optimization and process condition investigation. J. Membr. Sci. 470: 266–274.

He, X., Kim, T.-J. and Hägg, M.B. 2013a. CO_2 Capture with Membranes: Process Design and Feasibility Analysis, TCCS-7, Trondheim, Norway.

He, X., Kim, T.-J., Uddin, M.W. and Hägg, M.-B. 2013b. CO_2 Removal from High Pressure Natural Gas with Hybrid Fixed-site-carrier Membranes: Membrane Material Development, AIChE Annual Meeting 2013, San Francisco, USA.

He, X., Lie, J.A., Sheridan, E. and Hagg, M.-B. 2009. CO_2 Capture by hollow fibre carbon membranes: experiments and process simulations. Energy Procedia 1: 261–268.

He, X., Lindbråthen, A. and Hägg, M.-B. 2017a. Pilot Demonstration on Membranes for Post-combustion CO_2 Capture 9th Trondheim Conference on CO_2 Capture, Transport and Storage (TCCS-9), Trondheim.

He, X., Lindbråthen, A., Kim, T.-J. and Hägg, M.-B. 2017b. Pilot testing on fixed-site-carrier membranes for CO_2 capture from flue gas. IJGGC 64: 323–332.

He, X., Nieto, D.R., Lindbråthen, A. and Hägg, M.-B. 2017c. Membrane System Design for CO_2 Capture, Process Systems and Materials for CO_2 Capture. John Wiley & Sons, Ltd, pp. 249–281.

He, X., Yu, Q., Hägg and M.-B. 2013c. CO_2 Capture. *In*: Hoek, E.M.V. and Tarabara, V.V. (eds.). Encyclopedia of Membrane Science and Technology. John Wiley & Sons, Inc.

Hoffert, M.I., Caldeira, K., Benford, G., Criswell, D.R., Green, C., Herzog, H. et al. 2002. Advanced technology paths to global climate stability: Energy for a greenhouse planet. Science 298: 981–987.

Huang, J., Zou, J. and Ho, W.S.W. 2008. Carbon Dioxide Capture Using a CO_2-Selective Facilitated Transport Membrane. Ind. Eng. Chem. Res. 47: 1261–1267.

Hussain, A. and Hägg, M.-B. 2010. A feasibility study of CO_2 capture from flue gas by a facilitated transport membrane. J. Membr. Sci. 359: 140–148.

International Energy Outlook 2010—Highlights. [cited 2017 September 28]; Available from: http://www.eia.doe.gov/oiaf/ieo/highlights.html.

IPCC Special Report Carbon Dioxide Capture and Storage: Summary for Policymakers. 2005 [cited 2017 March 12th]; Available from: http://www.ipcc.ch/pdf/special-reports/srccs/srccs_summaryforpolicymakers.pdf.

Khan, M.M., Filiz, V., Bengtson, G., Shishatskiy, S., Rahman, M.M., Lillepaerg, J. et al. 2013. Enhanced gas permeability by fabricating mixed matrix membranes of functionalized multiwalled carbon nanotubes and polymers of intrinsic microporosity (PIM). J. Membr. Sci. 436: 109–120.

Kim, D.-H., Baek, I.-H., Hong, S.-U. and Lee, H.-K. 2011. Study on immobilized liquid membrane using ionic liquid and PVDF hollow fiber as a support for CO_2/N_2 separation. J. Membr. Sci. 372: 346–354.

Kim, S., Han, S.H. and Lee, Y.M. 2012. Thermally rearranged (TR) polybenzoxazole hollow fiber membranes for CO_2 capture. J. Membr. Sci. 403-404: 169–178.

Kim, S. and Lee, Y. 2012. Thermally rearranged (TR) polymer membranes with nanoengineered cavities tuned for CO_2 separation. J. Nanopart. Res. 14: 1–11.

Kim, T.-J., Li, B. and Hägg, M.-B. 2004. Novel fixed-site–carrier polyvinylamine membrane for carbon dioxide capture. J. Polym. Sci. Part B: Polym. Phys. 42: 4326–4336.

Kim, T.-J., Vrålstad, H., Sandru, M. and Hägg, M.-B. 2013. Separation performance of PVAm composite membrane for CO2 capture at various pH levels. J. Membr. Sci. 428: 218–224.

Kreuzer, F. and Hoofd, L. 2011. Facilitated Diffusion of Oxygen and Carbon Dioxide, Comprehensive Physiology. John Wiley & Sons, Inc.

Labreche, Y., Fan, Y., Rezaei, F., Lively, R.P., Jones, C.W. and Koros, W.J. 2014. Poly(amide-imide)/Silica Supported PEI Hollow Fiber Sorbents for Postcombustion CO_2 Capture by RTSA. ACS Applied Materials & Interfaces 6: 19336–19346.

Lie, J.A., Vassbotn, T., Hägg, M.-B., Grainger, D., Kim, T.-J. and Mejdell, T. 2007. Optimization of a membrane process for CO_2 capture in the steelmaking industry. IJGGC 1: 309–317.

Lin, H. and Freeman, B.D. 2005. Materials selection guidelines for membranes that remove CO_2 from gas mixtures. J. Mol. Struct. 739: 57–74.

Luis, P., Neves, L.A., Afonso, C.A.M., Coelhoso, I.M., Crespo, J.G., Garea, A. et al. 2009. Facilitated transport of CO_2 and SO_2 through Supported Ionic Liquid Membranes (SILMs). Desalination 245: 485–493.

McKeown, N.B 2012. Polymers of Intrinsic Microporosity. ISRN Materials Science 2012, 16.

McKeown, N.B., Budd, P.M., Msayib, K.J., Ghanem, B.S., Kingston, H.J., Tattershall, C.E. et al. 2005. Polymers of Intrinsic Microporosity (PIMs): Bridging the Void between Microporous and Polymeric Materials. Chemistry – A European Journal 11: 2610–2620.

Merkel, T. 2016. Pilot Testing of a Membrane System for Post-Combustion CO_2 Capture-Final report.

Merkel, T.C., Lin, H., Wei, X. and Baker, R. 2010. Power plant post-combustion carbon dioxide capture: An opportunity for membranes. J. Membr. Sci. 359: 126–139.

Merkel, T.C., Wei, X., He, Z., White, L.S., Wijmans, J.G., Baker and R.W. 2013. Selective Exhaust Gas Recycle with Membranes for CO_2 Capture from Natural Gas Combined Cycle Power Plants. Ind. Eng. Chem. Res. 52: 1150–1159.

Metz, S.J., Mulder, M.H.V. and Wessling, M. 2004. Gas-permeation properties of poly(ethylene oxide) Poly(butylene terephthalate) block copolymers. Macromolecules 37: 4590–4597.

Morisato, A., Shen, H.C., Sankar, S.S., Freeman, B.D., Pinnau, I. and Casillas, C.G. 1996. Polymer characterization and gas permeability of poly(1-trimethylsilyl-1-propyne) [PTMSP], poly(1-phenyl-1-propyne) [PPP], and PTMSP/PPP blends. J. Polym. Sci. Part B: Polym. Phys. 34: 2209–2222.

Myers, C., Pennline, H., Luebke, D., Ilconich, J., Dixon, J.K., Maginn, E.J. et al. 2008. High temperature separation of carbon dioxide/hydrogen mixtures using facilitated supported ionic liquid membranes. J. Membr. Sci. 322: 28–31.

Neves, L.A., Crespo, J.G. and Coelhoso, I.M. 2010. Gas permeation studies in supported ionic liquid membranes. J. Membr. Sci. 357: 160–170.

Neves, L.A., Nemestóthy, N., Alves, V.D., Cserjési, P., Bélafi-Bakó, K. and Coelhoso, I.M. 2009. Separation of biohydrogen by supported ionic liquid membranes. Desalination 240: 311–315.

Pan, X., Zhang, J., Xue, Q., Li, X., Ding, D., Zhu, L. et al. Mixed Matrix Membranes with Excellent CO2 Capture Induced by Nano-Carbon Hybrids. ChemNanoMat, n/a-n/a.

Park, H.B., Han, S.H., Jung, C.H., Lee, Y.M. and Hill, A.J. 2010. Thermally rearranged (TR) polymer membranes for CO_2 separation. J. Membr. Sci. 359: 11–24.

Park, H.B., Jung, C.H., Lee, Y.M., Hill, A.J., Pas, S.J., Mudie, S.T. et al. 2007. Polymers with cavities tuned for fast selective transport of small molecules and ions. Science 318: 254–258.

Paul, D.R. and Jampol'skij, J.P. 1994. Polymeric gas separation membranes. CRC Press, Boca Raton, Fla.

Pohlmann, J., Bram, M., Wilkner, K. and Brinkmann, T. 2016. Pilot scale separation of CO2 from power plant flue gases by membrane technology. IJGGC 53: 56–64.

Pope, D.S., Koros, W.J. and Hopfenberg, H.B. 1994. Sorption and dilation of poly(1-(trimethylsilyl)-1-propyne) by carbon dioxide and methane. Macromolecules 27: 5839–5844.

Quan, S., Li, S.W., Xiao, Y.C. and Shao, L. 2017. CO_2-selective mixed matrix membranes (MMMs) containing graphene oxide (GO) for enhancing sustainable CO_2 capture. IJGGC 56: 22–29.

Ramasubramanian, K. and Ho, W.S.W. 2011. Recent developments on membranes for post-combustion carbon capture. Current Opinion in Chemical Engineering 1: 47–54.

Reijerkerk, S.R. 2010. Polyether based block copolymer membranes for CO_2 separation [PhD]. University of Twente, Enschede, p. 245.

Reijerkerk, S.R., Knoef, M.H., Nijmeijer, K. and Wessling, M. 2010. Poly(ethylene glycol) and poly(dimethyl siloxane): Combining their advantages into efficient CO_2 gas separation membranes. J. Membr. Sci. 352: 126–135.

Reijerkerk, S.R., Wessling, M. and Nijmeijer, K. .2011. Pushing the limits of block copolymer membranes for CO_2 separation. J. Membr. Sci. 378: 479–484.

Robeson, L.M. 2008. The upper bound revisited. J. Membr. Sci. 320: 390–400.

Samanta, A., Zhao, A., Shimizu, G.K.H., Sarkar, P. and Gupta, R. 2011. Post-Combustion CO2 Capture Using Solid Sorbents: A Review. Ind. Eng. Chem. Res. 51: 1438–1463.

Sandru, M. 2009. Development of a FSC membrane for selective CO_2 capture. Norwegian University of Science and Technology, Trondheim, pp. IX, 176 s.

Sandru, M., Haukebø, S.H. and Hägg, M.-B. 2010. Composite hollow fiber membranes for CO_2 capture. J. Membr. Sci. 346: 172–186.

Sandru, M., Kim, T.-J., Capala, W., Huijbers, M. and Hägg, M.-B. 2013. Pilot Scale Testing of Polymeric Membranes for CO2 Capture from Coal Fired Power Plants. Energy Procedia 37: 6473–6480.

Scholes, C.A., Chen, G.Q., Lu, H.T. and Kentish, S.E. 2016. Crosslinked PEG and PEBAX Membranes for Concurrent Permeation of Water and Carbon Dioxide. Membranes 6, 0001.

Thomas, S., Pinnau, I., Du, N. and Guiver, M.D. 2009. Pure- and mixed-gas permeation properties of a microporous spirobisindane-based ladder polymer (PIM-1). J. Membr. Sci. 333: 125–131.

Tong, Z. and Ho, W.S.W. 2016. Facilitated transport membranes for CO_2 separation and capture. Separ Sci. Technol: 1–12.

Uchytil, P., Schauer, J., Petrychkovych, R., Setnickova, K. and Suen, S.Y. 2011. Ionic liquid membranes for carbon dioxide–methane separation. J. Membr. Sci. 383: 262–271.

Uddin, M.W. and Hägg, M.-B. 2012a. Effect of monoethylene glycol and triethylene glycol contamination on CO2/CH4 separation of a facilitated transport membrane for natural gas sweetening. J. Membr. Sci. 423-424: 150–158.

Uddin, M.W. and Hägg, M.-B. 2012b. Natural gas sweetening—the effect on CO_2–CH_4 separation after exposing a facilitated transport membrane to hydrogen sulfide and higher hydrocarbons. J. Membr. Sci. 423-424: 143–149.

W, Y., Car, A., Wind, J. and Peinemann, K. 2010. Nanometric thin film membranes manufactured on square meter scale: ultra-thin films for CO_2 capture. Nanotechnology 21.
Wang, C., Ryman, C. and Dahl, J. 2009. Potential CO_2 emission reduction for BF–BOF steelmaking based on optimised use of ferrous burden materials. IJGGC 3: 29–38.
Woo, K.T., Lee, J., Dong, G., Kim, J.S., Do, Y.S., Hung, W.-S. et al. 2015. Fabrication of thermally rearranged (TR) polybenzoxazole hollow fiber membranes with superior CO_2/N_2 separation performance. J. Membr. Sci. 490: 129–138.
Yang, H., Xu, Z., Fan, M., Gupta, R., Slimane, R.B., Bland, A.E. et al. 2008. Progress in carbon dioxide separation and capture: A review. J. Environ. Sci. 20: 14–27.
Yave, W., Car, A., Funari, S.S., Nunes, S.P. and Peinemann, K.-V. 2010. CO_2-philic polymer membrane with extremely high separation performance. Macromolecules 43: 326–333.
Yong, W.F., Li, F.Y., Xiao, Y.C., Chung, T.S. and Tong, Y.W. 2013. High performance PIM-1/Matrimid hollow fiber membranes for CO_2/CH_4, O_2/N_2 and CO_2/N_2 separation. J. Membr. Sci. 443: 156–169.
Yu, C.-H., Huang, C.-H. and Tan, C.-S. 2012. A review of CO_2 capture by absorption and adsorption. Aerosol and Air Quality Research 12: 745–769.
Zhang, X., Zhang, X., Dong, H., Zhao, Z., Zhang, S. and Huang, Y. 2012. Carbon capture with ionic liquids: overview and progress. Energy & Environmental Science 5: 6668–6681.
Zhao, W., He, G., Nie, F., Zhang, L., Feng, H. and Liu, H. 2012. Membrane liquid loss mechanism of supported ionic liquid membrane for gas separation. J. Membr. Sci. 411-412: 73–80.
Zou, J. and Ho, W.S.W. 2006. CO_2-selective polymeric membranes containing amines in crosslinked poly(vinyl alcohol). J. Membr. Sci. 286: 310–321.

6

Advanced Membrane Materials for CO_2 Separation from Natural Gas

A.K. Zulhairun, N. Yusof, W.N.W. Salleh, F. Aziz* and *A.F. Ismail**

INTRODUCTION

Methane (CH_4) is an important commodity for domestic uses, as the feedstock for industrial and energy generation sectors, and the fuel for transportation (Economides and Wood 2009). Raw natural gas from the underground reservoirs is composed of primarily CH_4. While carbon dioxide (CO_2) is often the major contaminant present with a fraction of other impurities such as hydrogen sulfide, helium, nitrogen, water, and heavier hydrocarbons (Adewole et al. 2013, Faramawy et al. 2016). The raw natural gas has to undergo a series of purification steps in order to remove the aforementioned contaminants. The removal of CO_2 will prevent pipeline corrosion due to reaction of CO_2 and water, increase the heating value of the sales gas, and minimize the transportation energy of the gas to the onshore plants (Rufford et al. 2012). Since natural gas holds an enormous market and profound usage in diverse sectors, substantial efforts have been invested to explore and establish the most efficient and economical method to purify the gas, especially for removing the CO_2. The most common technique applied for CO_2 removal is chemical absorption using amine solutions (e.g., mono-ethanolamine (MEA), Methyldiethanolamine (MDEA), etc.) (Sreedhar et al. 2017). CO_2 is removed by means of contacting the raw natural gas with aqueous amine in a huge column where CO_2 will be stripped away. The absorbent will be regenerated by boiling to release the captured CO_2, before being

Advanced Membrane Technology Research Centre (AMTEC), Universiti Teknologi Malaysia, 81310 UTM Skudai, Johor Darul Ta'zim, Malaysia.
* Corresponding authors: zulhairun@utm.my; afauzi@utm.my

pumped back to the absorption column to repeat the process. This technique practically achieves complete CO_2 removal due to high chemical affinity (Baker and Lokhandwala 2008). However, the system requires massive components comprising absorption and desorption columns, intense energy associated to reboiler for solvent regeneration, high capital (complex and large footprint system) and operational costs (energy and chemical absorbent) (Rufford et al. 2012). On top of that, this technique would be effective only for gas streams containing low CO_2 concentrations (no greater than 20 mol%) (Baker and Lokhandwala 2008).

Since the deployment of membrane technology for gas separation on an industrial scale in the 1980s, this relatively new separation technology has been proven competitive to that of absorption, adsorption, and cryogenic processes (Kundu et al. 2014). Membrane technology offers lower capital investment due to its simplicity of operation (no phase change, no moving or rotating parts, thus no need for intense monitoring or supervision); compact, lightweight modular system requires small footprint as well eases the transportation and installation in remote locations, i.e., offshore facilities; the process is considerably safer, imposing low environmental impact, where no hazardous/toxic wastes are produced. Furthermore, as opposed to absorption, better separation could be achieved by membrane system with increasing CO_2 concentration in the feed (Baker and Lokhandwala 2008).

In spite of its profound advantages, the key membrane performance i.e., permeability and selectivity are limited by a trade-off relationship where a membrane of high permeability exhibits low selectivity, and vice-versa (Robeson 1991, Robeson 2008). Other issues such as the membranes mechanical stability under high operating pressures, CO_2-induced swelling or plasticization, and physical ageing may lead to deteriorating performance over a period of operation (Adewole et al. 2013, George et al. 2016). To date, extreme efforts have been made to improve the separation performance and the robustness of the membranes by exploring new polymers as well molecular sieving nanoparticles. This chapter reviews some of the most recent and advanced polymeric and inorganic materials in the development of membranes aimed for the removal of CO_2 from natural gas mixtures. Special emphasis on the separation performance and the structure-property relationship of high free volume polymers such as structurally modified polyimides (PIs), Thermally-Rearranged (TR) polymers, and Polymers of Intrinsic Microporosity (PIMs) is made. Several types of microporous inorganic materials which include zeolites, carbon nanotubes (CNTs), graphene, and Metal Organic Frameworks (MOFs) are discussed in view of their important characteristics for CO_2 removal and their potential for membrane development. Recent efforts in combining the advantages of both polymeric and inorganic materials to form novel Mixed Matrix Membranes (MMMs) and their performance are also discussed. The key issues to be addressed in order to propel current researches towards industrial application are proposed considering the mass potential of these materials for removing CO_2 from the natural gas fields.

Polymeric Materials

Polymer has been the most attractive, and essentially the prime material for membrane-based gas separation applications. In contrast to other materials such as ceramics and metals, polymeric materials possess good processability, ease of handling, and very cost effective (George et al. 2016). The transport of gas molecules through a polymeric membrane is a complex mechanism. The separation is accomplished due to the differences in gases mobility across the dense polymer matrix which do not strictly based on their molecular sizes. This permeation mechanism is best described by the solution-diffusion theory (Barrer 1951). Diffusion is the basis of the solution-diffusion model in which the permeable components dissolve in the membrane materials and then transverse across the membrane propelled by partial pressure gradient across the membrane (Pandey and Chauhan 2001). The permeability can be defined as the product of diffusivity coefficient (D) and solubility coefficient (S). Solubility coefficient relates to the chemical interactions between the gas species with the membrane material. Meanwhile, the diffusivity factor mainly affected by the available free volume or voids between the polymer chains and the stiffness of the chains. This key concept has been the primary basis in the architectural design of novel polymers for gas separation membranes since the polymer structure and its properties will dictate the separation performance. In the following section, we will discuss three types of the most advanced, high performance polymers which have been developed for gas separation in the last decade: polyimides, thermally-rearranged polymers, and polymers of intrinsic microporosity. This chapter discusses the structural modifications involved to further enhance the performance of novel polymers including the introduction of CO_2-philic functional groups for higher solubility and tailoring the polymer backbone with bulky side group and rigid chemical bonds in order to increase the free volume and its rigidity.

Polyimides

Paramount efforts on developing high performance membrane materials for CO_2/CH_4 separation have been intensively reported over this decade. The latest trend shows growing interest in utilizing polyimide-based membranes primarily due to its good separation performance as well as good mechanical, thermal, and chemical stability (Xiao et al. 2009). Polyimide basically comprises of dianhydride and diamine segments. In the search for better membrane materials for CO_2/CH_4 separation, the combinations of diverse choice of aromatic dianhydrides and diamines (Fig. 6.1) have resulted in very interesting membrane properties and gas transport performances. The differences in polymer backbone structure such as chain rigidity, the chemical affinity of functional groups and its size strongly influence the spatial arrangement of polymer chain and hence their free volume elements. For instance, Matrimid® 5218, a commercial polyimide which has been primarily used for CO_2 separation, contains carbonyl linkage (from 3,3',4,4'-benzophenone tetracarboxylic dianhydride, BTDA) which provides decent gas permeability due to its chain flexibility. While, fluorine-containing polyimide particularly 4,4'-(hexafluoroisopropylidene) diphthalic anhydride (6FDA)-based membranes exhibit even higher permeability which is ascribed to enhanced free volume elements due to its bulky conformation despite its rigid structure. In essence, polymer chain rigidity determines the selectivity while the free volume elements govern the permeability. Polyimides have both rigid-

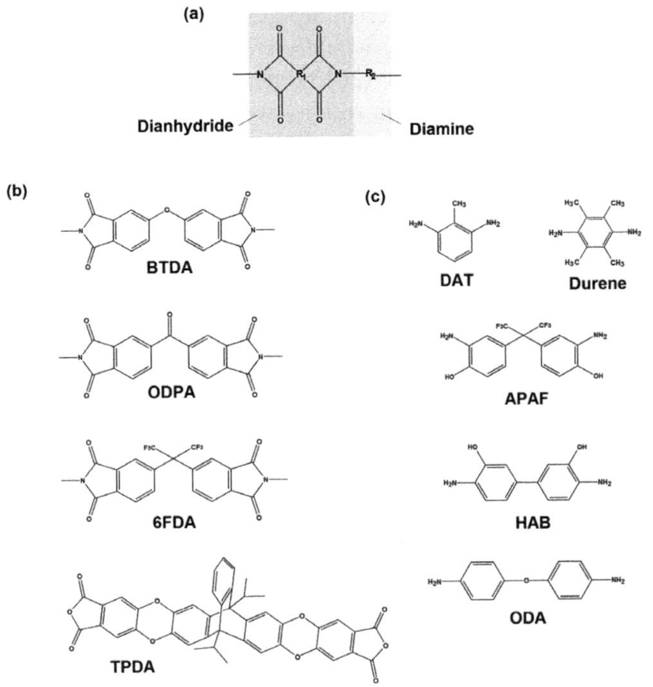

Figure 6.1. (a) Chemical structure of typical polyimide; (b) some examples of dianhydrides; 3,3',4,4'-benzophenone tetracarboxylic dianhydride (BTDA), 3,3',4,4'-tetracarboxylic diphenylether dianhydride (ODPA), 4,4'-(hexafluoroisopropylidene) diphthalic anhydride (6FDA), and 9,10-diisopropyl-triptycene (TPDA); and (c) some examples of diamines; 2,6-diaminotoluene (DAT), 2,3,5,6-tetramethyl benzene-1,4-diamine (Durene), 2,2-Bis(3-amino-4-hydroxylphenyl)-hexafluoropropane (APAF), 3,3'-dihydroxy-4,4'-diamino-biphenyl (HAB), 4,4'-diaminophenyl ether (ODA).

flexible chain and high free volume properties which endow good gas selectivity and permeability.

There have been various strategies for improving the performance of polyimide membranes, one of which is by introducing CO_2-philic side chains to the polymer backbone. Several reports on poly(ionic salt)s functionalization demonstrated astounding advantage on CO_2 selectivity due to significant increase in CO_2 solubility and reduction in CH_4 diffusivity (Kammakakam et al. 2015, Gye et al. 2017). Ionic-mediated crosslinked 6FDA-durene have shown overall improvement in CO_2 selectivity (from 13 to > 30) as well as excellent thermal, mechanical, and anti-plasticization pressure due to highly cross-linked polymer chain evidenced by significant reduction in *d*-spacing. However, the significant downturn was the loss in its permeability. Gye et al. (2017) reported that the introduction of PEG-ionic imidazolium group to the 6FDA-durene reduced the permeability from 495 barrer to 72 barrer while the CO_2/CH_4 selectivity was notably improved from 13 to 42. Other structural modification approaches include the addition of pendant functional groups to the reacting monomer. These pendant functional groups are responsible in increasing the free volume and hence the permeability. Guo et al. (2013b) replaced the *ortho*-positioned hydroxyl groups on the 3,3'-dihydroxy-4,4'-diamino- biphenyl

(HAB) attached to 6FDA with different acetates by reacting acetic anhydride and pivalic anhydride. The permeability would increase with increasing the size of the pendant group. However, Guo et al. (2013a) demonstrated that the increase in free volume without increasing the chain rigidity would lead to enhanced permeability (3 barrer to 57 barrer) but poor selectivity (48 reduced to 18).

In different study, Zhang et al. (2017) synthesized 6FDA-based polyimides with spirobichroman-based (SBC) diamines with different substituted groups (-CF_3, -H, and -CH_3). The incorporation of SBC backbone with the spiro and non-coplanar nature had increased the free volume and reduced the polymer chain flexibility. Fluoro-SBC diamine contained the bulkiest CF_3 group hence exhibited the highest free volume, and higher permeability than other SBC-polyimides. Several emerging studies have reported another approach to significantly enhance the polymer free volume by utilizing a bridged-bicyclic contorted dianhydride centre, namely 9,10-diisopropyl-triptycene (Ghanem et al. 2014, Swaidan et al. 2015, Alaslai et al. 2016). By integrating both rigid and contorted moieties into the polyimide backbone, intra-chain microporosity is developed within the polymer matrix giving rise to both high permeability and selectivity.

Swaidan et al. (2015) studied the effect of thermal annealing on several polyimide membranes as a mean to suppress membrane plasticization. They showed that 6FDA-APAF polyimide experienced a 30% reduction in permeability and 40% enhancement in CO_2/CH_4 selectivity when the membrane was annealed from 120 to 250°C for 24 hours. Annealing procedure would result in chain densification and the collapsing of the free volume which explained the changes in the gas transport properties of polyimide membranes. Densification-induced annealing is useful to suppress CO_2 plasticization effect under high partial pressure feed conditions, which is a typical drawback of common polyimides. Thermal annealing of "certain polyimides" would result in a new class of highly permeable polymers, known as thermally rearranged polymers which is discussed in the next section. Table 6.1 is provided to portray a broad performance comparison among different classes of polyimides previously discussed to that of others reported in the literature.

Thermally Rearranged Polymers

For typical amorphous polymers, gas molecules are diffused through free volume cavities which are caused by random thermal motion. These free volumes are non-uniform; the size and distributions vary widely with the environment. Large cavities are beneficial for high permeability of gases but lead to poor selectivity. For instance, poly(1-trimethylsilyl-1-propyne) (PTMSP), possesses extremely high CO_2 permeability exceeding 10,000 barrer but has very low selectivity, that is less than 2 for CO_2/CH_4 separation (Gomes et al. 2005, Matteucci et al. 2008). Sheer interest in polyimides has led to the development of novel, high free volume polymers with narrower cavity size and distributions. It was discovered that some polyimides undergo intramolecular reactions upon heat treatment above their glass temperature

Table 6.1. Polyimide membranes for CO_2/CH_4 separation.

Polyimides	Tg (°C)	FFV	P_{CO_2} (Barrer)	α_{CO_2/CH_4}	References
Matrimid® 5218	313	0.110	9	36	Guiver et al. (2002)
Matrimid® 5218	308	-	8	40	Rangel et al. (2013)
6FDA-ODA	294	0.169	14	44	Nik et al. (2012)
6FDA-HAB	-	0.150	12	38	Sanders et al. (2012)
6FDA-HAB	314	-	3	48	Guo et al. (2013b)
6FDA-HAB-Ac	280	-	11	42	Guo et al. (2013b)
6FDA-HAB-Pac	288	-	57	18	Guo et al. (2013b)
6FAP–BisADA	216	-	3	32	Guo et al. (2013a)
6FDA-MPD	295	0.178	8	76	Comesaña-Gándara et al. (2015)
6FDA-DAP	313	0.176	11	100	Comesaña-Gándara et al. (2015)
6FDA-DAR	338	0.171	8	102	Comesaña-Gándara et al. (2015)
6FDA-APAF	315	-	7	96	Swaidan et al. (2015)
TPDA-APAF	-	-	46	53	Swaidan et al. (2015)
TPDA-ATAF	-	-	325	30	Swaidan et al. (2015)
TPDA-mPDA	-	-	349	32	Alaslai et al. (2016)
TPDA-DAR	-	-	215	46	Alaslai et al. (2016)
6FDA-DAT-TB	395	0.200	218	33	Zhuang et al. (2016)
ODPA-TMPDA	260	-	25	29	Duan et al. (2014)
ODPA-DAT-TB	402	0.181	14	14	Zhuang et al. (2016)
BTDA-DAT-TB	419	0.170	20	12	Zhuang et al. (2016)
6FDA-durene	424	0.191	495	13	Kammakakam et al. (2015)
6FDA-durene-HmXPip	335	-	476	35	Kammakakam et al. (2015)
6FDA-durene-HtXPip	310	-	73	32	Kammakakam et al. (2015)
6FDA-durene-PEG-Im	328	0.168	72	42	Gye et al. (2017)
6FDA-durene-Butyl-Im	-	0.189	219	28	Gye et al. (2017)
6FDA-FSBC	261	0.160	66	26	Zhang et al. (2017)
6FDA-SBC	267	0.142	32	25	Zhang et al. (2017)
6FDA-MSBC	281	0.133	21	28	Zhang et al. (2017)
6FDA-DAM	383	-	426	16	Mao and Zhang (2017)
6FDA-DAM	354	-	21	33	Boroglu and Yumru (2017)
6FDA-3MPA	-	-	570	21	Hasebe et al. (2017)

Unit: P (Barrer) 10^{-10} $cm^3(STP)/cm^2$ s cmHg; Acronyms: Ac (Acetate); Pac (Pivalic Acetate); TB (Tröger's Base); Im (Imidazole); SBC (Spiro-bichroman), HmXPip (Homogeneous crosslinked piperazinium), HtXPip (Heterogenous crosslinked piperazinium) which lead to structural rearrangement while in their solution or solid state (Calle et al. 2013). Thermal rearrangement process may also occur between neighboring polymer chains, resulting in a more stable, cross-linked structural framework (Liu et al. 2016). High free volume, microporous polybenzoxazoles (PBO), polybenzimidazoles (PBI), and polybenzothiazoles (PBT) can be formed upon polyimide rearrangement under high temperatures; hence they are called Thermally-Rearranged (TR) polymers. These polymers composed of rigid-rod structures with high-torsional energy barriers to rotation between phenylene-heterocyclic rings. Its rigid main chain possesses low cohesive energy which leads to relatively higher molecular free volume than that of typical glassy polymers i.e., polysulfones, polycarbonates, etc. (Meier et al. 1994). TR polymers possess better thermal stability than its corresponding precursor but may have less mechanical stability due to its amorphous structure compared to crystalline polyimide (Chang et al. 2000).

TR polymer precursors are basically polyimides or co-polyimides, but not all polyimides are TR-able. TR-able polyimides should contain an *ortho*-positioned functionality in imide ring to form either, PBO, PBI, or PBT (Fig. 6.2). Polyimide precursors can be synthesized via polycondensation of monomers through chemical imidization such as azeotropic imidization (Han et al. 2010b, Brunetti et al. 2017, Scholes et al. 2017), ester acid (Guo et al. 2013a,b, Liu et al. 2016), or thermal imidization (Han et al. 2010b). TR polymers are insoluble in common organic solvents, thus cannot be fabricated directly from its solution form (Tullos et al. 1999, Chang et al. 2000). Therefore, typical TR membranes are fabricated by casting the polyimide precursor, followed by controlled solvent evaporation before being gradually annealed at high temperatures in inert, air, or vacuum atmosphere. During annealing, stepwise inter- and intra-molecular cyclization occurs by decarboxylation (Chang et al. 2000) transforming the imide to benzoxazole, benzimidazole, or benzothiazole. Polyimide thermal conversions are accompanied by quantitative loss of carbon dioxide. The degree of thermal rearrangement can be evaluated with a thermo-gravimetric analyzer, by determining the actual weight loss due to the release of CO_2 during the thermal rearrangement (Tullos et al. 1999).

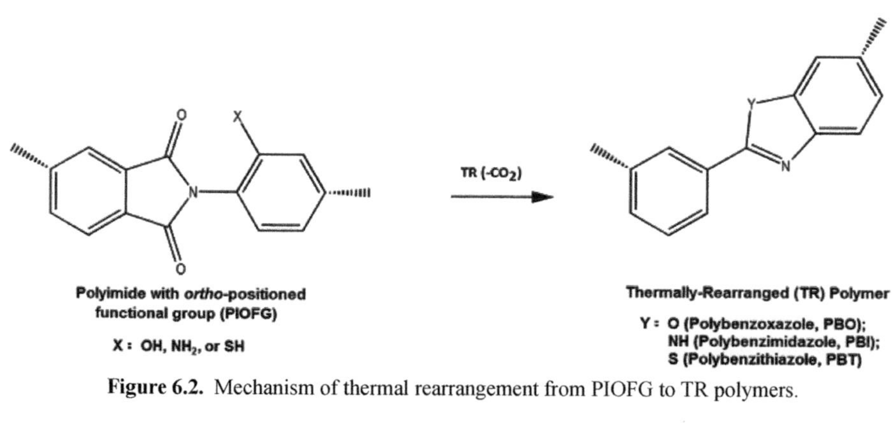

Figure 6.2. Mechanism of thermal rearrangement from PIOFG to TR polymers.

The gas separation performance of TR membranes is highly tunable by varying the precursor, fabrication and treatment protocols. Various combinations of the polyimide precursor basic structure have been reported to be significant factors in determining the free volume elements and its distributions (Smith et al. 2015). Jung et al. (2010) varied the TR-able to non-TR-able domain in their co-polyimide precursors and proved that the highest CO_2 permeability would be attained by maximizing the TR-able domain. A 120% FFV increment was observed when the membrane consisted of purely hydroxyl polyimide (fully TR-able). Chain flexibility of TR precursor and thermal conversion dictates the degree of rearrangement and hence the free volume topologies (size of cavities, distributions, and the fractional free volume). Calle and Lee (2011) prepared a novel poly(ether-benzoxazole) (PEBO) membranes by thermal rearrangement of an ether containing fluoro poly(hydroxyl-imide) (2,2-Bis(4-(4-amino-3-hydroxyphenoxy)phenyl)-hexafluoropropane) (6FBAHPP). The ether linkage is highly flexible, thus has higher rotational freedom which permits faster rearrangement rate. Furthermore, the rearrangement could possibly take place at lower temperatures.

Substitution of hydroxyl group of the diamino-biphenyl (HAB) segment with bulky pendant groups (Fig. 6.3) were reported to be effective for improving the gas permeation properties (Guo et al. 2013b). Liu et al. (2016) prepared TR-PBO based from 6FDA-HAB-Ac precursor similar to that of Guo et al. (2013a) by employing higher conversion temperature which resulted in much superior CO_2 permeability (443 vs. 174 barrer). According to Park et al. (2007) higher conversion temperature would increase the cavity size as well as the free volume elements, which is in agreement with the later studies (Calle and Lee 2011, Sanders et al. 2012).

Han et al. (2010b) demonstrated that the performance of TR membranes was also influenced by the methods of the precursor imidization. It was revealed that TR membranes following the route of thermal imidization possessed higher free volume elements than that of azeotropically imidized TR membranes. Thermally imidized PBO and acetate containing pendant group had shown astonishing permeability,

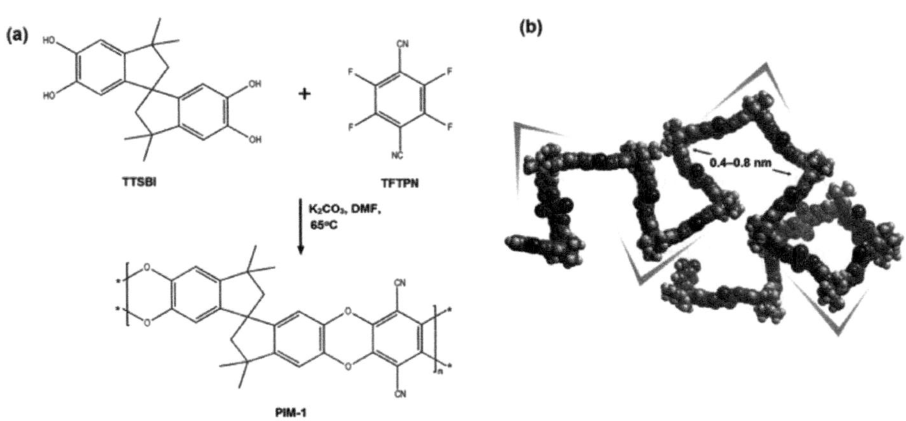

Figure 6.3. Thermal rearrangement (TR) of the poly(hydroxyimide) and its derivatized polyimides to polybenzoxazole (PBO) (Guo et al. 2013b).

ranging from 4,000 to almost 6,000 barrer. In their other study, high performance TR membrane was also synthesized based on thermal imidization method (Han et al. 2010a). Thermally imidized PBI possessed excellent CO_2 permeability exceeding 1,600 barrer and CO_2/CH_4 selectivity of 46. Calle et al. (2013) demonstrated a novel strategy for tuning the cavity size and free volume of TR-PBO copolymer membranes by trans-esterification cross-linking reaction of *ortho*-positioned hydroxyl polyimide precursors with 1,4-butylene glycol in the solid state by thermal annealing at 250°C prior to structural rearrangement. Upward-and-downward trend in permeability was observed with increasing crosslinker concentrations which was ascribed by the increased amount of micropores but decreasing free volumes. It was also revealed that cross-linked hydroxyl-polyimides converted to PBO at slower rates (and slower with increasing cross-linker concentrations) due to hindered chain mobility. Scholes et al. (2017) observed the reduction in membrane density and the formation of larger cavity sizes (which is indicative of increasing free volume elements) for cross-linked TR-PBOI membranes.

The performances of TR membranes discussed here are summarized in Table 6.2. Overall, TR membranes possess outstanding gas separation performance, surpassing the 2008 upper bound limit for CO_2/CH_4 separation, as well as other notable gas pairs such as O_2/N_2 and H_2/N_2. Gas transport through the TR membranes is strongly influenced by a diffusivity contribution due to a high free volume that is typical with regard to microporous structures. For more condensable gases like CO_2, SO_2 and H_2S, the transport would be dominated by solubility factor which lead to reduction of permeability with increasing operating temperature and pressure. In principle, larger cavities contribute to the higher rate of gas transport, whereas the presence of small micropores works for the improved gas-pair selectivity. Park et al. (2007) described that TR polymers comprise of markedly large free volumes, with hourglass, constricted cavities which resemble carbon molecular sieves. Park et al. (2007) also revealed that, unlike conventional glassy polymers, TR-based membranes are not susceptible to plasticization even at high CO_2 concentration (~ 80 mol%) and high CO_2 fugacity (~ 15 atm). Furthermore, TR membranes are also found to be resistant to plasticization by SO_2 (Scholes et al. 2017). Fabrication of TR-based membranes in hollow fiber configuration have been recently reported (Brunetti et al. 2016, Woo et al. 2016a, 2016b), exhibiting astounding CO_2 flux up to 2000 GPU; three-orders of magnitude higher than commercial membranes (Kim et al. 2012, Dong et al. 2015). Advancement in the development of TR-membranes thus far signifies its potential and viability for industrial-scale implementation not in so distant future.

Polymers of Intrinsic Microporosity

Polymers of Intrinsic Microporosity (PIMs), is a new class of amorphous, high FFV polymer comprised of stiff ladder-type structures with contorted sites which lead to a permanent microporosity within its backbone. The first work introducing PIM derivatives was reported by Budd and his co-workers (2004). In their pioneering work, the PIMs numerals (PIM-1 to 6) were used to denote six different combinations of monomers used in the synthesis. Since then, every work with PIMs has used the same numeral notations. Among them, PIM-1 has been the most extensively studied due to its

Table 6.2. TR membranes for CO_2/CH_4 separation.

TR Polymers (Precursors)	TR Temp. (°C)	FFV	P_{CO_2} (Barrer)	$\alpha_{\frac{CO_2}{CH_4}}$	References
PBO (6FDA-o-bisAPAF)	450	-	1,610	48	Park et al. (2007)
PBI (6FDA-DAB)	450	0.267	1,624	46	Han et al. (2010a)
PBO-PI (BPDA-bisAPAF-ODA)	450	-	1,014	24	Jung et al. (2010)
tPBO (6FDA-bisAPAF)	450	0.280	4,201	28	Han et al. (2010b)
aPBO (6FDA-bisAPAF)	450	0.220	398	34	Han et al. (2010b)
cPBO (6FDA-bisAPAF-Ac)	450	0.350	5,568	22	Han et al. (2010b)
sPBO (6FDA-bisAPAF-Ac)	450	0.330	5,903	23	Han et al. (2010b)
PEBO (6FDA-6FBAHPP)	450	0.207	485	29	Calle and Lee (2011)
PBO (6FDA-*p*-HAB-Ac)	450	0.196	410	24	Sanders et al. (2012)
PBO (6FDA-o-bisAPAF)	450	-	261	35	Calle et al. (2013)
X-PBOI (6FDA-o-bisAPAF-DABA)	450	-	980	30	Calle et al. (2013)
PBO (6FDA-*p*-HAB)	400	-	51	36	Guo et al. (2013b)
PBO (6FDA-*p*-HAB)	450	-	95	31	Guo et al. (2013a)
PBO (6FDA-*p*-HAB-Ac)	400	-	174	34	Guo et al. (2013b)
PBO (6FDA-*p*-HAB-Pac)	400	-	211	18	Guo et al. (2013b)
PBO (6FAP-bisADA)	400	-	13	33	Guo et al. (2013a)
PBO (6FDA-DAP)	450	0.194	142	38	Comesaña-Gándara et al. (2015)
PBO (6FDA-DAR)	450	0.210	354	40	Comesaña-Gándara et al. (2015)
PBO (6FDA-*p*-HAB-Ac)	450	0.194	443	21	Liu et al. (2016)
PBO (6FDA-*p*-HAB)	400	0.176	75	33	Liu et al. (2016)
PBOI (6FDA-HAB-DAM)	375	-	105	24	Brunetti et al. (2017)
PBOI (6FDA-HAB-DAM)	400	0.182	196	-	Scholes et al. (2017)
X-PBOI (6FDA-HAB-DAM-DABA)	400	0.185	218	-	Scholes et al. (2017)

Unit: P (Barrer) 10^{-10} cm^3(STP)/cm^2 s cmHg; Acronyms: PBO (polybenzoxazole), PBI (polybenzimidazole), PEBO (poly(ether–benzoxazole), PBOI (polybenzoxazole-co-imide), X (cross-linked).

astoundingly high FFV and gas permeability. Figure 6.4(a) illustrates the polymerization of PIM-1 by condensation of 5,5',6,6'-tetrahydroxy-3,3,3',3'-tetramethyl-1,1'-spirobisindane (TTSBI) and tetrafluoroterephthalonitrile (TFTPN), which is basically the formation of polybenzodioxane. Note that within PIM-1 backbone, there is a spiro-center where a single tetrahedral C atom shared by two aromatic rings. As can be seen from the molecular model in Fig. 6.4(b) the spiro-center introduces a sharp bend into the chain which resulted in a substantial amount of trapped free volume (FFV = 0.26) (Thomas et al. 2009). The backbone is fully aromatic; no single bonds in the backbone about which rotation can occur which resulted in considerably high glass transition temperature $T_g > 400$°C (Staiger et al. 2008).

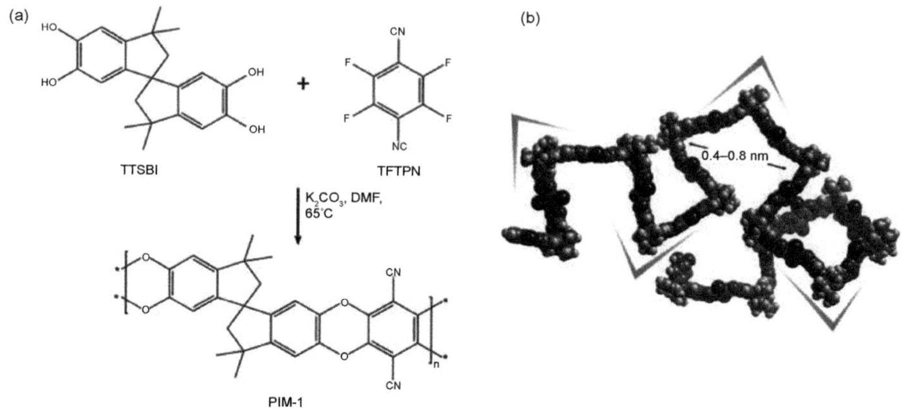

Figure 6.4. (a) PIM-1 synthesis route, and; (b) molecular model of a segment of PIM-1 showing its rigid, contorted structure (with sharp bends indicated by the arrows) (Budd et al. 2005).

Table 6.3. PIM-1 membranes for CO_2/CH_4 separation.

Casting solvent	Post-treatment	P_{CO_2} (Barrer)	$\alpha_{\frac{CO_2}{CH_4}}$	References
THF	-	2,300	18	Budd et al. (2005)
DCM	-	3,496	10	Staiger et al. (2008)
DCM	-	3,815	14	Yong et al. (2012)
$CHCl_3$	-	3,488	11	Hao et al. (2014)
$CHCl_3$	-	3,799	12	Wu et al. (2015)
$CHCl_3$	MeOH	12,600	17	Budd et al. (2008)
$CHCl_3$	MeOH	6,500	15	Thomas et al. (2009)
$CHCl_3$	MeOH	6,000	-	Ahn et al. (2010)
$CHCl_3$	MeOH	8,461	12	Du et al. (2012)
$CHCl_3$	MeOH	5,919	16	Swaidan et al. (2014)
$CHCl_3$	MeOH	10,683	8	Zhao et al. (2016)

Owing to its contorted structure PIM-based membranes exhibit very attractive separation performances, showing ultra-high permeability (easily surpasses thousands of barrers) while having a fairly good selectivity as shown in Table 6.3. However, the free volume elements could collapse due to polymer chain relaxation in a passage of time hence diminishing the permeability (Tiwari et al. 2017). Budd et al. (2008) revealed that post-treating the PIM membranes by means of soaking in methanol successfully prevent the collapse of free volume which resulted in much higher permeability than that of the pristine membrane. This procedure has been accustomed to conventional high free volume polymers to delay or reverse the effects of physical ageing (Merkel et al. 2000) or to prevent the pore from collapsing as in asymmetric membranes (Clausi and Koros 2000). Although a similar protocol has been applied by different researchers, wide variation in the performances can be observed, signifying that there is a reproducibility issue. For instance, Staiger et al. (2008), Yong et al. (2012), and Hao et al. (2014) reported the CO_2 permeability of PIM-1 membranes around 3,400 to 3,800 barrer, while Thomas et al. (2009), Ahn et al. (2010), and Swaidan et al. (2014) showed that the CO_2 permeability of PIM-1 after post-treatment with methanol raised to around 6,000 to 6,500 barrer. Budd et al. (2008) and Zhao et al. (2016) reported even staggering CO_2 permeability, exceeding 10,000 barrer. This variability might come from different chemical grades, synthesis conditions, and techniques which were not clearly disclosed in these previous publications. This suggests the need to further investigate the effects of different casting solvents and post-treatments, and the mechanisms of how these solvents affect the free volume formation in PIMs in order to identify the intrinsic performance of this polymer.

Owing to its high free volume and it large solubility coefficient, PIMs are essentially located above the Robeson's upper-bound. In order to further transcend to the upper-right corner of the diagram, the gas discriminating ability, i.e., the selectivity should be further enhanced. The most convenient strategy of obtaining high performance membranes would be through polymer blending. Blend membranes based on PIMs with different compositions of commodity polymers such as Matrimid (Yong et al. 2012, Yong et al. 2013), polyetherimide (Hao et al. 2014), polyamideimide (Torlon®) (Yong et al. 2014), and co-polyimide (P84) (Salehian et al. 2016) have been reported. Asymmetric hollow fiber membrane based on PIM-1 and its blends have also been published in recent years showing that blending PIM material with other polymers greatly increase its fabrication scalability (Yong et al. 2013, Jue et al. 2017).

As can be seen from Table 6.4, although the selectivity of the PIM-blend membranes is significantly higher than the pristine PIMs (CO_2/CH_4 selectivity < 20), the ultra-high permeability of PIM had also been sacrificed especially when the composition of PIM is too low. Nevertheless, from another perspective, the permeability of low permeable polymers can be considerably improved even with the addition of small amount of PIM (5 to 10 wt%) into the blend (Hao et al. 2014). Ideally, the permeability increment should follow a linear trend (on a logarithmic scale) as a function of PIM composition in the blend due to the reduction in density and increase in FFV. However, Yong et al. (2012) observed that PIM-1 and Matrimid blends were partially miscible when PIM constituted 20 to 80% from the blend resulted in large deviation in density-FFV relationship. The experimental permeability was found to be lower than the predicted value due to the unaccounted strong Charge Transfer Complexes (CTCs) between the ether groups in PIM-1 and carbonyl groups in the Matrimid. Hao et al. (2014) speculated that substantial drop in PIM-1 permeability from 2,877 barrer (PIM-1/PEI 90:10) to 477 barrer (PIM-1/PEI 70:30) possibly due to the filling of Ultem molecules into PIM-1 pores hence reduced the FFV. Similar to Yong et al. (2012), they also found that Ultem was only partially miscible with PIM-1 when the PIM-1 concentration was between 30 and 70%. This is ascribed to PIM-1 insolubility in common polar aprotic solvents which also could limit its full potential and possible industrial applications. In order to address this issue, PIM-1 have been structurally modified as carboxylated-PIM-1 (cPIM-1) by substituting the cyano- with carboxylic group; making it soluble in polar aprotic

Table 6.4. PIM-blend membranes for CO_2/CH_4 separation.

PIM-blends	P_{CO_2} (Barrer)	$\alpha_{\frac{CO_2}{CH_4}}$	References
Matrimid	9.6	36	Yong et al. (2012)
PIM-1/Matrimid (10:90)	17	34	
PIM-1/Matrimid (30:70)	56	31	
PIM-1/Matrimid (50:50)	155	28	
PIM-1/Matrimid (70:30)	558	25	
PIM-1/Matrimid (90:10)	1,953	16	
PIM-1	3,815	14	
PEI	1.48	37	Hao et al. (2014)
PIM-1/PEI (10:90)	3.95	34	
PIM-1/PEI (30:70)	9.27	35	
PIM-1/PEI (50:50)	52	23	
PIM-1/PEI (70:30)	477	17	
PIM-1/PEI (90:10)	2,877	12	
PIM-1	3,488	11	
Torlon	0.54	42	Yong et al. (2014)*
cPIM-1/Torlon (90:10)	1,013	23	
cPIM	2,654	13	
P84	1.30	44.8	Salehian et al. (2016)*
cPIM-1/P84 (90:10)	2,061	23.4	

* Carboxylated PIM is denoted as cPIM.

solvents, e.g., *N*-methyl-2-pyrrolidone (NMP) or *N,N*-dimethylformamide (DMF), and assisting the formation of hydrogen bonding (Yong et al. 2014, Salehian et al. 2016). Yong et al. (2014) and Salehian et al. (2016) blended cPIM-1 with Torlon and P48 polyimide, respectively. Although the intrinsic permeability for cPIM-1 is lower than PIM-1, due to its better solubility and miscibility with the secondary polymer, cPIM-1 90:10 blends were able to retain its high permeability (> 1,000 barrer) with enhanced selectivity (> 20).

Several studies have also reported the gas separation performances for other cyano-substituted PIMs, namely thioamide (Mason et al. 2011), methylated tetrazole (Du et al. 2012), amidoxime (Swaidan et al. 2014), and amine (Mason et al. 2014) as summarized in Table 6.5. Some modification efforts did result in profound enhancement in CO_2 selectivity of PIM membranes (CO_2/CH_4 > 20) (Mason et al. 2011, Du et al. 2012, Swaidan et al. 2014). Mason et al. (2014) demonstrated that the incorporation of primary amines in the PIM backbone strongly enhanced the CO_2

Table 6.5. Gas separation performances of structurally-modified PIMs.

Structurally-modified PIMs	P_{CO_2} (Barrer)	$\alpha_{\frac{CO_2}{CH_4}}$	References
Thio-PIM-1	1,120	20	Mason et al. (2011)
Methylated Tetrazole-PIM	1,391	22	Du et al. (2012)
Amidoxime-PIM-1	1,153	34	Swaidan et al. (2014)
Amine-PIM-1	1,890	6	Mason et al. (2014)
PIM-$COO^{-}Al^{3+}$	907	31	Zhao et al. (2016)
PIM-Triptycene-TB	9,709	11	Carta et al. (2014)
PIM-EA-TB	7,696	10	Carta et al. (2014)
Spirobifluorene-PIM	13,900	13	Bezzu et al. (2012)
6FDA-PIM-OH	208	35	Ma et al. (2012)
PMDA-PIM-OH	198	26	Ma et al. (2012)
PBO-PIM	635	13	Swaidan et al. (2013)

affinity which in turn reduced the membrane permeability and selectivity due to the reduction in CO_2 diffusivity. However, non-interacting gases like H_2 and He was not much affected, hence shifting the permselectiveness to H_2/CO_2 and He/CO_2. While recently, Zhao et al. (2016) employed a post-synthesis modification on PIM-1 with multivalent ions after converting the cyano group to carboxyl group. The presence of polar functional groups and cross-linkers leads to microstructural tightening due to enhanced intermolecular interactions which in turn improve the diffusivity selectivity significantly. Besides changing the cyano group from the PIM backbone, some researchers have replaced the spirobisindane segment with more rigid segment, e.g., 9,9′-spirobifluorene (SBF) (Bezzu et al. 2012) and triptycene (Carta et al. 2014). The rigid segment contributed to higher FFV hence improved the permeability notably. However, the replacement of spirobisindane with dianhydride segment lead to a tighter structure with lower FFV similar to polyimides. Therefore, the permeability would be much lower to typical PIMs, but with much higher selectivity. Furthermore, thermally-rearrangeable polyimide of intrinsic microporosity (TR-PI-PIMs) could be attained by polymerizing 6FDA-dianhydride with 3,3,3′,3′-tetramethyl-1,1′-spirobisindane-5,5′-diamino-6,6′-diol to form TR-able PI-PIM precursor (Ma et al. 2012, Swaidan et al. 2013). The PI-PIM is later subjected to thermal treatment, forming novel PBO-PIM as shown in Fig. 6.5. Swaidan et al. (2013) revealed that these membranes exhibited better stability under mixed-gas testing, where CH_4 permeability was suppressed hence negating the plasticization phenomenon which was often observed from conventional polyimides. It was also reported that PIM-1 backbone could be rearranged upon UV irradiation through hemolytic cleavage at the spiro-carbon center forming more stable backbone arrangement (Li et al. 2012, Li and Chung 2013).

The advancement in polymer backbone architecturing promises the birth of novel polymers which could possibly impose as game-changing material for membrane applications in CO_2 removal. In addition, due to its inherent microporosity and excellent chemical stability, PIMs were also found to be a promising material for adsorbents

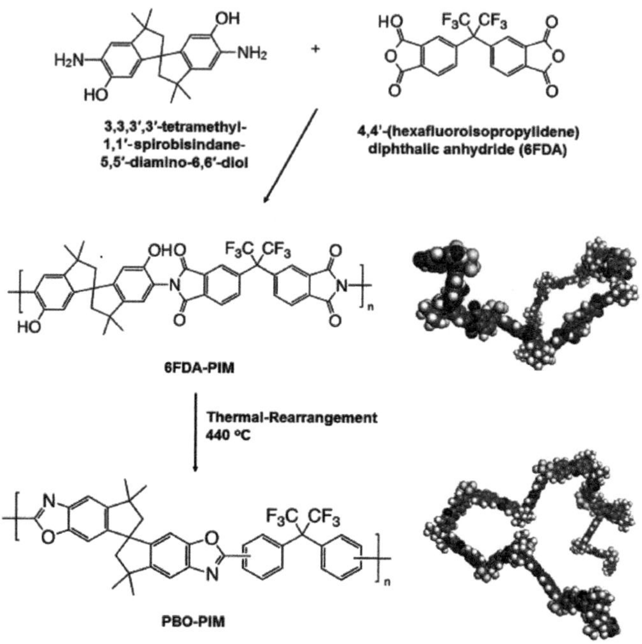

Figure 6.5. Illustration of 6FDA-PIM synthesis followed by thermal rearrangement to form PBO-PIM. The molecular model depicts the microporosity of contorted polymer structure (Swaidan et al. 2013).

(Pang et al. 2015), nanofiltration membranes for organic solvent separation (Fritsch et al. 2012, Satilmis et al. 2015) and other vapor/liquid phase membrane separations (Adymkanov et al. 2008, Thomas et al. 2009, Weng et al. 2015).

Inorganic Materials

As previously discussed for polymer-based membranes, gas separation is achieved by solution-diffusion model due to its dense structure. In contrast, gas molecules cannot penetrate the densely packed structure of inorganic materials. Around the 1940s, porous glass material comprising homogeneous pores of 20 – 40 Å in diameter was fabricated by Corning Incorporation (Pandey and Chauhan 2001). Such porous material separates gas molecules based on a sieving action; the size of the pore is smaller than the gas to be excluded while large enough to permit the permeation of the desired gas component. Since CO_2 kinetic diameter is very small (3.3 Å) and there is only a small size difference with CH_4 (3.8 Å), microporous molecular sieve of narrow pore size distribution has to be tailored. The following section reviews some of the most attractive microporous materials recently studied for CO_2 separation such as zeolite, carbon molecular sieves, graphene, carbon nanotubes (CNTs), and Metal Organic Frameworks (MOFs).

Zeolite

Zeolites are highly crystalline, hydrated aluminosilicate with opened three-dimensional framework structures, with regular intra-crystalline cavities and channels

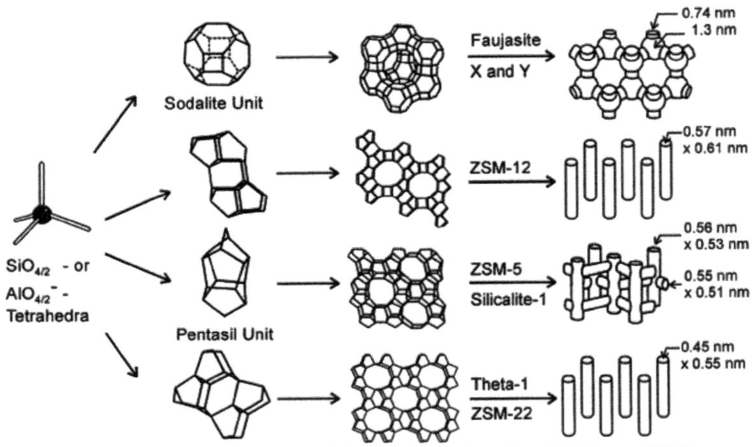

Figure 6.6. Structure of four different types of zeolites (Weitkamp 2000).

of molecular dimensions. Due to their narrow pore size distribution approaching the size of gas molecules, zeolites have the ability to accurately discriminate gas species primarily on the basis of molecular sizes. Most commonly used zeolites for gas adsorption and separation possess pores or interconnected cages of precisely sized apertures ranging between 2 and 10 Å. There are various types of zeolites such as Faujasite-type zeolites, SAPO-34 zeolites, MFI-type zeolite, ZSM-5, etc. Figure 6.6 shows the structure of four selected zeolites. Faujasite-type (FAU) zeolites comprised of silicon, aluminium and oxygen with varying Si/Al ratio. FAU-type zeolites consist of b-cages and hexagonal prisms, connected in such a manner that large internal super-cages are created. They had been produced commercially and widely applied as industrial adsorbents (Villarreal et al. 2017). Their open three-dimensional pore system results in fast intra-crystalline diffusion rates which makes them a promising CO_2 adsorbing material (Pillai et al. 2012, Prats et al. 2017).

Silicalite-1 is an MFI-type zeolite which composed of pure SiO_2 (Zhao et al. 2015, Wang et al. 2017). Silicalite-1 has the same crystalline structure as ZSM-5, but contains higher Si/Al ratio (> 1,000) while ZSM-5 ranges from 10 to infinity (Arudra et al. 2014, Zito et al. 2017). Silicalite-1 is favorable for the separation of gas mixtures containing non-polar molecules with similar physical properties due to the "smooth" adsorption surface containing few equilibrium ions (Wang et al. 2017). Despite its excellent selectivity and adsorption properties, the production of silicalite-1 is relatively costly due to exhaustive synthesis steps. The average time taken for synthesizing silicalite-1 is 4 to 12 days depending on the methods employed (Hu et al. 2017).

There are several available methods for preparing different types of zeolites (Table 6.6) such as hydrothermal method (Wang et al. 2010, Hu et al. 2017, Sun et al. 2017), steam assisted crystallization (Alfaro et al. 2001), crystal seed method (Wang et al. 2017) and others. Recently, Wang et al. (2017) successfully crystallized highly uniform microporous silicalite-1 (20 nm) in a relatively shorter period (20 hours) using a crystal-seeds method (Wang et al. 2017). The produced microporous silicalite-1 exhibited higher adsorption capacity and adsorption selectivity for CO_2/

Table 6.6. Methods to synthesize several types of zeolites.

Zeolite	Synthesis Method	References
Silicalite-1	Hydrothermal method	Wang et al. (2010), Hu et al. (2017)
	Embedded-seeding method	Sun et al. (2017)
	Steam assisted crystallization	Alfaro et al. (2001)
	Crystal seed method	Wang et al. (2017)
Zeolite 13X	Alkaline-fusion method	Kongnoo et al. (2017)
Faujasite	*In situ* hydrothermal method	Liu et al. (2016)
Zeolite Y	Rapid crystallization method	Wang et al. (2015)
ZSM-5	Facile synthesis method	Li et al. (2017)
Zeolite-4A	Hydrothermal method	Wynnyk et al. (2017), Vieira et al. (2015)

CH_4, CO_2/N_2, CH_4/N_2, and C_2H_6/CH_4 compared to mesoporous silicalite-1. Kongnoo et al. (2017) synthesized zeolite 13X from palm oil mill fly ash (POMFA) via alkaline fusion method. By employing acid treatment with 4–8 M HCl for 4 hours, a remarkable increase in the mesopore and total pore volume of the zeolite 13X were observed. The increased in the total pore volume lead to increment in CO_2 adsorption capacities compared to those of the untreated zeolite 13X (Kongnoo et al. 2017).

There have been a lot of studies reported on the use of silicalite as the adsorbents, as well as composites membranes for gas separations applications (Sun et al. 2017). Wang et al. (2017) synthesized microporous and mesoporous silicalite-1 and studied their adsorption properties towards various gas mixtures. High adsorption capacity and adsorption selectivity was observed in microporous silicalite-1 but higher selectivity was demonstrated by mesoporous silicalite-1 for all the gases tested (CH_4/N_2, CO_2/CH_4, CO_2/N_2 and C_2H_6/CH_4). This is attributed to the 'smooth' adsorption surface of silicalite-1 that is ideal for the separation of gas mixtures containing non-polar molecules with similar physical properties, such as CH_4/N_2. This suggested that the pore size only affects the adsorption capacity but not affecting their selectivity. Sjöberg et al. (2015) fabricated three different types of MFI zeolites; silicalite-1, NaZSM-5 and BaZSM-5. Micrometer-thick MFI film were grown on graded alumina supports. Higher CO_2 flux was observed in silicalite-1 membrane compared with the NaZSM-5 membrane and the BaZSM-5 membrane at all investigated conditions. They observed that silicalite-1 has the highest selectivity of CO_2/H_2 due to more favorable adsorption sites, which resulted in larger difference in fractional surface loading between feed and permeate side of the membrane. While recently, Sun et al. (2017) embedded silicalite-1 on the external surface of ceramic hollow fibers by a novel embedded-seeding method. They only studied the H_2/N_2 selectivity and found that membrane embedded with silicalite-1 displayed high H_2/N_2 selectivity because a continuous silicalite-1 membrane was formed on the Si-Al support even after one growth cycle. Hernandez et al. (2015) study the influence of the system geometry, which is tubular configuration using the Wicke–Kallenbach (WK) method. They concluded that the feed flow rates and the volume of the tubular separation system is the significant parameter in the tubular configurations. However, further studies are needed to understand the transport mechanisms in such structures.

Zeolite 4A is another types of zeolite with considerably high cation exchange capacity with the largest number of acidity sites (Ni et al. 2014). It has been extensively used in ion exchange processes such as drying gases and liquids and adsorption (Ni et al. 2014). From Vieira et al. (2015) study, adsorption capacity of CO_2 as high as 78.4% was recorded. Hauchhum and Mahanta (2014) compared the CO_2 adsorption capacity of zeolite 13X and zeolite 4A by pressure swing adsorption in fixed bed. Their findings showed that zeolite 13X offers higher adsorption capacity than that of zeolite 4A mainly due to higher surface area of zeolite 13X (720 m^2/g) compared to zeolite 4A (434 m^2/g), respectively.

Carbon Nanomaterials

Emerging carbon-based materials such as CNTs and graphene (specifically graphene oxide, GO) have been extensively employed in CO_2 capture as well as other separation processes since its first detail physical interpretation by Iijima (1991) and Novoselov et al. (2004), for CNTs and graphene, respectively. Graphene is an atom-thick, two-dimensional sheet of carbon arranged in honeycomb-like structure. It is considered as the world's thinnest material. The term "Graphene" combines the information of "graph" because of the graphite and "ene" for organic compounds with carbon-carbon double bonds, as in "alkene". Graphene can be regarded as the basic building block for all other carbonaceous graphitic nanomaterials of different structures. Carbon nanotube is the cylindrical arrangement of graphene sheet with a diameter as small as 0.4 nm and stretches in one-dimensional projection up to few microns (Labropoulus et al. 2015). CNTs, whether single-walled (SWCNT) or multi-walled (MWCNTs) can be synthesized from carbon precursors through various methods such as arc discharged, laser ablation and chemical vapor deposition (CVD) (Arora and Sharma 2014, Shah and Tali 2016). Furthermore, graphene could be obtained by exfoliation from graphite (Novoselov et al. 2004) which is considered as a top-down approach . Other than that, graphene can be grown directly from carbon precursor through CVD process, similar to CNTs (Zhang et al. 2013).

These materials are known for their unique nano-scale geometries with very high aspect ratio, astounding mechanical strengths and other remarkable features attributed to the carbon atom arrangements and hybridizations which make them excellent material in various modern applications ranging from microelectronics, sensors, reinforced materials, energy storage, etc. In gas separation applications, inherently smooth interior of CNTs channel act as a super-highway for molecular transport. Noy (2013) described the gas transport across the CNT channel by a simplified model of kinetic theory and postulated that it is of great importance that the channel is defect- and adsorbate-free for swift gas diffusion across the channel and along the nanotube surface. Controlling the orientation of the nanotubes parallel to the gas feed would allow for optimal gas permeation across the membrane. Figure 6.7 shows the morphology of the vertically aligned (VA) CNT membrane. Zhang et al. (2014a,b) reported the permeability of various gases 30–60 fold faster than the theoretical Knudsen diffusion for paralyne-VACNTs composite membrane which

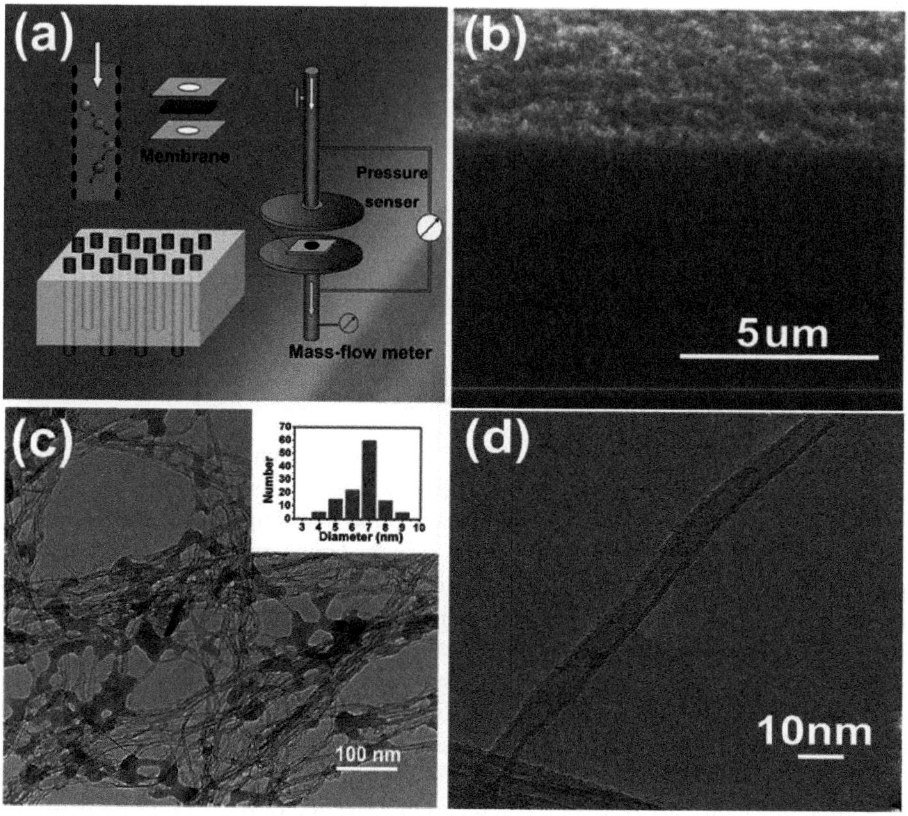

Figure 6.7. (a) Illustration of the VACNT membrane structure, assembly, and gas flow measurement; (b) SEM image of the as-grown VACNTs; (c) TEM image of the CNTs. The inset shows the histogram of inner diameters of the CNTs; (d) HRTEM image clearly shows the tube structure of the CNTs (Zhang et al. 2014a).

was ascribed to the small membrane thickness (about 10 μm). Successful fabrication of VACNT membrane with subnanometer channel (~ 0.4 nm) has also been reported by Labropoulus et al. (2015) by growing the CNTs based on zeolite seed.

For graphene, its oxidized derivative, known as Graphene Oxide (GO) is more favorable in gas separation and capture due to the abundance of oxygen functional groups on its plane which increases its solubility. In addition, the presence of intrinsic defects on its plane could act as molecular sieves. Pristine graphene on the other hand is perfectly impermeable, not even helium can pass through the atomically thin sheet (Bunch et al. 2008). Due to its atomically thin structure, the gas diffusion over the graphene nanopore is predicted to be extremely fast, even faster than in CNTs (Sun and Bai 2017). Recent studies have shown that pores passivated with nitrogen or fluorine atom would increase the selectivity of CO_2 over CH_4 by reducing the energy barrier for CO_2 permeation across the ultrathin nanoporous graphene (Wang et al. 2017). Although very high gas permeance was predicted for these membranes, CO_2/CH_4 selectivity is often in the range of Knudsen selectivity.

CNTs and graphene are amenably functionalized with various chemical compounds. Functionalization with CO_2-philic groups would be the most attractive strategy for improving their selectivity either as free-standing membrane or upon the incorporation into polymer membrane matrix (Ansaloni et al. 2015, Peng et al. 2017). Amine containing compound such as 3-aminopropyl-triethoxysilane (APTS) and tetraethylenepentamine (TEPA) has been used to modify CNTs for enhanced adsorption capacity (Cinke et al. 2003, Su et al. 2009, Irani et al. 2017). CO_2 is adsorbed on the particle surface by means of physical force thus enabling adsorbent regeneration at low temperature (Su et al. 2009). CNTs and graphene typically possess high specific surface area which is highly advantageous for CO_2 capture and storage (Szczęśniak et al. 2017). Ganesan and Shaijumon (2016) have designed graphene-based absorbent with ultra-high surface area (~ 3240 m^2/g) from thermally exfoliated graphite oxide by controlled KOH chemical activation process. Its high surface area and total pore volume (2.23 cm^3/g) provided large number of available sites for adsorption hence resulted in high CO_2 adsorption capacity of 21.1 mmol/g at 20 bar pressure. The interaction and adsorption properties of these carbon-based nanomaterials clearly suggest its prominent importance in the application of CO_2 separation and capture.

Furthermore, Carbon Molecular Sieves (CMS) is one of the carbon-based materials that have also been extensively synthesized for CO_2/CH_4 separation. CMS can attain high selectivity without losing the productivity and thus surpass the Robeson's upper bound. Unlike polymeric membranes, CMS membranes are usually durable and capable of withstanding adverse and rigorous environment because of their high thermal and chemical resistance (Park and Lee 2003). CMS demonstrates attractive characteristics among other molecular sieving materials for having excellent shape selectivity for planar molecule (Kyotani 2000). An ideal bimodal pore distribution is commonly used to represent CMS pore properties, which comprises of micropores (6–20 Å) connected with ultramicropores (< 6 Å). During gas separation process, micropores would provide sorption sites, while ultramicropores would enable molecular sieving, making CMS membrane both highly permeable and selective (Das et al. 2010). Among the polymer precursors that have been used in literatures, aromatic polyimide-type polymer appears to be one of the most promising candidates to yield CMSMs with superior separation properties. This is mainly due to the ability of this material to withstand the high temperature treatments without softening and decomposing rapidly, good physical properties, and high chemical stability (Fu et al. 2017). These materials can produce CMSM with high carbon yields and sustains their structural shapes after high temperature treatment. The great potential and advantages offered by CMS will definitely lead to their wide applications over the coming years.

Metal Organic Frameworks

Metal organic framework or MOFs are crystalline organic-inorganic hybrid compounds. They are made up of metal ions in the form of clusters, chains, layers or 3D arrangement, bonded to organic ligands which comprises of complexing segments such as carboxylates, phosphonates, and N-containing compounds. The organic and

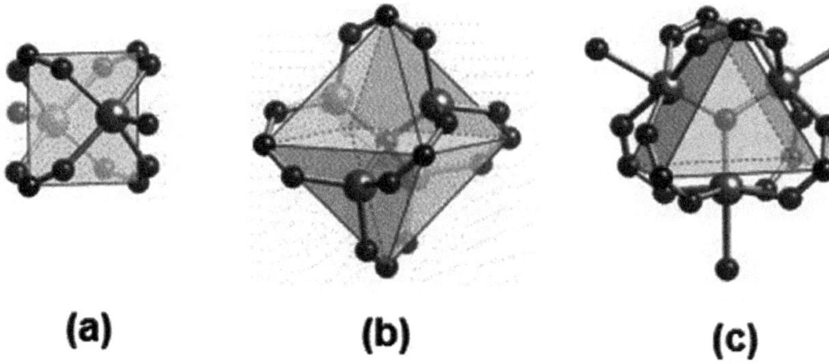

Figure 6.8. The Secondary Building Units (SBU) metal cluster bonded with organic linkers in the form of (a) square "paddlewheel", with two terminal ligand sites, (b) the octahedral "basic zinc acetate" cluster, and (c) the trigonal prismatic oxo-centered trimer, with three terminal ligand sites (Rowsell and Yaghi 2004).

inorganic phases are bonded via strong iono-covalent or dative bonds leading to various possible building units as illustrated in Fig. 6.8. There are innumerable types of MOFs since these materials can be engineered from almost all known elements in the periodic table. Structurally, MOF usually consisted of three-dimensional inorganic-organic hybrid networks originally formed from a cluster of metal ions (e.g., Al, Cu, Zn, and Cr). These elements could be strongly bonded by carboxylate or pyridyl organic groups. As a result of this strong bond, MOFs possessed a well-defined framework structure: rigid, dynamic, flexible and open metal sites. In comparison to other solid materials, one of the main advantage of MOF is its pore size and functionality. These properties can be tailored by selecting appropriate metal salts, organic linkers, functional groups as well as the synthesis and activation methods (Sabouni et al. 2014). The desired properties are: high specific surface area (> 1,000 m^2/g), high chemical and thermal stability, low densities (0.21 to 1 g/cm^3), and high volume of voids (55–90%) (Zhu and Xu 2014).

The structures of MOFs can be classified into four structures, which are open metal sites, surface functionalized framework, rigid framework, and dynamic framework. The structural differences result in different CO_2 capturing mechanisms. For MOFs with open metal sites, the selective separation of CO_2 occurs by means of the coordination of CO_2 at the end of the network. One example of MOFs with open site structures are Cu_3 (BTC_2) encompassing Cu(COO) metal connected by BTC linker (D'Alessandro et al. 2010). At elevated temperature, the open bound solvent molecules located at axial end of each Cu metal center will be removed (Venna and Carreon 2015). Based on the polarity and dipole moment, these open sites possessed charges that attract certain gases molecules to be attached. On the other hand, surface functionalized framework works by grafting different functional groups of high affinity towards CO_2 such as arylamine, akylamine, and hydroxyl groups to the surface of MOFs (Murray et al. 2010). As a result of introduction of the functional groups, the selective interaction between the functional groups and the frameworks themselves is enhanced. In contrast to non-functionalized framework, the functionalized MIL-53 showed an excellent CO_2 uptake and adsorption enthalpy

(Couck et al. 2009). For MOFs with rigid and flexible frameworks, the mechanism of adsorption is typically occurring due to molecular sieving of adsorbate-adsorbent interactions (Horike et al. 2009). The flexible framework of MOFs is more favorable for gas separation as this kind of structure weakens the size-exclusion separation ability of MOF-based molecular sieve membranes and also beneficial for adsorption-based separation such as pervaporation (Li et al. 2015).

Typically, MOFs are synthesized via solvothermal techniques. In this method, the solvent will be heated above its boiling point where this process allows better solubility of the metal salts (Li et al. 1999). This method provides consistent control over the synthesis conditions and able to produce MOFs of high crystallinity. Recently, a three-dimensional pillared-layer metal-organic framework has been synthesized using a partially methylated linear rigid linker (LH2) in the presence of the co-linker DPB (1,4-dipyridylbenzene) under solvothermal condition (Sahu et al. 2015). The desolvated framework shows significant CO_2 adsorption over N_2 and CH_4 at 273 K with an isosteric heat of adsorption of ~ 21 kJ/mol, suggesting a strong interaction of the CO_2 molecules with the framework walls. Besides solvothermal method, new methods have been reported for synthesizing MOFs such as microwave synthesis and mechanochemical synthesis (Butova et al. 2016). The microwave-assisted method requires electrical radiation to facilitate the reaction between the organic linker and free inorganic ions which will be aligned within the alternating fields (Lee et al. 2013). This method is able to provide higher energy to overcome the activation energy barriers during the synthesis. As a result, the synthesis duration could be shortened. Cho et al. (2015) successfully synthesized high-quality Co-MOF-74 crystals desired for CO_2 adsorption and catalytic application only in an hour. As for mechanochemical synthesis, this method does not require any electrical current, instead, it basically relies on mechanical energy, e.g., ball milling (James et al. 2012). For this method, none or a very small amount of solvent is used (Bowmaker 2013). The mechanochemical approaches can be divided into two: hot-spot model and magma-plasma model. The former approach involves the contact of two slid surfaces. The friction between these two surfaces causes a sharp increase of temperature and thus yields MOFs as the product. Masoomi et al. (2014) reported two new MOFs, TMU-4 and TMU-5 with dicarboxylate and linear N- donor pyridyl-based ligands prepared by mechanosynthesis which involved 15 minutes of hand-grinding. For the case of magma-plasma model, diffusion and enhanced reactivity is achieved as a result of instant release of large energy, hence releasing plasma. This state-of-the-art method is able to yield highly crystalline MOFs with higher product yield. However, very limited theories has been provided to explain the reaction occurred during the plasma-induced process (Butova et al. 2016).

The advancement of MOFs synthesis methods and performances has gained significant attention during recent years for gas adsorption and separation applications. Conventional porous adsorbents (e.g., zeolites, porous carbon and porous silica) although possess good affinity towards CO_2, they have low internal surface area hence low CO_2 uptakes (Yong et al. 2002). There three important criteria of a good adsorbent: (1) the presence of functional groups or open metal sites; (2) high specific surface area; and (3) narrow pore size distribution (Howarth et al. 2017, Xiang et al. 2012). The presence of functional groups or open metal sites during dry condition will allow the polarizability of CO_2 and thus boost the CO_2 capture (Venna and Carreon 2015). The high surface area of MOFs with open metal sites act as a large platform for adsorption while selective towards specific gases.

It was reported that $Cu_3(BTC_2)$ able to adsorb 12.7 mmol CO_2/g at 15 bar, and still did not reach it saturation point due to a large surface area (1571 m^2/g) and pore volume (0.79 cm^3/g) (Liang et al. 2009). A MOF named as Mg-MOF-74 possessed a surface area of 1174 m^2/g and has open sites has been reported to has the highest CO_2 uptake which was 37.9 wt% at room temperature with pressure of 1 bar (Rosi et al. 2005). The structure of this MOFs is similar to the MOF-74 (Zn) but the metal cluster can be changed to either Mg, Ni, Cu and Zn while the linker is 2,5-dihydroxyterephthalic acid (Rosi et al. 2005). Cu-TDPAT with specific surface area of 1938 m^2/g had also showed a good CO_2 uptake which was 26.0 wt% (Li et al. 2012). Cu-TDPAT composed of open metal sites of Cu_2(-COO_4) and organic linker of 2,4,6-tris(3,5-dicarboxylphenylamino)-1,3,5-triazine. Zeolitic MOFs such as Zn(BTZ) composed of Zn^{2+} and organic linker of 1,5-bis(5-tetrazolo)-3-oxapentane) has been reported to capture CO_2 up to 8.1%, and the introduction of amine groups has successfully enhance the overall uptake (Cui et al. 2012). Other zeolitic MOFs such as ZIF-300, ZIF-301 and ZIF-302 that consist of tetrahedral Zn imidazole and functionalized imidazolate linker normally capture CO_2 based on physisorption (Poloni et al. 2014). Even with the presence of humidity up to 80%, no performance loss was observed when these zeolitic MOFs were used. SIFSIX-3-Zn MOF composed of Zn^{2+} and pyrazine displayed a very good affinity towards CO_2 as it possess narrow pore size, allowing capture of the gas molecules at confined space (Nugent et al. 2013). Similar to zeolitic MOFs, these MOFs can also be used with the presence of high humidity, approximately 74% without causing any uptake loss. The narrowly distributed pores sizes with the dimensions equivalent to that of the gas molecular diameter are some important criteria of MOFs for gas separation. From the kinetic perspective, an adsorbent with a well-defined pore size between the kinetic diameters of the gas molecules would provide an infinitely high selectivity. Additionally, the polarity of the framework enables weak interactions between the porous materials and the guest molecules, which further facilitate the separation of the mixture (Yao et al. 2017).

Advanced Mixed Matrix Membranes

Although microporous inorganic materials might offer excellent permeability and selectivity transcending the Robeson's upper bound, the intricacy of fabrication would cause the material to cost 1 to 3 orders of magnitude more expensive than the conventional polymer-based membranes (Koros and Mahajan 2000). Inorganic membranes also face reproducibility issues and it is difficult to obtain defect-free membranes with a large quantity or as a module with large surface-to-volume ratio (Lai 2018). The applicability of inorganic-based membranes has also been hindered due to its fragile nature. On the other hand, while currently commercially available polymeric membranes are a cheaper solution, their performance are no match to that of inorganic membranes (Pandey and Chauhan 2001). The advent in membrane science has shown that the individual deficiencies of polymeric and inorganic membranes could be overcome by combining both phases giving birth to a composite material which will possess the synergistic properties of each phase: high separation performance of the microporous inorganic phase as the dispersed filler and the ease of processability of the polymer as the membrane matrix. This approach can be regarded as a potential mean to overcome the trade-off relationship and the performance upper boundary on the Robeson's diagram. Furthermore, the state-of-the-art organic–inorganic hybrid membranes, or Mixed Matrix Membranes (MMMs) would also possess enhanced physical, thermal and mechanical properties which could withstand the aggressive and harsh environments (high pressure, temperature, corrosive, etc.) in natural gas processing (Cheng et al. 2018). To date, MMMs have been applied to various membrane applications such as pervaporation (Garg et al. 2011, Khosravi et al. 2012), fuel cell (Jaafar et al. 2011), photocatalysis (Li et at. 2014), and water purification application (Das et al. 2014, Zinadini et al. 2014), to name a few. Studies analogous to gas separation MMMs have been evolving at an unprecedented pace. Some researchers have also demonstrated that one may incorporate more than one type of fillers into the membrane matrix to further enhance the gas separation performance (Zornoza et al. 2011, Zornoza et al. 2012, Galve et al. 2013, Valero et al. 2014, Li et al. 2015, Castarlenas et al. 2017). The MMM concept is best portrayed by Fig. 6.9.

As postulated for MMMs based on common polymer-filler systems, the addition of porous fillers in those advanced microporous polymers may also result in the increase of free volume due to the combination of nanoparticles' cavities and loosen polymer packing at the filler-polymer interface. While, good filler and polymer interactions would restrict the chain motion about its rotation thus increase its rigidity. For instance, the incorporation of 1wt% zeolite-T into 6FDA-durene resulted in 25% increase in FFV and 4% in chain rigidity (on account to the increment in glass transition temperature); hence resulted in a profound improvement for both CO_2 permeability and CO_2/CH_4 selectivity (Jusoh et al. 2017). Additional free volume as a result of filler perturbation in between the polymer chains was observed to enhance the diffusion coefficient of ZIF-8-filled PIM-1 (Bushell et al. 2013). Li et

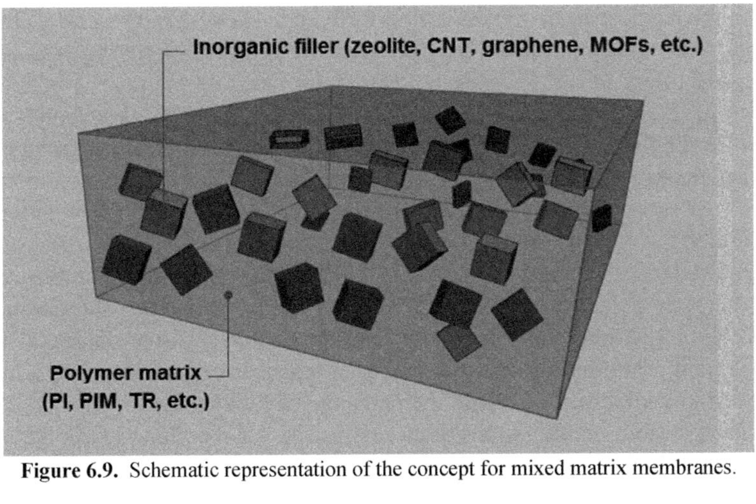

Figure 6.9. Schematic representation of the concept for mixed matrix membranes.

al. (2015) demonstrated that the addition of CNTs resulted in increasing permeability but lower selectivity, while the incorporation of GO increased the selectivity but lower permeability. Subsequently, the addition of CNT and GO led to significant enhancement in both permeability and selectivity suggesting the synergistical effect from both filler.

While more recently novel covalent organic framework based on Schiff-base network (SNW-1) was blended with PIM-1 membrane exhibited total augmentation on the membrane's properties; 20% increase in tensile strength, 40% in Young's modulus, and 65% in FFV which contributed to almost 70% enhancement in diffusivity coefficient with 10 wt% of filler loading (Wu et al. 2017). NH_2-containing filler such as SNW-1 or other NH_2-functionalized fillers (Nik et al. 2012, Ansaloni et al. 2015, Khdhayyer et al. 2017) would form favorable hydrogen bonds with cyano group and ether linkage in PIMs as well as in polyimides which ensure exceptional compatibility. Utilization of various silane agents have been reported for modifying the surface of fillers, which acts as the bridge between the filler and the polymer phase (Chen et al. 2012, Ansaloni et al. 2015). Stronger and more stable bonding between the polymer chain and the inorganic filler could be attained by covalent cross-linking. Hydroxyl-functionalized MOF-74 was covalently cross-linked at the PIM-1 terminal fluoride hence synergistically forming an inter-connected microporous network throughout the MMMs (Tien-Binh et al. 2016). The membranes exhibited astounding CO_2 permeability and CO_2/CH_4 selectivity of 21,000 barrer and 19, respectively. The cross-linked membranes were also proven stable under mixed gas and high CO_2 partial pressure conditions, displaying unprecedented anti-plasticization properties.

Meanwhile, it is typical for high free volume polymers that they are prone to physical ageing (compaction of the free volume) which could result in loss of permeability up to 50% compared to as-fabricated membranes (Starannikova et al. 2004, Budd et al. 2008, Staiger et al. 2008, Carta et al. 2014). Brunetti et al. (2017) hypothesized that the addition of inorganic fillers into high free volume polymers might be efficient for suppressing the physical ageing phenomenon. They fabricated TR-able 6FDA-based co-polyimide with oxidized-MWCNTs as the dispersed phase. Heat treatment at 375°C yielded TR-PBOI-MWCNT MMMs which demonstrated notable improvement in its separation performance. However, no significant impact on ageing was observed. They did find something more interesting; the presence of inorganic fillers in the TR-able polymer matrix would result in a phenomenon exclusive for this type of polymers. The presence of MWCNTs was believed to be responsible for the enhancement of the thermal rearrangement to a greater degree (11.3%) compared to the pristine TR membrane (6.7%). Thermal conductivity of MWCNTs is substantially higher that the matrix, hence it may allow greater heat conduction to the polymer network and accelerates the rearrangement process. It was also suggested that the acid functionalities on the CNTs would catalyze the acid-base and/or charge transfer complexes with the polymer, reducing the activation energy of the decarboxylation-cyclization reaction, hence enhances the degree of thermal rearrangement. This finding may also suggest that rearrangement process might be achieved at lower temperature with the presence of catalyst.

To date, various combinations of inorganic fillers and commodity polymers have been reported, but it was rather limited for MMMs-based on structurally modified

Table 6.7. Mixed matrix membranes for CO_2/CH_4 separation.

Mixed Matrix Membranes	Filler Loading (wt%)	P_{CO_2} (Barrer)	α_{CO_2/CH_4}	Operating conditions (°C/bar)		References
Matrimid/SAPO-34	20	7 (+55%)	67 (+97%)	25	10	Peydayesh et al. (2013)
Matrimid/ZIF-12	40	13 (+122%)	67 (+133%)	35	4	Boroglu et al. (2017)
Matrimid/GO-UiO-66	24	21 (+200%)	51 (+46%)	35	1	Castarlenas et al. (2017)
Matrimid/CuBTC	20	25 (+226%)	38 (UC)	35	2	Duan et al. (2014)
Matrimid/GO-CNT 50:50	10	38 (+330%)	85 (+150%)	30	2	Li et al. (2015)
ODPA-TMPDA/CuBTC	20	260 (+940%)	29 (UC)	35	2	Duan et al. (2014)
6FDA-ODA/UiO-66	25	50 (+250%)	46 (UC)	35	10	Nik et al. (2012)
6FDA-ODA/NH_2-UiO-66	25	14 (-5%)	52 (+17%)	35	10	Nik et al. (2012)
6FDA-ODA/CuBTC	25	22 (+51%)	51 (+16%)	35	10	Nik et al. (2012)
6FDA-ODA/NH_2-CuBTC	25	27 (+85%)	60 (+35%)	35	10	Nik et al. (2012)
6FDA-ODA/FAU-EMT	25	14 (-20%)	55 (+30%)	35	10	Chen et al. (2012)
6FDA-durene/ZIF-8	5	693 (+48%)	17 (+136%)	30	3.5	Jusoh et al. (2016)
6FDA-durene/zeolite-T	1	844 (+80%)	19 (+173%)	30	3.5	Jusoh et al. (2017)
6FDA-DAM/Noria	9.4	289 (-47%)	21 (+28%)	25	1	Mao and Zhang (2017)
6FDA-DAM/Noria-CO^{t}Bu	20	543 (+27%)	15 (UC)	25	1	Mao and Zhang (2017)
6FDA-DAM/ZIF-11	20	258 (+1148%)	31 (UC)	30	4	Boroglu and Yumru (2017)
6FDA-3MPA/APTES-Silica	25	1,920 (+237%)	28 (+33%)	35	1	Hasebe et al. (2017)
BTDA-DMMDA/O-MWCNT	3	9 (+292%)	25 (+143%)	15	1	Sun et al. (2017)
TR-PBOI/O-MWCNT	0.5	126 (+20%)	26 (UC)	35	10	Brunetti et al. (2017)
PIM-1/FS	23.5[a]	13,400 (+123%)	-	23	3.5	Ahn et al. (2010)
PIM-1/MFI-3	35.5[a]	2,530 (–74%)	14 (UC)	25	1	Mason et al. (2013)

Table 6.7 contd. ...

...Table 6.7 contd.

Mixed Matrix Membranes	Filler Loading (wt%)	P_{CO_2} (Barrer)	α_{CO_2/CH_4}	Operating conditions (°C/bar)		References
PIM-1/SWCNT	1	15,721 (+154%)	9 (–80%)	30	2	Khan et al. (2013)
PIM-1/MWCNT	1	7,813 (+26%)	10 (–57%)	30	2	Khan et al. (2013)
PIM-1/g-C_3N_4	1	10,528 (+62%)	8 (–37%)	30	2	Tian et al. (2016)
PIM-1/ZIF-8	28[a]	17,050 (+35%)	12 (–44%)	21	1	Bushell et al. (2013)
PIM-1/ZIF-8	43[a]	19,350 (+54%)	7 (–133%)	21	1	Bushell et al. (2013)
PIM-1/Mg-MOF-74	20	21,269 (+223%)	19 (+60%)	25	2	Tien-Binh et al. (2016)
PIM-1/UiO-66	30	11,000 (+33%)	14 (–14%)	25	1	Khdhayyer et al. (2017)
PIM-1/UiO-66-NH_2	30	9,000 (+8%)	13 (–20%)	25	1	Khdhayyer et al. (2017)
PIM-1/UiO-66-$(COOH)_2$	30	9,000 (+8%)	14 (–12%)	25	1	Khdhayyer et al. (2017)
PIM-1/SNW-1	10	7,553 (+106%)	14 (+27%)	30	2	Wu et al. (2017)

[a] Filler loading based on volume percent (vol%).
Parentheses indicate the percentage improvement over the neat membrane, UC indicates nearly unchanged value.
Acronyms: O-MWCNT (oxidized MWCNT), FS (fumed-silica), Noria-Co^{t}Bu (Noria containing t-butyl ester groups), SNW-1 (covalent organic framework).

polyimides, PIMs, and especially TR polymers. Despite its limited publications, performance summary in Table 6.7 proved that the incorporation of advanced inorganic fillers such as microporous zeolite, CNTs, graphene, and various MOFs have helped to improve the performance of these advanced polymeric membranes to another level. Although in several studies, the permeability-selectivity trade-off "curse" still exists after the incorporation of the discrete particle phase, it could be overcome by ensuring the fine control of filler-polymer interfacial interactions, its morphology, and the distribution throughout the polymer matrix. Improvements in membrane fabrication protocols as well as the modifiability of the fillers' surface chemistry through various functionalization techniques have resulted in MMMs with far more superior performances as previously discussed.

To gain a better insight on the current development of membrane materials for CO_2 removal, CO_2 separation performances of advanced polymers summarized in Tables 6.1-6.5 and their corresponding MMMs (Table 6.7) incorporated with different type of porous inorganic fillers discussed in the previous section are charted on the Robeson's permeability-selectivity diagram for performance analysis. It can be clearly seen in Fig. 6.10 that each class of polymer scattered in a similar region: polyimides having high selectivity but relatively lower permeability positioned mostly

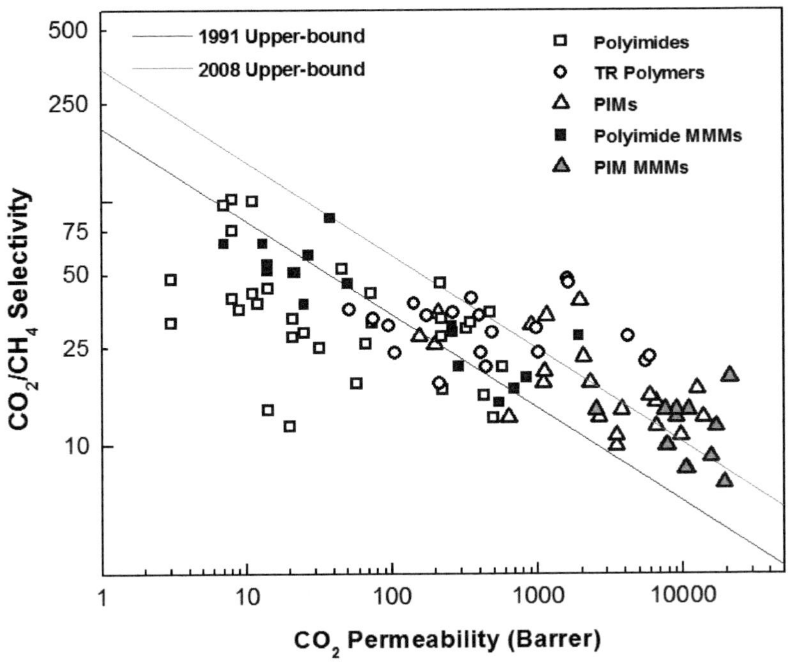

Figure 6.10. Robeson's upper-bound analysis of polyimides, TR polymers, PIMs and their respective MMM derivatives.

below the upper-bound; TR polymers are in between the 1991's and 2008's upper-bound for having good permeability and selectivity, and; PIMs are located further to the right due to its super-permeability, hence surpassing both the upper-bounds. Note that, most MMMs, incorporated with various fillers possess improved performance moving to the upper-right direction. One may also realize that, the upper-bound should be renewed, but with less steep slope. Witnessing such positive progresses in material development, it is not long before these materials could substitute the conventional membranes commonly made of cellulose acetate and polysulfone to be used for sweetening the natural gas as well as in other gas separation processes.

Conclusion and Future Directions

The advances in chemistry of polymer synthesis have led to assortment of selective and more permeable polymeric membranes. Pseudo-microporous polymers such as TR and PIMs have shown performance commensurate with the most promising candidates for CO_2 separation processes. Polymer molecular architecturing consisting stiff chain and non-collapsing free volumes as in TR polymers is highly desirable. For TR-based membranes to be more attractive, the formation route should be further simplified in order to achieve the molecular rearrangement. Instead of thermally-induced molecular rearrangement, catalytic-rearrangement at low temperature, or even, at room temperature might be possible to produce high free volume, PBOs or PBIs with similar microporosity to that of TR counterparts.

Despite the attractive gas transport properties of most microporous inorganic materials such that zeolites, MOFs, CNTs and graphene, free standing membranes derived from its constituent material could only be formed in small scale. We have witnessed that with the addition of rigid and permanently microporous fillers, the overall properties of polymeric membranes could be synergistically enhanced; improving the separation performance through molecular sieving action as well as reinforcing its physical stability. Their strong polymer-inorganic structure in MMMs allows for stable performance even under harsh conditions as in natural gas sweetening process. Evidenced by increasing number of publications related to combinations of these advanced materials, we may witness an upward shift of the upper bound boundary in the near future.

One of the key issues is transforming these advanced membranes into a reliable design of industrial membrane modules with high surface area-to-volume ratio. Most studies have almost exclusively focused on dense flat sheet or thin film composite membrane morphologies. To put industrial applicability into perspective, consider a conventional CTA membrane with intrinsic permeability ca. 10 barrer gives of CO_2 permeance of 100GPU when the membrane's thickness is 100 nm. Imagine that if PIM-1 membrane of the same thickness could possibly be fabricated the CO_2 permeance would reach up to 10,000 GPU. Therefore, it is of great interest to see the exciting development of these materials having ultra-thin skinned morphology which would revolutionize the technology for gas separation. Effort in fabricating asymmetric membrane comprising the advanced materials discussed in this chapter has been seen underway but may need to be further optimized and streamlined. Great challenge lies in the appropriate scaling-up from laboratory-scale hollow fiber membrane modules to industrial-scale modules, which requires a thorough understanding on the impacts of operating conditions (pressure, temperature, and concentration profiles) as well as fluid flow behavior at the membrane interfaces in the module. Scientific research conducted thus far has set exciting paths for unimaginable innovation and might lead to fruition of novel materials with astounding properties for addressing our industrial needs as well as mitigating global warning issues related to the emission of greenhouse gases into our environment.

Acknowledgements

The authors would like to acknowledge research management and financial supports from Universiti Teknologi Malaysia and the financial supports from the Ministry of Higher Education under research grants 02E65, 02E26, and 02K04.

References

Adewole, J.K., Ahmad, A.L., Ismail, S. and Leo, C.P. 2013. Current challenges in membrane separation of CO_2 from natural gas: A review. Int. J. Greenhouse Gas Control. 17: 46–65.

Adymkanov, S.V., Yampolskii, Y.P., Polyakov, A.M., Budd, P.M., Reynolds, K.J., McKeown, N.B. et al. 2008. Pervaporation of alcohols through highly permeable PIM-1 polymer films. Polym. Sci. Ser. A. 50: 444–450.

Ahn, J., Chung, W.J., Pinnau, I., Song, J., Du, N., Robertson, G.P. et al. 2010. Gas transport behavior of mixed-matrix membranes composed of silica nanoparticles in a polymer of intrinsic microporosity (PIM-1). J. Membr. Sci. 346: 280–287.

Alaslai, N., Ghanem, B., Alghunaimi, F. and Pinnau, I. 2016. High-performance intrinsically microporous dihydroxyl-functionalized triptycene-based polyimide for natural gas separation. Polymer 91: 128–135.

Alfaro, S., Arruebo, M., Coronas, J.N., Menéndez, M. and Santamaría, J. 2001. Preparation of MFI type tubular membranes by steam-assisted crystallization. Microporous Mesoporous Mater. 50: 195–200.

Ansaloni, L., Zhao, Y., Jung, B.T., Ramasubramaniam, K., Baschetti, M.G. and Ho, W.S.W. 2015. Facilitated transport membranes containing amino-functionalized multi-walled carbon nanotubes for high-pressure CO_2 separations. J. Membr. Sci. 490: 18–28.

Arora, N. and Sharma, N.N. 2014. Arc discharge synthesis of carbon nanotubes: Comprehensive review. Diamond & Related Materials 50: 135–150.

Arudra, P., Bhuiyan, T.I., Akhtar, M.N., Aitani, A.M., Al-Khattaf, S.S. and Hattori, H. 2014. Silicalite-1 As Efficient Catalyst for Production of Propene from 1-Butene. ACS Catalysis 4: 4205–4214.

Baiyan, L., Zhang, Z., Li, Y., Yao, K., Zhu, Y., Deng, Z. et al. 2012. Enhanced Binding Affinity, Remarkable Selectivity, and High Capacity of CO_2 by Dual Functionalization of a Rht-Type Metal-Organic Framework. Angew. Chem. Int. Ed. 51(6): 1412–15.

Baker, R.W. and K. Lokhandwala. 2008. Natural gas processing with membranes: An Overview. Ind. Eng. Chem. Res. 47: 2109–2121.

Barrer, R.M. 1951. Diffusion in and through solids. Cambridge University Press, London.

Bezzu, C.G., Carta, M., Tonkins, A., Jansen, J.C., Bernardo, P., Bazzarelli, F. et al. 2012. A spirobifluorene-based polymer of intrinsic microporosity with improved performance for gas separation. Adv. Mater. 24: 5930–5933.

Boroglu, M.S. and A.B. Yumru. 2017. Gas separation performance of 6FDA-DAM-ZIF-11 mixed-matrix membranes for H_2/CH_4 and CO_2/CH_4 separation. Sep. Purif. Technol. 173: 269–279.

Boroglu, M.S., M. Ugur and I. Boz. 2017. Enhanced gas transport properties of mixed matrix membranes consisting of Matrimid and RHO type ZIF-12 particles. Chem. Eng. Res. Des. 123: 201–213.

Bowmaker, G.A. 2013. Solvent-Assisted Mechanochemistry. Chem. Commun. 49 (4): 334–48.

Brunetti, A., Cersosimo, M., Dong, G., Woo, K.T., Lee, J., Kim, J.S. et al. 2016. *In situ* restoring of aged thermally rearranged gas separation membranes. J. Membr. Sci. 520: 671–678.

Brunetti, A., Cersosimo, M., Kim, J.S., Dong, G., Fontananova, E., Lee, Y.M. et al. 2017. Thermally rearranged mixed matrix membranes for CO2separation: An aging study. Int. J. Greenhouse Gas Control 61: 16–26.

Budd, P.M., Ghanem, B.S., Makhseed, S., McKeown, N.B., Msayib, K.J. and Tattershall, C.E. 2004. Polymers of intrinsic microporosity (PIMs): robust, solution-processable, organic nanoporous materials. Chem. Commun. 230: 230–231.

Budd, P.M., Msayib, K.J., Tattershall, C.E., Ghanem, B.S., Reynolds, K.J., McKeown, N.B. et al. 2005. Gas separation membranes from polymers of intrinsic microporosity. J. Membr. Sci. 251: 263–269.

Budd, P.M., McKeown, N.B., Ghanem, B.S., Msayib, K.J., D. Fritsch Starannikova, L. et al. 2008. Gas permeation parameters and other physicochemical properties of a polymer of intrinsic microporosity: Polybenzodioxane PIM-1. J. Membr. Sci. 325: 851–860.

Bunch, J.S., Verbridge, S.S., Alden, J.S., van der Zande, A.M., Parpia, J.M., Craighead, H.G. et al. 2008. Impermeable atomic membranes from graphene sheets. Nano Lett. 8(8): 2458–2462.

Bushell, A.F., Attfield, M.P., Mason, C.R., Budd, P.M., Tampolskii, Y., Starannikova, L. et al. 2013. Gas permeation parameters of mixed matrix membranes based on the polymer of intrinsic microporosity PIM-1 and the zeolitic imidazolate framework ZIF-8. J. Membr. Sci. 427: 48–62.

Butova, V.V., Soldatov, M.A., Guda, A.A., Lomachenko, K.A. and Lamberti, C. 2016. Metal-Organic Frameworks: Structure, Properties, Methods of Synthesis and Characterization. Russ. Chem. Rev. 85(3): 280–307.

Calle, M. and Lee, Y.M. 2011. Thermally Rearranged (TR) Poly(ether-benzoxazole) Membranes for Gas Separation. Macromolecules 44: 1156–1165.

Calle, M., Doherty, C.M., Hill, A.J. and Lee, Y.M. 2013. Cross-Linked Thermally Rearranged Poly(benzoxazole-co-imide) Membranes for Gas Separation. Macromolecules 46(20): 8179–8189.

Carta, M., Croad, M., Malpass-Evans, R., Jansen, J.C., Bernardo, P., Clarizia, G. et al. 2014. Triptycene induced enhancement of membrane gas selectivity for microporous Troger's base polymers. Adv. Mater. 26: 3526–3531.

Castarlenas, S., Tellez, C. and Coronas, J. 2017. Gas separation with mixed matrix membranes obtained from MOF UiO-66-graphite oxide hybrids. J. Membr. Sci. 526: 205–211.

Chang, J.H., Park, K.M., Lee, S.M. and Oh, J.B. 2000. Two-Step Thermal Conversion from Poly(amic acid) to Polybenzoxazole via Polyimide: Their Thermal and Mechanical Properties. J. Polym. Sci.: Part B: Polymer. Physics. 38: 2537–2545.

Chen, X.Y., Nik, O.G., Rodrigue, D. and Kaliaguine, S. 2012. Mixed matrix membranes of aminosilanes grafted FAU/EMT zeolite and cross-linked polyimide for CO_2/CH_4 separation. Polymer. 53: 3269–3280.

Cheng, Y., Wang, Z. and Zhao, D. 2018. Mixed matrix membranes for natural gas upgrading: current status and opportunities. Ind. Eng. Chem. Res. 57(12): 4139–4169.

Cho, H.-Y., Yang, D.-A., Kim, J., Jeong, S.-Y. and Ahn, W.-S. 2012. CO2 adsorption and catalytic application of Co-MOF-74 synthesized by microwave heating. Cat. Today. 185 (1): 35-40.

Cinke, M., Li, J., Bauschlicher, Jr., Ricca, A. and Meyyappan, M. 2003. CO_2 adsorption in singlewalled carbon nanotubes. Chem. Phys. Lett. 376: 761–766.

Clausi, D.T. and Koros, W.J. 2000.Formation of defect-free polyimide hollow fiber membranes for gas separations. J. Membr. Sci. 167(1): 79–89.

Comesaña-Gándara, B., Hernandez, A., de la Campa, J.G., de Abajo, J., Lozano, A.E. and Lee, Y.M. 2015. Thermally rearranged polybenzoxazoles and poly(benzoxazole-co-imide)s from ortho-hydroxyamine monomers for high performance gas separation membranes. J. Membr. Sci. 493: 329–339.

Couck, S., Denayer, J.F.M., Baron, G.V., Rémy, T., Gascon, J. and Kaptejin, F. 2009. An Amine-Functionalized MIL-53 Metal-Organic Framework with Large Separation Power for CO_2 and CH_4. J. Am. Chem. Soc. 131(18): 6326–27.

Cui, P., Ma, Y.G., Li, H.H., Zhao, B., Li, J.R., Cheng, P. et al. 2012. Multipoint Interactions Enhanced CO_2 Uptake: A Zeolite-like Zinc-Tetrazole Framework with 24-Nuclear Zinc Cages. J. Am. Chem. Soc. 134(46): 18892–95.

D'Alessandro, D.M., Smit,B. and Long, J.R. 2010. Carbon Dioxide Capture: Prospects for New Materials. Angew. Chem. Int. Ed. 49(35). 6058–6082.

Das, M., Perry, J.D. and Koros, W.J. 2010. Gas-transport-property performance of hybrid carbon molecular sieve-polymer materials. Ind. Eng. Chem. Res. 49(19): 9310–9321.

Das, R., Ali, M.E., Hamid, S.B.A., Ramakrishna, S. and Chowdhury, Z.Z. 2014. Carbon nanotube membranes for water purification: A bright future in water desalination. Desalination. 336: 97–109.

Dong, G., Woo, K.T., Kim, J., Kim, J.S. and Lee, Y.M. 2015. Simulation and feasibility study of using thermally rearranged polymeric hollow fiber membranes for various industrial gas separation applications. J. Membr. Sci. 496: 229–241.

Du, N., Dal-Cin, M.M., Pinnau, I., Nicalek, A., Robertson, G.P. and Guiver, M.D. 2011. Azide-based Cross-Linking of Polymers of Intrinsic Microporosity (PIMs) for Condensable Gas Separation. Macromolecular Rapid Comm. 32(8): 631–636.

Du, N., Robertson, G.P., Dal-Cin, M.M., Scoles, L. and Guiver, M.D. 2012. Polymers of intrinsic microporosity (PIMs) substituted with methyl tetrazole. Polymer 53: 4367–4372.

Duan, C., Jie, X., Liu, D., Cao, Y. and Yuan, Q. 2014. Post-treatment effect on gas separation property of mixed matrix membranes containing metal organic frameworks. J. Membr. Sci. 92–102.

Economides, M.J. and Wood, D.A. 2009. The state of natural gas. J. Nat. Gas Sci. Eng. 1: 1–13.

Faramawy, S., Zaki, T. and Sakr, A.A.-E. 2016. Natural gas origin, composition, and processing: A review. J. Nat. Gas Sci. Eng. 34: 34–54.

Fritsch, D., Merten, P., Heinrich, K., Lazar, M. and Priske, M. 2012. High performance organic solvent nanofiltration membranes: development and thorough testing of thin film composite membranes made of polymers of intrinsic microporosity (PIMs), J. Membr. Sci. 401–402: 222–231.

Fu, Y.J., Hu, C.C., Lin, D.W., Tsai, H.A., Huang, S.H., Hung, W.S. et al. 2017. Adjustable microstructure carbon molecular sieve membranes derived from thermally stable polyetherimide/polyimide blends for gas separation. Carbon 13: 10–17.

Galve, A., Sieffert, D., Staudt, C., Ferrando, M., Guell, C., Tellez, C. et al. 2013. Combination of ordered mesoporous silica MCM-41 and layered titanosilicate JDF-L1 fillers for 6FDA-based copolyimide mixed matrix membranes. J. Membr. Sci. 431: 163–170.

Ganesan, A. and M.M. Shaijumon. 2016. Activated graphene-derived porous carbon with exceptional gas adsorption properties. Microporous Mesoporous Mater 220: 21–27.

Garg, P., Singh, R.P. and Choudhary, V. 2011. Pervaporation separation of organic azeotrope using poly(dimethyl siloxane)/clay nanocomposite membranes. Sep. Purif. Technol. 80: 435–444.

George, G., Bhoria, N., AlHallaq, S., Abdala, A. and Mittal, V. 2016. Polymer membranes for acid gas removal from natural gas. Sep. Purif. Technol. 158: 333–356

Ghanem, B.S., R. Swaidan, E. Litwiller and I. Pinnau. 2014. Ultra-microporous triptycene-based polyimide membranes for high-performance gas separation. Adv. Mater. 26: 3688–3692.

Gomes, D., Nunes, S.P. and Peinemann, K.V. 2005. Membranes for gas separation based on poly(1-trimethylsilyl-1-propyne)–silica nanocomposites. J. Membr. Sci. 246: 13–25.

Guiver, M.D., Robertson, G.P., Dai, Y., Bilodeau, F., Kang, Y.S., Lee, K.J. et al. 2002. Structural Characterization and Gas-Transport Properties of Brominated Matrimid Polyimide. J. Polym. Sci.: Part A: Polymer Chemistry. 40: 4193–4204.

Guo, R., Sanders, D.F., Smith, Z.P., Freeman, B.D., Paul, D.R. and McGrath, J.E. 2013a. Synthesis and characterization of thermally rearranged (TR) polymers: effect of glass transition temperature of aromatic poly(hydroxyimide) precursors on TR process and gas permeation properties. J. Mater. Chem. A 1: 6063–6072.

Guo, R., Sanders, D.F., Smith, Z.P., Freeman, B.D., Paul, D.R. and McGrath, J.E. 2013b. Synthesis and characterization of Thermally Rearranged (TR) polymers: influence of ortho-positioned functional groups of polyimide precursors on TR process and gas transport properties. J. Mater. Chem. A. 1: 262–272.

Gye, B., Kammakakam, I., You, H., Nam, S.Y. and Kim, T.H. 2017. PEG-imidazolium-incorporated polyimides as hig3h-performance CO2- selective polymer membranes: The effects of PEG-imidazolium content. Sep. Purif. Technol. 179: 283–290.

Han, S.H., Lee, J.E., Lee, K.J., Park, H.B. and Lee, Y.M. 2010a. Highly gas permeable and microporous polybenzimidazole membrane by thermal rearrangement. J. Membr. Sci. 357: 143–151.

Han, S.H., Misdan, N., Kim, S., Doherty, C.M., Hill, A.J. and Lee, Y.M. 2010b. Thermally Rearranged (TR) Polybenzoxazole: Effects of Diverse Imidization Routes on Physical Properties and Gas Transport Behaviors. Macromolecules 43: 7657–7667.

Hao, L., Li, P. and Chung, T.S. 2014. PIM-1 as an organic filler to enhance the gas separation performance of Ultem polyetherimide. J. Membr. Sci. 453: 614–623.

Hasebe, S., Aoyama, S., Tanaka, M. and Kawakami, H. 2017. CO_2 separation of polymer membranes containing silica nanoparticles with gas permeable nano-space. J. Membr. Sci. 536: 148–155.

Hauchhum, L. and Mahanta, P. 2014. CO2 capture onto zeolite 13X and zeolite 4a by pressure swing adsorption in a fixed Bed. Appl. Mech. and Mater. 592-594: 1456–1460.

Hernández, M.G., Salinas-Rodríguez, E., Gómez, S.A., Roa-Neri, J.A.E., Alfaro, S. and Valdés-Parada, F.J. 2015. Helium permeation through a silicalite-1 tubular membrane. Heat and Mass Transfer 51: 847–857.

Horike, S., Shimomura, S. and Kitagawa, S. 2009. Soft Porous Crystals. Nat. Chem. 1(9): 695–704.

Howarth, A.J., Peters, A.W., Vermeulen, N.A., Wang, T.C., Hupp, J.T. and Farha, O.K. 2017. Best practices for the synthesis, activation, and characterization of metal–organic frameworks. Chem. Mater. 29(1): 26–39.

Hu, S., Ren, W., Cai, D., Hughes, T.C., Qin, P. and Tan, T. 2017. A mixed matrix membrane for butanol pervaporation based on micron-sized silicalite-1 as macro-crosslinkers. J. Membr. Sci. 533: 270–278.

Iijima, S. 1991. Helical microtubules of graphitic carbon. Nature 354: 56–58.

Irani, M., Jacobson, A.T., Gasem, K.A.M. and Fan, M. 2017. Modified carbon nanotubes/tetraethylenepentamine for CO_2 capture. Fuel. 206: 10–18.

Jaafar, J., Ismail, A.F., Matsuura, T. and Nagai, K. 2011. Performance of SPEEK based polymer–nanoclay inorganic membrane for DMFC. J. Membr. Sci. 382: 202–211.

James, S.L., Adams, C.J., Bolm, C., Braga, D., Collier, P., Friščić, T. et al. 2012. Mechanochemistry: Opportunities for New and Cleaner Synthesis. Chem. Soc. Rev. 41(1): 413–47.

Jue, M.L., Breedveld, V. and Lively, R.P. 2017. Defect-free PIM-1 hollow fiber membranes. J. Membr. Sci. 530: 33–41.

Jung, C.H., Lee, J.E., Han, S.H., Park, H.B. and Lee, Y.M. 2010. Highly permeable and selective poly(benzoxazole-co-imide) membranes for gas separation. J. Membr. Sci. 350: 301–309.

Jusoh, N., Yeong, Y.F., Cheong, W.L., Lau, K.K. and Shariff, A.M. 2016. Facile fabrication of mixed matrix membranes containing 6FDA-durene polyimide and ZIF-8 nanofillers for CO_2 capture. J. Ind. Eng. Chem. 44: 164–173.

Jusoh, N., Yeong, Y.F., Lau, K.K. and Shariff, A.M. 2017. Enhanced gas separation performance using mixed matrix membranes containing zeolite T and 6FDA-durene polyimide. J. Membr. Sci. 525: 175–186.

Kammakakam, I., Yoon, H.W., Nam, S.Y., Park, H.B. and Kim, T.H. 2015. Novel piperazinium-mediated crosslinked polyimide membranes for high performance CO_2 separation. J. Membr. Sci. 487: 90–98.

Khan, M.M., Filiz, V., Bengtson, G., Shishatskiy, S., Rahman, M.M., Lillepaerg, J. and Abetz, V. 2013. Enhanced gas permeability by fabricating mixed matrix membranes of functionalized multiwalled carbon nanotubes and polymers of intrinsic microporosity (PIM). J. Membr. Sci. 436: 109–120.

Khdhayyer, M.R., Esposito, E., Fuoco, A., Monteleone, M., Giorno, L., Jansen, J.C. et al. 2017. Mixed matrix membranes based on UiO-66 MOFs in the polymer of intrinsic microporosity PIM-1. Sep. Purif. Technol. 173: 304–313.

Khosravi, T., Mosleh, S., Bakhtiari, O. and Mohammad, T. 2012. Mixed matrix membranes of Matrimid 5218 loaded with zeolite 4A for pervaporation separation of water–isopropanol mixtures. Chem. Eng. Res. Des. 90: 2353–2363.

Kim, S., Han, S.H. and Lee, Y.M. 2012. Thermally rearranged (TR) polybenzoxazole hollow fiber membranes for CO_2 capture. J. Membr. Sci. 403–404: 169–178.

Kongnoo, A., Tontisirin, S., Worathanakul, P. and Phalakornkule, C. 2017. Surface characteristics and CO_2 adsorption capacities of acid-activated zeolite 13X prepared from palm oil mill fly ash. Fuel 193: 385–394.

Koros, W.J. and Mahajan, R. 2000. Pushing the limits on possibilities for large scale gas separation: which strategies? J. Membr. Sci. 175: 181–196.

Kundu, P.K., Chakma, A. and Feng, X. 2014. Effectiveness of membranes and hybrid membrane processes in comparison with absorption using amines for post-combustion CO2 capture. Int. J. Greenh. Gas Con. 28: 248–256.

Kyotani, T. 2000. Control of pore structure in carbon. Carbon. 38(2): 269–286.

Lai, Z. 2018. Development of ZIF-8 membranes: opportunities and challenges for commercial applications. Curr. Opin. Chem. Eng. 20: 78–85.

Labropoulos, A., Veziri, C., Kapsi, M., Pilatos, G., Likodimos, V., Tsapatsis, M. et al. 2015. Carbon Nanotube Selective Membranes with Sub-Nanometer, Vertically Aligned Pores, and Enhanced Gas Transport Properties. Chem. Mater. 27(24): 8198–8210.

Lee, Y., Kim, R.J. and Ahn, W.S. 2013. Synthesis of Metal-Organic Frameworks: A Mini Review. Korean J. Chem. Eng. 30(9): 1667–1680.

Li, C., Ren, Y., Gou, J., Liu, B. and Xi, H. 2017. Facile synthesis of mesostructured ZSM-5 zeolite with enhanced mass transport and catalytic performances. Appl. Surf. Sci. 392: 785–794.

Li, F.Y., Xiao, Y., Ong, Y.K. and Chung, T.S. 2012. UV-rearranged PIM-1 polymeric membranes for advanced hydrogen purification and production. Adv. Energy Mater. 2: 1456–1466.

Li, F.Y. and Chung, T.S. 2013. Physical aging, high temperature and water vapour permeation studies of UV-rearranged PIM-1 membranes for advanced hydrogen purification and production. Int. J. Hydrog. Energy. 38: 9786–9793.

Li, H., Eddaoudi, M., Keeffe, M.O. and Yaghi, O.M. 1999. Design and synthesis of an exceptionally stable and highly porous metal-organic framework. Nature 402: 276–79.

Li, X., Wu, X., He, G., Sun, J., Xiao, W. and Tan, Y. 2014. Microspheroidization treatment of macroporous TiO2 to enhance its recycling and prevent membrane fouling of photocatalysis–membrane system. Chem. Eng. J. 251: 58–68.

Li, X., Ma, L., Zhang, H., Wang, S., Jiang, Z., Guo, R. et al. 2015. Synergistic effect of combining carbon nanotubes and graphene oxide in mixed matrix membranes for efficient CO_2 separation. J. Membr. Sci. 479: 1–10.

Li, Y. and Yang, W. 2015. Molecular sieve membranes: From 3D zeolites to 2D MOFs. Chin. J. Catal. 36(5): 692–697.

Liang, Z., Marshall, M. and Chaffee, A.L. 2009. Comparison of Cu-BTC and zeolite 13X for adsorbent based CO_2 separation. Energy Procedia 1: 1265–1271.

Liu, Q., Borjigin, H., Paul, D.R., Riffle, J.S., McGrath, J.E. and Freeman, B.D. 2016. Gas permeation properties of thermally rearranged (TR) isomers and their aromatic polyimide precursors. J. Membr. Sci. 518: 88–99.

Liu, Y., Yan, C., Qiu, X., Li, D., Wang, H. and Alshameri, A. 2016. Preparation of faujasite block from fly ash-based geopolymer via in-situ hydrothermal method. J. Taiwan Inst. Chem. Eng. 59: 433–439.

Ma, X., Swaidan, R., Belmabkhout, Y., Zhu, Y., Litwiller, E., Jouiad, M. et al. 2012. Synthesis and Gas Transport Properties of Hydroxyl-Functionalized Polyimides with Intrinsic Microporosity. Macromolecules 45(9): 3841–3849.

Mao, H. and Zhang, S. 2017. Mixed-matrix membranes incorporated with porous shape-persistent organic cages for gas separation. J. Colloid Interf. Sci. 490: 29–36.

Mason, C.R., Maynard-Atem, L., Al-Harbi, N.M., Budd, P.M., Bernado, P., Bazzarelli, F. et al. 2011. Polymer of Intrinsic Microporosity Incorporating Thioamide Functionality: Preparation and Gas Transport Properties. Macromolecules 44: 6471–6479.

Mason, C.R., Buonomenna, M.G., Golemme, G., Budd, P.M., Galiano, F., Figoli, A. et al. 2013. New organophilic mixed matrix membranes derived from a polymer of intrinsic microporosity and silicalite-1. Polymer 54: 2222–2230.

Mason, C.R., Maynard-Atem, L., Heard, K.W., Satilmis, B., Budd, P.M., Friess, K. et al. 2014. Enhancement of CO2 affinity in a polymer of intrinsic microporosity by amine modification. Macromolecules 47(3): 1021–1029.

Masoomi, M.Y., Stylianoe, K.C., Morsali, A., Retaileau, P. and Maspoch, D. 2014. Selective CO_2 Capture in Metal–Organic Frameworks with Azine-Functionalized Pores Generated by Mechanosynthesis. Cryst. Growth Des. 14 (5): 2092–2096.

Matteucci, S., Kusuma, V.A., Sanders, D., Swinnea, S. and Freeman, B.D. 2008. Gas transport in TiO_2 nanoparticle-filled poly (1-trimethylsilyl-1-propyne). J. Membr. Sci. 307: 196–217.

Meier, I.K., Langsam, M. and Klotz, H.C. 1994. Selectivity enhancement via photooxidative surface modification of polyimide air separation membranes. J. Membr. Sci. 94: 195–212.

Merkel, T.C., Bondar, V., Nagai, K. and Freeman, B.D. 2000. Sorption and transport of hydrocarbons and perfluorocarbon gases in poly(1-trimethylsilyl-1-propyne). J. Polym. Sci. Part B: Polym. Phys. 38: 273–296.

Murray, L.J., Dinca, M., J. Yano, Chavan, S., Bordiga, S., Brown, C.M. et al. 2010. Highly-Selective and Reversible O2 Binding in Cr 3(1,3,5-benzenetricarboxylate)2. J. Am. Chem. Soc. 132 (23): 7856–57.

Ni, X., Zheng, Z., Wang, X., Zhang, S. and Zhao, M. 2014. Fabrication of hierarchical zeolite 4A microspheres with improved adsorption capacity to bromofluoropropene and their fire suppression performance. J. Alloys Compd. 592: 135–139.

Nik, O.G., Chen, X.Y. and Kaliaguine, S. 2012. Functionalized metal organic framework-polyimide mixed matrix membranes for CO_2/CH_4 separation. J. Membr. Sci. 413-414: 48–61.

Novoselov, K.S., Geim, A.K., Morozov, S.V., Jiang, D., Zhang, Y., Dubonos, S.V. et al. 2004. Electric field effect in atomically thin carbon films. Science 306(5696): 666–669.

Noy, A. 2013. Kinetic model of gas transport in carbon nanotubes. J. Phys. Chem. C. 117(15): 7656–7660.

Nugent P., Belmabkhout, Y., Burd, S.D., Cairns, A.J., Luebke, R., Forrest, K. et al. 2013. Porous materials with optimal adsorption thermodynamics and kinetics for CO2 separation. Nature 495(7439): 80–84.

Pandey, P. and Chauhan. R.S. 2001. Membranes for gas separation. Prog. Polym. Sci. 26: 853–893.

Pang, S.H., Jue, M.L., Leisen, J., Jones, C.W. and Lively, R.P. 2015. PIM-1 as a Solution-Processable "Molecular Basket" for CO_2 Capture from Dilute Sources. ACS Macro Lett. 4(12): 1415–1419.

Park, H.B. and Lee, Y.M. 2003. Pyrolytic carbon-silica membrane: A promising membrane material for improved gas separation. J. Memb. Sci. 213(1-2): 263–272.

Park, H.B., Jung, C.H., Lee, Y.M., Hill, A.J., Pas, S.J., Mudie, S.T. et al. 2007. Polymers with cavities tuned for fast selective transport of small molecules and ions. Science 318(5848): 254–258.

Peng, D., Wang, S., Tian, Z., Wu, X., Wu, Y., Wu, H. et al. 2017. Facilitated transport membranes by incorporating graphene nanosheets with high zinc ion loading for enhanced CO_2 separation. J. Membr. Sci. 522: 351–362.

Peydayesh, M., Asarehpour, S., Mohammadi, T. and Bakhtiari, O. 2013. Preparation and characterization of SAPO-34 – Matrimid® 5218 mixed matrix membranes for CO_2/CH_4 separation. Chem. Eng. Res. Des. 91: 1335–1342.

Pillai, R.S., Peter, S.A. and Jasra, R.V. 2012. CO_2 and N_2 adsorption in alkali metal ion exchanged X-Faujasite: Grand canonical Monte Carlo simulation and equilibrium adsorption studies. Microporous Mesoporous Mater. 162: 143–151.

Poloni, R., Lee, K., Berger, R.F., Smit, B. and Neaton, J.B. 2014. Understanding Trends in CO_2 Adsorption in Metal-Organic Frameworks with Open-Metal Sites. J. Phys. Chem. Let. 5(5): 861–65.

Prats, H., Bahamon, D., Alonso, G., Giménez, X., Gamallo, P. and Sayós, R. 2017. Optimal Faujasite structures for post combustion CO2 capture and separation in different swing adsorption processes. J. CO_2 Util. 19: 100–111.

Qiu, L.-G., Li, Z.-Q., Wu, Y., Wang, W., Xu, T. and Jiang, X. 2008. Facile synthesis of nanocrystals of a microporous metal-organic framework by an ultrasonic method and selective sensing of organoamines. Chem. Commun. 31: 3642–44.

Rangel, E.R., Maya, E.M., Sanchez, F., de Abajo, J. and J.G. de la Campa. 2013. Gas separation properties of mixed-matrix membranes containing porous polyimides fillers. J. Membr. Sci. 447: 403–412.

Robeson, L.M. 1991. Correlation of separation factor versus permeability for polymeric membranes. J. Membr. Sci. 62: 165–185.

Robeson, L.M. 2008. The upper bound revisited. J. Membr. Sci. 320: 390–400.

Rosi, N.L., Kim, J., Eddaoudi, M., Chen, B., O'Keeffe, M. and Yaghi, O.M. 2005. Rod Packings and Metal-Organic Frameworks Constructed from Rod-Shaped Secondary Building Units. J. Am. Chem. Soc. 127(5): 1504–18.

Rostamnia, S. and Morsali, A. 2014. Basic Isoreticular Nanoporous Metal-Organic Framework for Biginelli and Hantzsch Coupling: IRMOF-3 as a Green and Recoverable Heterogeneous Catalyst in Solvent-Free Conditions. RSC Adv. 4(21): 10514–18.

Rowsell, J.L.C. and Yaghi, O.M. 2004. Metal–Organic Frameworks : A New Class of Porous Materials. Microporous and Mesoporous Materials 73: 3–14.

Rufford, T.E., Smart, S., Watson, G.C.Y., Graham, B.F., Boxall, J., Diniz da Costa, J.C. et al. 2012. The removal of CO2 and N2 from natural gas: A review of conventional and emerging process technologies. J. Pet. Sci. Eng. 94-95: 123–154.

Sabouni, R., Kazemian, H. and Rohani, S. 2014. Carbon Dioxide Capturing Technologies: A Review Focusing on Metal Organic Framework Materials (MOFs). Environ. Sci. Pollut. Res. 21(8): 5427–5449.

Sahu, J., Aijaz, A., Xu, Q. and Bharadwaj, P.K. 2015. A three-dimensional pillared-layer metal-organic framework: Synthesis, structure and gas adsorption studies. Inorg. Chim. Acta. 430: 193–198

Salehian, P., Yong, W.F. and Chung, T.S. 2016. Development of high performance carboxylated PIM-1/P84 blend membranes for pervaporation dehydration of isopropanol and CO_2/CH_4 separation. J. Membr. Sci. 518: 110–119.

Sanders, D.F., Smith, Z.P., Ribeiro, Jr. C.P., Guo, R., McGrath, J.E., Paul, D.R. et al. 2012. Gas permeability, diffusivity, and free volume of thermally rearranged polymers based on 3,3-dihydroxy-4,4-diamino-biphenyl (HAB) and 2,2-bis-(3,4-dicarboxyphenyl) hexafluoropropane dianhydride (6FDA). J. Membr. Sci. 409–410: 232–241.

Satilmis, B., M.N. Alnajrani and P.M. Budd. 2015. Hydroxyalkylaminoalkylamide PIMs: selective adsorption by ethanolamine-and diethanolamine-modified PIM-1. Macromolecules 48(16): 5663–5669.

Scholes, C.A., Stevens, G.W. and Kentish, S.E. 2012. Membrane gas separation applications in natural gas processing. Fuel. 96: 15–28.

Scholes, C.A., Dong, G., Kim, J.S., Jo, H.J., Lee, J. and Lee, Y.M. 2017. Permeation and separation of SO_2, H_2S and CO_2 through thermally rearranged (TR) polymeric membranes. Sep. Purif. Technol. 179: 449–454.

Shah, K.A. and Tali, B.A. 2016. Synthesis of carbon nanotubes by catalytic chemical vapour deposition: A review on carbon sources, catalysts and substrates. Mat. Sci. Semicon. Proc. 41: 67–82.

Sjöberg, E., Barnes, S., Korelskiy, D. and Hedlund, J. 2015. MFI membranes for separation of carbon dioxide from synthesis gas at high pressures. J. Membr. Sci. 486: 132–137.

Smith, Z.P., Hernandez, G., Gleason, K.L., Anand, A., Doherty, C.M., Konstas, K. et al. 2015. Effect of polymer structure on gas transport properties of selected aromatic polyimides, polyamides and TR polymers. J. Membr. Sci. 493: 766–781.

Sreedhar, I., Nahar, T., Venugopal, A. and Srinivas, B. 2017. Carbon capture by absorption – Path covered and ahead. Renew. Sust. Energ. Rev. 76: 1080–1107.

Staiger, C.L., Pas, S.J., Hill, A.J. and Cornelius, C.J. 2008. Gas separation, free volume distribution, and physical aging of a highly microporous spirobisindane polymer. Chem. Mater. 20: 2606–2608.

Starannikova, L., Khodzhaeva, V. and Yampolskii, Yu. 2004. Mechanism of aging of poly[1-(trimethylsilyl)-1-propyne] and its effect on gas permeability. J. Membr. Sci. 244: 183–191.

Su, F., Lu, C., Cnen, W., Bai, H. and Hwang, J.F. 2009. Capture of CO2 from flue gas via multiwalled carbon nanotubes. Sci. Total Environ. 407: 3017–3023.

Sun, C. and Bai, B. 2017. Fast mass transport across two-dimensional graphene nanopores: Nonlinear pressure-dependent gas permeation flux. Chem. Eng. Sci. 165: 186–191.

Sun, H., Wang, T., Xu, Y., Gao, W., Li, P. and Niu, J. 2017. Fabrication of polyimide and functionalized multi-walled carbon nanotubes mixed matrix membranes by *in-situ* polymerization for CO_2 separation. Sep. Purif. Technol. 177: 327–336.

Sun, M., Wang, X., Meng, B., Tan, X. and Liu, S. 2017. A simple embedded-seeding method to prepare silicalite-1 membrane on porous α-Al2O3 hollow fibers. Mater. Lett. 194: 122–125.

Swaidan, R., Ma, X., Liwiller, E. and Pinnau, I. 2013. High pressure pure-and mixed-gas separation of CO2/CH4 by thermally-rearranged and carbon molecular sieve membranes derived from a polyimide of intrinsic microporosity. J. Membr. Sci. 447: 387–394.

Swaidan, R., Ghanem, B.S., Litwillwer, E. and Pinnau, I. 2014. Pure-and mixed-gas CO_2/CH_4 separation properties of PIM-1 and an amidoxime-functionalized PIM-1. J. Membr. Sci. 457: 95–102.

Swaidan, R., Ghanem, B., Litwiller, E. and Pinnau, I. 2015. Effects of hydroxyl-functionalization and sub-Tg thermal annealing on high pressure pure- and mixed-gas CO_2/CH_4 separation by polyimide membranes based on 6FDA and triptycene-containing dianhydrides. J. Membr. Sci. 475: 571–581.

Szczęśniak, B., Choma, J. and Jaroniec, M. 2017. Gas adsorption properties of graphene-based materials. Adv. Colloid Interf. Sci. 243: 46–59.

Taylor-Pashow, K.M.L., Rocca, J.D., Xie, Z., Tran, S. and Lin, W. 2009. Postsynthetic Modifications of Iron-Carboxylate Nanoscale Metal-Organic Frameworks for Imaging and Drug Delivery. J. Am. Chem. Soc. 131(40): 14261–63.

Thomas, S., Pinnau, I., Du, N. and Guiver, M.D. 2009. Pure- and mixed-gas permeation properties of a microporous spirobisindane-based ladder polymer (PIM-1). J. Membr. Sci. 2009: 125–131.

Tian, Z., Wang, S., Wang, Y., Ma, X., Cao, K., Peng, D. et al. 2016. Enhanced gas separation performance of mixed matrix membranes from graphitic carbon nitride nanosheets and polymers of intrinsic microporosity. J. Membr. Sci. 514: 15–24.

Tiem-Binh, N., Vinh-Thang, H., Chen, X.Y., Rodrigue, D. and Kaliaguine, S. 2016. Crosslinked MOF-polymer to enhance gas separation of mixed matrix membranes. J. Membr. Sci. 520: 941–950.

Tiwari, R.R., Jin, J., Freeman, B.D. and Paul, D.R. 2017. Physical aging, CO_2 sorption and plasticization in thin films of polymer with intrinsic microporosity (PIM-1). J. Membr. Sci. 537: 362–371.

Tranchemontagne, D.J., Hunt, J.R. and Yaghi, O.M. 2008. Room Temperature Synthesis of Metal-Organic Frameworks: MOF-5, MOF-74, MOF-177, MOF-199, and IRMOF-0. Tetrahedron. 64(36): 8553–57.

Tullos, G.L., Powers, J.M., Jeskey, S.J. and Mathias, L.J. 1999. Thermal conversion of hydroxy-containing imides to benzoxazoles: polymer and model compound study. Macromolecules 32: 3598–3612.

Valero, M., Zornoza, B., Tellez, C. and Coronas, J. 2014. Mixed matrix membranes for gas separation by combination of silica MCM-41 and MOF NH2-MIL-53(Al) in glassy polymers. Micropor Mesopor Mater. 192: 23–28.

Venna, S.R. and Carreon, M.A. 2015. Metal Organic Framework Membranes for Carbon Dioxide Separation. Chem. Engineering Sci. 124: 3–19.

Vieira, L.O., Madeira, A.C., Merlini, A., Melo, C.R., Mendes, E., Santos, M.G. et al. 2015. Synthesis of 4A-Zeolite for Adsorption of CO_2. Materials Science Forum 805: 632–637.

Villarreal, A., Garbarino, G., Riani, P., Finocchio, E., Bosio, B., Ramírez, J. et al. 2017. Adsorption and separation of CO_2 from N2-rich gas on zeolites: Na-X faujasite vs. Na-mordenite. J. CO_2 Util. 19: 266–275.

Wang, B., Sun, C., Li, Y., Zhao, L., Ho, W.S.W. and Dutta, P.K. 2015. Rapid synthesis of faujasite/polyethersulfone composite membrane and application for CO_2/N_2 separation. Micropor. Mesopor. Mater. 208: 72–82.

Wang, C., J. Liu, J. Yang and J. Li. 2017. A crystal seeds-assisted synthesis of microporous and mesoporous silicalite-1 and their $CO_2/N_2/CH_4/C_2H_6$ adsorption properties. Micropor. Mesopor. Mater. 242: 231–237.

Wang, X., Zhang, X., Liu, H., Yeung, K.L. and Wang, J. 2010. Preparation of titanium silicalite-1 catalytic films and application as catalytic membrane reactors. Chem. Eng. J. 156: 562–570.

Wang, Y., Yang, Q., Zhong, C. and Li, J. 2017. Theoretical investigation of gas separation in functionalized nanoporous graphene membranes. App. Surf. Sci. 407: 532–539.

Weitkamp, J. 2000. Zeolites and catalysis. Solid State Ionics 131: 175–188.

Weng, X., Baez, J.E., Khiterer, M., Hoe, M.Y., Bao, Z. and Shea, K.J. 2015. Chiral polymers of intrinsic microporosity: selective membrane permeation of enantiomers. Angew. Chem. Int. Ed. 54: 11214–11218.

Woo, K.T., Dong, G., Lee, J., Kim, J.S., Do, Y.S., Lee, W.H. et al. 2016a. Ternary mixed-gas separation for flue gas CO2 capture using high performance thermally rearranged (TR) hollow fiber membranes. J. Membr. Sci. 510: 472–480.

Woo, K.T., Lee, J., Dong, G., Kim, J.S., Do, Y.S., Jo, H.J. et al. 2016b. Thermally rearranged poly(benzoxazole-co-imide) hollow fiber membranes for CO_2 capture. J. Membr. Sci. 498: 125–134.

Woo, K.T., Lee, J., Dong, G., Kim, J.S., Do, Y.S., Hung, W.S. et al. 2015. Fabrication of thermally rearranged (TR) polybenzoxazole hollow fiber membranes with superior CO_2/N_2 separation performance. J. Membr. Sci. 490: 129–138.

Wu, X., Tian, Z., Wang, S., Peng, D., Yang, L., Wu, Y. et al. 2017. Mixed matrix membranes comprising polymers of intrinsic microporosity and covalent organic framework for gas separation. J. Membr. Sci. 528: 273–283.

Wu, X.M., Zhang, Q.G., Lin, P.J., Y.Q., Zhu, A.M. and Liu, Q.L. 2015. Towards enhanced CO_2 selectivity of the PIM-1 membrane by blending with polyethylene glycol. J. Membr. Sci. 493: 147–155.

Wynnyk, K.G., Hojjati, B., Pirzadeh, P. and Marriott, R.A. 2017. High-pressure sour gas adsorption on zeolite 4A. Adsorption 23: 149–162.

Xiang, S., He, Y., Zhang, Z., Wu, H., Zhou, W., Krishna, R. et al. 2012. Microporous Metal-Organic Framework with Potential for Carbon Dioxide Capture at Ambient Conditions. Nature. Commun. 3: 954.

Xiao, Y., Low, B.T., Hosseini, S.S., Chung, T.S. and Paul, D.R. 2009. The strategies of molecular architecture and modification of polyimide-based membranes for CO_2 removal from natural gas—A review. Prog. Polym. Sci. 34: 561–580.

Yanaranop, P., Santoso, B., Etzion, R. and Jin, J. 2016. Facile conversion of nitrile to amide on polymers of intrinsic microporosity (PIM-1). Polymer 98: 244–251.

Yao, K.X., Chen, Y., Lu, Y., Zhao, Y. and Ding, Y. 2017. Ultramicroporous carbon with extremely narrow pore distribution and very high nitrogen doping for efficient methane mixture gases upgrading. Carbon 122: 258–265.

Yong, Z., Mata, V. and Rodrigues, A.E. 2002. Adsorption of carbon dioxide at high temperature—a Review. Sep. Purif. Technol. 26 (2): 195–205.

Yong, W.F., Li, F.Y., Xiao, Y.C., Li, P., Pramoda, K.P., Tong, Y.W. et al. 2012. Molecular engineering of PIM-1/Matrimid blend membranes for gas separation. J. Membr. Sci. 407-408: 47– 57.

Yong, W.F., Li, F.Y., Xiao, Y.C., Chung, T.S. and Tong, Y.W. 2013. High performance PIM-1/Matrimid hollow fiber membranes for CO_2/CH_4, O_2/N_2 and CO_2/N_2 separation. J. Membr. Sci. 443: 156–169.

Yong, W.F., Li, F.Y., Chung, T.S. and Tong, Y.W. 2014. Molecular interaction, gas transport properties and plasticization behaviour of cPIM-1/Torlon blend membranes. J. Membr. Sci. 462: 119–130.

Zhang, C., Li, P. and Cao, B. 2017. Effects of the side groups of the spirobichroman-based diamines on the chain packing and gas separation properties of the polyimides. J. Membr. Sci. 530: 176–184.

Zhang, L., Zhao, B., Wang, X., Liang, Y., Qiu, H., Zheng, G. et al. 2014a. Gas transport in vertically-aligned carbon nanotube/parylene composite membranes. Carbon 66: 11–17.

Zhang, L., Yang, J., Wang, X., Zhao, B. and Zheng, G. 2014b. Temperature-dependent gas transport performance of vertically aligned carbon nanotube/parylene composite membranes. Nano. Res. Lett. 9: 448(8).

Zhang, Y., Zhang, L. and Zhou, C. 2013. Review of chemical vapor deposition of graphene and related applications. Acc. Chem. Res. 46(10): 2329–2339.

Zhao, H., Xie, Q., Ding, X., Chen, J., Hua, M., Tan, X. et al. 2016. High performance post-modified polymers of intrinsic microporosity (PIM-1) membranes based on multivalent metal ions for gas separation. J. Membr. Sci. 514: 305–312.

Zhao, R., Chen, J., Liu, J., Fan, J. and Du, J. 2015. Morphologies-controlling synthesis of silicalite-1 and its adsorption property. Mater. Lett. 139: 494–497.

Zhu, Q.-L. and Xu, Q. 2014. Metal–organic Framework Composites. Chem. Soc. Rev. 43(16): 5468–5512.

Zhuang, Y., Seong, J.G., Do, Y.S., Lee, W.H., Lee, M.J., Guiver, M.D. et al. 2016. High-strength, soluble polyimide membranes incorporating Tröger's Base for gas separation. J. Membr. Sci. 504: 55–65.

Zinadini, S., Zinatizadeh, A.A., Rahimi, M., Vatanpour, V. and Zangeneh, H. 2014. Preparation of a novel antifouling mixed matrix PES membrane by embedding graphene oxide nanoplates. J. Membr. Sci. 453: 292–301.

Zito, P.F., Caravella, A., Brunetti, A., Drioli, E. and Barbieri, G. 2017. Knudsen and surface diffusion competing for gas permeation inside silicalite membranes. J. Membr. Sci. 523: 456–469.

Zornoza, B., Seoane, B., Zamaro, J.M., Tellez, C. and Coronas, J. 2011. Combination of MOFs and zeolites for mixed-matrix membranes. ChemPhysChem. 12: 2781–2785.

Zornoza, B., Seoane, B., Zamaro, J.M., Tellez, C. and Coronas, J. 2012. Synergy gas separation effects when using fillers of different natures (MOFs and zeolites) in the same mixed matrix membrane. Procedia Eng. 44: 2118–2120.

7

Role of Aromatic Polyimide Membrane in CO_2 Separations

From Monomer to Polymer Perspective

Peng Chee TAN,[1] *Pei Ching OH*[2] and *Siew Chun LOW*[1,*]

INTRODUCTION

The past century has seen rapidly increasing carbon dioxide (CO_2) emission attributed to the continuous urbanization and industrialization process. In fact, the ever growing industries are responsible for approximately 40% of global energy-related CO_2 emissions (Ramírez-Santos et al. 2017). This has resulted in special attention for CO_2 removal technology especially in the treatment of industrial flue gas and natural gas purification. In this context, an improved separation efficiency of CO_2 from N_2 (constitutes the largest composition of flue gas) or CH_4 (a major component of natural gas) has become increasingly important in order to mitigate global warming effectively.

Among those available CO_2 removal technologies, for instance amine absorption, pressure swing adsorption, cryogenic distillation and membrane separation (Wang et al. 2005, Powell and Qiao 2006, Chua et al. 2014), polyimide membrane stands out as a viable and promising separation tool for industrial CO_2 removal. The industrial

[1] Universiti Sains Malaysia, School of Chemical Engineering, Engineering Campus, Seri Ampangan, Nibong Tebal, Pulau Pinang, Malaysia, 14300.
Email: peng_chee91@hotmail.com

[2] Universiti Teknologi PETRONAS, Department of Chemical Engineering, Bandar Seri Iskandar, Tronoh Perak Darul Ridzuan, Malaysia, 31750.
Email: peiching.oh@utp.edu.my

* Corresponding author: chsclow@usm.my

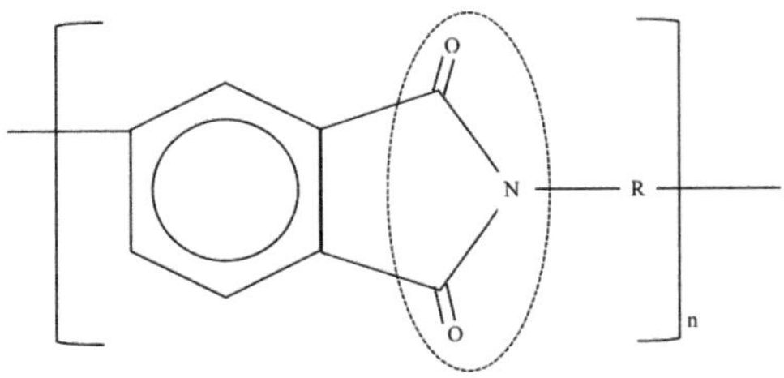

Figure 7.1. General structure of aromatic polyimide.

application of polyimide membrane was pioneered by Du Pont Co. (USA) in 1962, in which the polyimide membrane was employed for the separation of helium from natural gas (Xiao et al. 2009, Chen et al. 2015). Later in 1986, Ube Industries (Japan) also launched an ammonia production pilot plant that utilized asymmetric polyimide hollow fiber membrane for the recovery of hydrogen gas from the synthesis gas (Ohya et al. 1996). The efforts of Du Pont Co. and Ube Industries thus justify the potential of polyimide membrane in gas separation.

Indeed, polyimide possesses high thermal stability, a feature essential for industrial CO_2 removal. The aromatic structure in the polyimide backbone (Fig. 7.1) contributes strongly to the polyimide chain rigidity, which increases its glass transition temperature, T_g (Vanherck et al. 2013). A high T_g is beneficial for a CO_2 separation membrane because it can avoid the glassy-rubbery transformation of a polyimide membrane under high CO_2 concentration, which cause a drastic loss in the membrane selectivity and hence deteriorate the membrane gas separation performances (Wind et al. 2002, Qiu et al. 2011). Generally, the T_g of a polyimide membrane ranges from 280 to 400°C (Vanherck et al. 2013). In view of this, polyimide membrane is highly preferable for industrial CO_2 removal as the industrial processes typically operate at high temperature. Since a gas separation membrane should also be able to withstand high pressure and cope with aggressive feed streams, the excellent mechanical strength and chemical resistance (Kratochvil and Koros 2008, Tong et al. 2015) offered by the polyimide membrane further evidence its capability for industrial CO_2 removal.

Due to the attractive features of polyimide, intense research interests are focused towards polyimide gas separation membrane particularly in the field of CO_2 removal (Hillock and Koros 2007, Kraftschik et al. 2013, Heck et al. 2017, Yoshioka et al. 2017). This can be proven from the significant increment in publications related to polyimide membrane over the past decades, as depicted in Fig. 7.2. With such extensive literature, the present chapter summarizes the potential of polyimide membrane in CO_2 removal with strong emphasis on its structural flexibility. The key challenges that constraint the development of polyimide membrane, including the trade-off between membrane permeability and selectivity as well as the plasticization issue, and the available approaches to overcome those limitations are also highlighted. Besides, the simulation study and feasibility of polyimide membrane for CO_2 separation under real industrial conditions are also discussed.

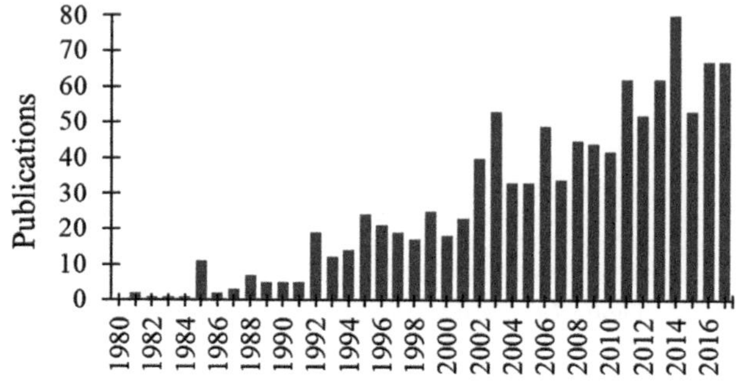

Figure 7.2. Publications related to polyimide gas separation membrane over the years. Data are obtained from Scopus.

Implementation of Polyimide Membrane in Gas Separation

Permeability-selectivity trade-off

In principle, a gas separation membrane should acquire both high permeability and high selectivity simultaneously in order to reduce the membrane area requirement and to ensure high purity of the separated gas streams, respectively (Ayala et al. 2003). However, the permeability and selectivity performances of a polymeric membrane, which includes the polyimide, are contradictive by nature. As demonstrated by the Robeson upper bound plot (Fig. 7.3), there is always a trade-off between permeability and selectivity, i.e., a high selectivity membrane concomitantly possesses low permeability and vice versa (Robeson 2008). This trade-off phenomenon can be explained by the gas transport mechanism across a polymeric membrane. For instance, a densely packed membrane with low free volume offers less diffusion pathways for the gas penetrants. At the same time, the small free volume sizes act as molecular sieves that allow slower penetration of bigger gas molecules across the membrane in relation to the smaller gas molecules. Therefore, the membrane selectivity is greatly improved with the sacrifice of the membrane permeability.

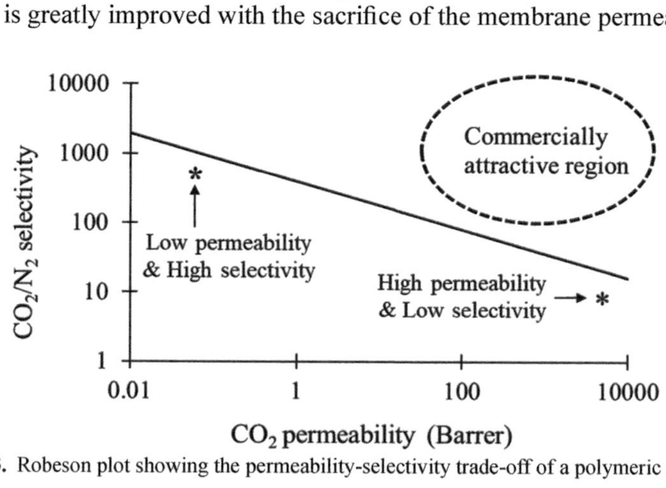

Figure 7.3. Robeson plot showing the permeability-selectivity trade-off of a polymeric membrane.

The permeability-selectivity trade-off is one of the major concerns that restricts the widespread application of polyimide membrane for industrial CO_2 removal. Substantial research efforts have been directed to surpass this upper bound limit to produce a polyimide membrane with separation performances in the commercially attractive region, as illustrated in Fig. 7.3. The two most common approaches employed by researchers to enhance the membrane separation performances are modification of polyimide backbone structure and fabrication of polyimide mixed matrix membrane. In this chapter, these two techniques are comprehensively discussed while major findings are reported.

Structural flexibility of polyimide

Structural flexibility is one of the most interesting features of polyimide. In general, polyimide is synthesized from the reaction between dianhydride and diamine monomers. Owing to the diversity of the monomers' chemical structures, a wide variety of polyimides with different physical properties and chemical characteristics can be obtained by varying the monomer combination. With proper structural alteration, it is hence possible to produce a polyimide membrane with better gas separation performances that surpass the Robeson upper bound plot. In this section, the structural modification of polyimide in past researches and their gas separation performances are highlighted with the monomers' structures illustrated in Table 7.1.

As a fundamental guideline, any structural alteration that inhibits efficient polymer packing and chain mobility tends to increase the gas permeability without significant loss in selectivity (Coleman and Koros 1990). In view of this, fluorinated polyimide with -$C(CF_3)_2$- linkage gains much research interest nowadays (Coleman and Koros 1990, Tanaka et al. 1992a, Kim et al. 2000). In particular, the bulky -$C(CF_3)_2$- group in the polyimide backbone tends to disrupt efficient polymer packing, hence yielding a polyimide membrane with high Fractional Free Volume (FFV) and thus high gas permeability (Cui et al. 2011, Chua et al. 2014). Despite the great improvement of gas permeability, the membrane selectivity is not much compromised as the introduction of -$C(CF_3)_2$- linkage also leads to stiffer polyimide chains due to the restricted torsional motion of the neighboring phenyl rings (Qiu et al. 2011, Chua et al. 2014).

The gas transport properties of a series of fluorinated and non-fluorinated polyimides membrane were investigated by Tanaka et al. (1992a). The -$C(CF_3)_2$- group was incorporated into the polyimide backbone via the introduction of 6FDA or 6FpDA (Table 7.1). It was found that the polyimide membrane with higher fluorine content had inhibited chain packing and higher gas permeability. For example, the CO_2 permeability of 6FDA-ODA (fluorine content = 14×10^{-3} mol/cm^3) was 4.7 times higher than that of PMDA-ODA (fluorine content = 0 mol/cm^3). When the fluorine content of the polyimide membrane was further increased by substituting ODA with 6FpDA, the 6FDA-6FpDA membrane (fluorine content = 24×10^{-3} mol/cm^3) showed even higher CO_2 permeability, which was 207% higher than that of 6FDA-ODA. However, only 22% reduction in CO_2/CH_4 selectivity was observed.

Besides the commonly reported -$C(CF_3)_2$- containing polyimide, Kim et al. (2000) introduced another polyimide with fluorinated alkyl side group by

Table 7.1. Structure of the commonly used monomers.

Monomer		Structure
6FDA	4,4′(hexafluoroisopropylidene) diphthalic anhydride	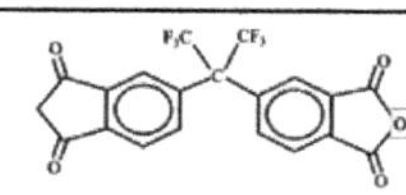
6FmDA	3,3′(hexafluoroisopropylidene) dianiline	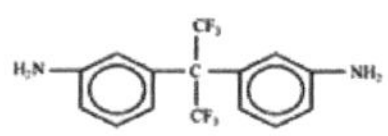
6FpDA	4,4′(hexafluoroisopropylidene) dianiline	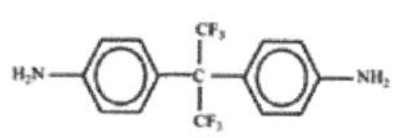
BAPAF	2,2-bis(3-amino-4-hydroxyphenyl) hexafluoropropane	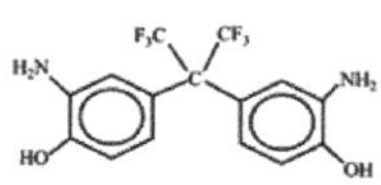
BAPM	α,α-bis(4-amino-3,5-dimethyphenyl)-1-phenylmethane	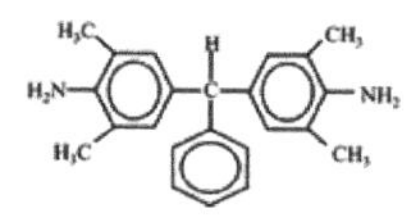
BAT	1,4-bis(4-aminophenoxy) triptycene	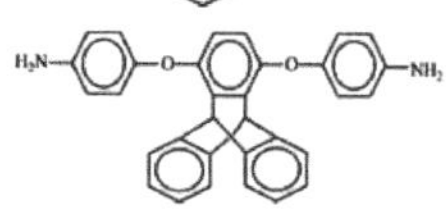
BDA	Benzidine	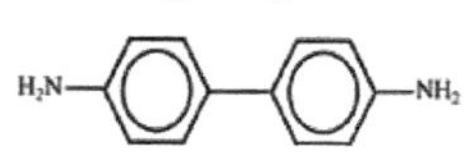
DABA	3,5-diaminobenzoic acid	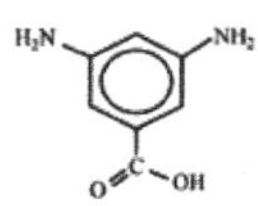
DAM	2,4,6-trimethyl-*m*-phenylenediamine	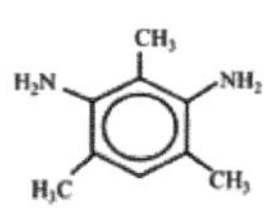
DAP	2,4-diaminophenol dihydrochloride	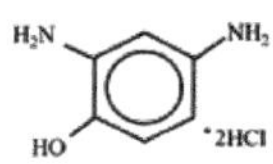
*m*MPD	2-methyl-1,3-phenylenediamine	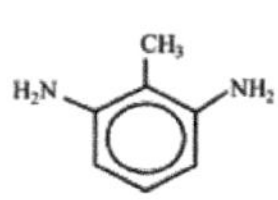
*m*PD	1,3-phenylenediamine	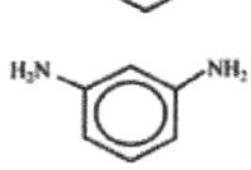
ODA	4,4′-oxydianiline	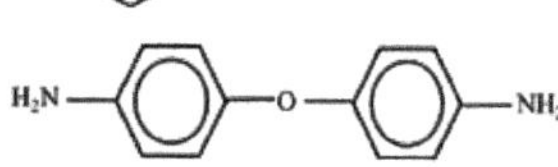
PFDAB	2-(perfluorohexyl)ethyl-3,5-diaminobenzoate	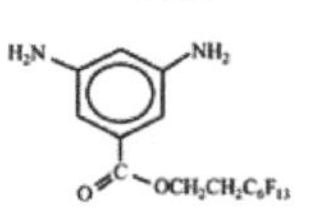
PMDA	Pyromellitic dianhydride	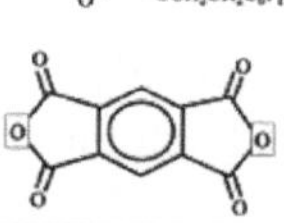

incorporating a fluorinated dianhydride, namely PFDAB (see Table 7.1), into the polyimide backbone. As the fluorinated alkyl side group reduced the cohesive energy and molecular interaction, a polyimide membrane with increased FFV was obtained. Hence, the PFDAB-containing polyimide membrane achieved higher CO_2 permeability as compared to the non-fluorinated one. Similarly, only slight selectivity loss was noticed.

Apart from the popularly researched fluorinated polyimide, the effect of bulky methyl substituent on gas permeability and selectivity has also been investigated by Tanaka et al. (1992b). It was reported that the incorporation of the methyl substituent restricted the internal rotation around the bonds between the phenyl rings and the imide rings, thus yielding stiffer polyimide chains. Besides, the steric hindrance caused by the nonplanar polymer structure and the bulky methyl substituent prevented the dense packing of the rigid polyimide chains. Hence, the FFV of the resulting polyimide membrane increased with the number of methyl substituents on the phenyl rings of diamine, i.e., DAM > *m*MPD > *m*PD (Table 7.1). Unsurprisingly, the gas permeability of the DAM-containing polyimide membrane was higher than that of the other two polyimide membrane albeit with certain degree of selectivity loss.

Although the incorporation of bulky groups into the polyimide backbone generally leads to significant improvement in gas permeability, it is often accompanied with selectivity loss. Hence, recent studies attempt to develop a new macromolecular structure with high FFV but narrow free volume size distribution to increase the gas permeability while at the same time maintaining high selectivity. Tong et al. (2015) prepared polyimides from 6FDA and diamine with bulky triptycene (BAT) or phenyl moiety (BAPM). Owing to the higher FFV, the BAPM-containing polyimide membrane demonstrated 10.3 times higher CO_2 permeability as compared to the BAT-containing polyimide membrane. However, enhancement in CO_2/N_2 selectivity was also observed, in which the selectivity was 3.3 times higher than that of the BAT-containing membrane. The higher selectivity of the BAPM-containing membrane was attributed to its favorable cavity size (R = 3.42–3.45 Å) for CO_2 separation, which is between the dynamic radius of CO_2 (3.3 Å) and other gas molecules (3.45–3.8 Å).

The integration of flexible linkage into the polyimide structure is another effective strategy in enhancing the gas permeability. For instance, Xuesong and Fengcai (1995) synthesized a polyimide with flexible ester linkage by introducing ODA diamine into the polymer backbone. It was expected that the flexible ester linkage in ODA could promote the intra-segmental motion of the polyimide chains, hence increased the FFV of the membrane. Therefore, higher gas permeability was achieved by the ODA-containing polyimide membrane as compared to the more rigid BDA-containing polyimide membrane.

Based on the above discussion, it can be summarized that the gas permeability of a polyimide membrane increases with the incorporation of bulky groups and flexible linkages into the polyimide structure. Nevertheless, the enhancement of gas selectivity is also equally important as aforementioned. Previous researches suggested that the gas selectivity can be improved by incorporating *meta*-linkage in the polyimide structure rather than *para*-linkage. Coleman and Koros (1990) demonstrated that the CO_2/CH_4 selectivity increased by 60% by replacing the 6FpDA moiety (*para*-

linkage) with 6FmDA moiety (*meta*-linkage) in the polyimide structure, presumably due to the lower sub-T_g motion of the *meta*-connected polyimide. A similar result was also reported by Tanaka et al. (1992a), in which the *meta*-linked polyimide membrane showed higher selectivity compared to the *para*-linked one particularly in H_2/CO and H_2/CH_4 separations.

Besides the monomer linkage position, the potential of polar groups in improving the gas selectivity was evidenced by Park et al. (2003). A series of 6FDA-based polyimide membranes with polar hydroxyl or carboxyl groups were synthesized. It was disclosed that the polyimide membrane containing polar moiety such as BAPAF, DAP and DABA (Table 7.1) possessed much higher selectivity (CO_2/CH_4 = 48–89 and CO_2/N_2 = 18–27) in comparison to the non-polar 6FDA-DAM membrane (CO_2/CH_4 = 20, CO_2/N_2 = 13). The high gas selectivity was attributed to the greater molecular interaction in the polar group-containing polyimide as reflected by the higher cohesive energy density value. However, the polar group-containing polyimide membrane showed much lower permeability than the non-polar polyimide membrane.

From the previous findings, it is obviously difficult to enhance both the gas permeability and selectivity at the same time by just using single polyimide with favorable characteristic (high permeability or high selectivity only) in membrane preparation. Therefore, researchers attempt to balance the membrane separation performances by developing copolyimide membrane consisting of both highly permeable moiety and strongly selective moiety (Kim et al. 2006a, Hillock et al. 2008, Qiu et al. 2013). For instance, Qiu et al. (2013) synthesized 6FDA-DAM:DABA copolyimide membrane for gas separation. Indeed, DAM, which constitutes bulky methyl groups, contributes to the high gas permeability while DABA is responsible for the high gas selectivity. Hence, good permeability and selectivity was achieved simultaneously by the 6FDA-DAM:DABA copolyimide membrane, in which its permeability and selectivity value was between those of 6FDA-DAM and 6FDA:DABA polyimide membranes. Therefore, it is possible to acquire a copolyimide membrane with balanced performances between permeability and selectivity through the proper adjustment of monomer combination.

The effectiveness of monomer combination adjustment in overcoming the Robeson's 2008 upper bound plot is of ultimate interest. Henceforth, the gas separation performances of various structurally different polyimides obtained from literature are assessed in detailed, as shown in Fig. 7.4. Among all the polyimide structures, the 6FDA-containing polyimide membrane demonstrates better gas separation efficiency. Unfortunately, its gas separation performances are still below the Robeson's 2008 trade-off plot. Obviously, it is rather difficult to overcome the trade-off between the gas permeability and selectivity with simply adjustment of monomer combination. Indeed, the alteration of monomer combination only serves as the primary screening in obtaining a polyimide structure with better gas separation performance. Afterward, a more sophisticated modification on that particular polyimide structure is necessary to push the gas separation performances beyond the trade-off plot. This includes the fabrication of polyimide mixed matrix membrane, which will be discussed in detail next.

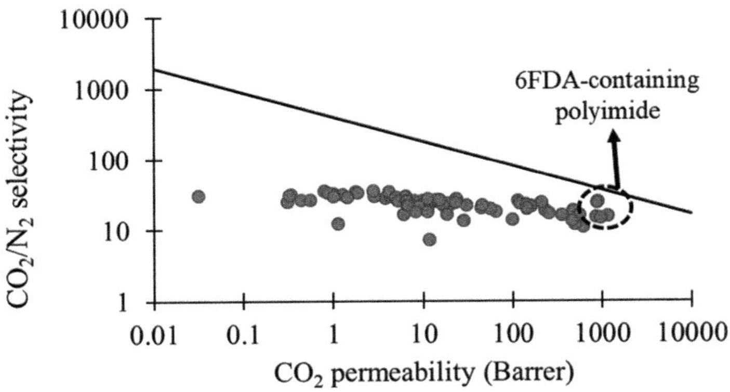

Figure 7.4. Robeson's 2008 upper bound plot showing the gas separation performances of various structurally different polyimides obtained from literature (Kim et al. 2006a, Qiu et al. 2013, Tong et al. 2015, Calle et al. 2013, Ayala et al. 2003, Kim et al. 2000, Coleman and Koros 1990, Hirayama et al. 1996, Shao et al. 2005, Kammakakam et al. 2015, Chua et al. 2014, Qiu et al. 2011, Kanehashi et al. 2013, Matsui et al. 1998, Fu et al. 2015, Japip et al. 2014, Wang et al. 2015, Lua and Shen 2013, Maya et al. 2010, Staudt-Bickel and Koros 1999, Liu et al. 2001, Shao et al. 2008, Zhang et al. 2017).

Advancement of Polyimide Mixed Matrix Membrane

As discussed in the previous section, polyimide (PI) is the most expansively studied polymer material for membranes because they have significantly better gas permeability and selectivity than typical glassy polymers (Bos et al. 1998). Some of the polyimides such as Matrimid and Kapton are widely available commercially. However, since the inception of Robeson upper bound limit in 1991(Robeson 1991) and then updated in 2008 (Robeson 2008), polymeric membrane, including PI have seemingly reached a limit in the trade-off between permeability and selectivity despite extensive efforts were done to improve the materials (Kim et al. 2006a, Hillock et al. 2008, Qiu et al. 2013, Kraftschik et al. 2013), as illustrated by the Robeson plot shown in Fig. 7.5. Plasticization and aging remain as two common causes that deteriorate membrane performances in extreme environment typically in high CO_2 concentration or high pressure operating conditions. In another words, highly permeable polymers such as PI have the potential for large scale separations only if the selectivity can be enhanced.

Notably, nanoparticles incorporation into membrane matrix, dubbed as Mixed Matrix Membrane (MMM), has been the latest and prominent membrane technology with capability for future applications. In MMMs, nanomaterials (mixtures of organic or inorganic fillers such as zeolite, carbon nanotube, MOF, organic cage and, etc.) are added to the continuous polymer phase (Fig. 7.6). The synergistic combination is expected to overcome the respective limitations of polymeric and inorganic membranes in gas separation. The PI serves as the main matrix, whereas some varieties of nano-fillers are embedded in the polymer matrix in trying to overcome the Robeson's upper bound.

The incorporation of nano-fillers is based on the state of the nano-fillers before membrane formation (Xiao et al. 2009). The first method involves the direct physical dispersion of nano-filler into PI matrix, which is accomplished by mechanical

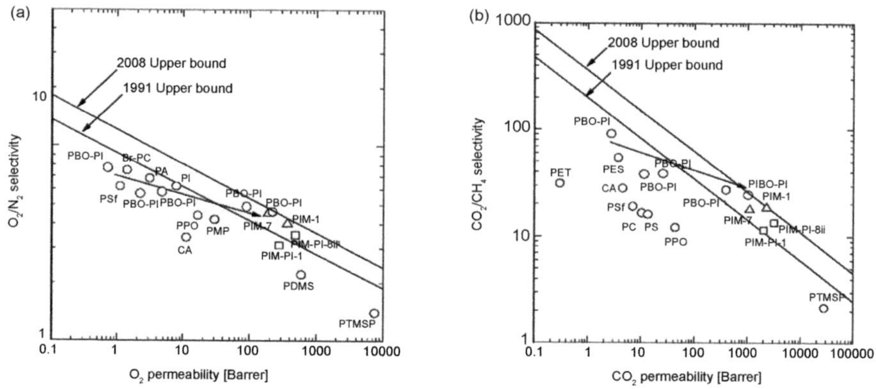

Figure 7.5. Upper bound limit for O_2/N_2 and CO_2/CH_4 separation by various type polymeric materials (Jung et al. 2010). [PET = poly(ethylene terephthalate); PSf = polysulfone; CA = cellulose acetate; PC = polycarbonate; PS = polystyrene; PPO = poly(phenylene oxide); PTMSP = poly(1-trimethylsilyl-1-propyne); PA = polyamide; PI = polyimide; PMP = poly(4-methyl-2-pentyne); PDMS = polydimethylsiloxane; PIM–PI = polymers with intrinsic microporosity–polyimide] Reprinted with permission from Jung et al. (2010). Copyright 2010 Elsevier.

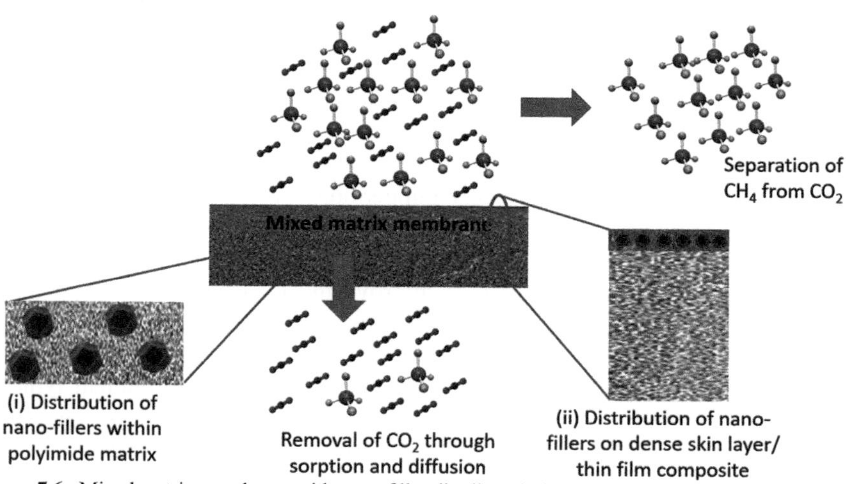

Figure 7.6. Mixed matrix membrane with nano-filler distributed: (i) within the PI matrix (ii) within dense skin selective layer (thin film composite).

stirring or ultrasonication to reduce particle agglomeration. Nik et al. (2011) had successfully fabricated defects free PI MMM by a dense film casting method. Low amount of 6FDA-ODA polymer was added to the amine-grafted zeolitic FAU/EMT suspension before incorporating the nano-filler into the polymer dope to make the particles more compatible with and homogeneously distributed within the bulk film. MMMs with higher loading of grafted amine zeolitic fillers showed improved mixed gas permeability and selectivity compared to the neat 6FDA-ODA membrane, in which the CO_2/CH_4 selectivity increased by more than 118%. The advantage of the direct physical dispersion of nano-filler into PI matrix is that the synthesis of inorganic fillers will be completed prior to the membrane fabrication, regardless of the membrane fabrication conditions. The second approach involves the use nano-filler precursor, which is added to the polymer solution to allow nano-filler formation during membrane formation (Cong et al. 2007, Xiao et al. 2009). For example, Lua and Shen (2013) prepared polyimide-silica composite membranes using the sol-gel technique. The silica precursor and silane coupling agent, which were tetraethyl orthosilica (TEOS) and (3-aminopropyl)triethoxysilane (APTES) respectively, were mixed into the polyimide solution prior to the membrane casting. This method promises a more homogeneous dispersion of inorganic particles in the polyimide matrix. With the incorporation of 7.2 wt% of silica into the polyimide membrane, the CO_2 permeability decreased as a result of the reduced solubility. Meanwhile, the CO_2/N_2 selectivity increased by 2.3 times as compared to the pure polyimide membrane.

The idea behind MMMs is to combine the advantages of PI processability and the good selective adsorption and/or diffusion properties of nano-fillers to achieve high permeability, selectivity, or both. Gas permeation through MMMs mainly occurs via solution diffusion mechanism. Two additional selective gas transport mechanism may exist with the addition of nano-fillers into polymer matrix (Sunarso et al. 2017): (1) physical and/or chemical surface interactions between nano-fillers and gas penetrants which may increase the diffusivity of gas molecules across the membrane and (2) size exclusion of gas penetrant by the framework pores of the nano-fillers. Yong et al. (2001) studied the gas separation performance of MMMs containing zeolite 13X and 4A with different pore diameters. PI/13X MMM (zeolite pore diameter of 7.4 Å) exhibited higher gas permeability while PI/4A MMM achieved higher permselectivity due to stronger sieving effect as the zeolite pore size of 3.8 Å is closer to the kinetic diameter of the gas molecules. Besides, the fabrication of poly(imide siloxane)-carbon nanotube MMM (6FDA-6FpDA-PDMS/CNTs MMM) by Kim et al. (2006b) demonstrated increased permeability of O_2, N_2 and CH_4 in proportion to the amount of open-ended CNTs in the polymer matrix. The presence of CNTs as inorganic phase induced higher FFV in the mixed matrix, which offered more high diffusivity tunnels for gas penetrant thus enhancing the gas permeability.

Various inorganic particles have been employed as fillers in the synthesis of polyimide MMM. The commonly evaluated fillers are zeolite (e.g., zeolite A, SAPO-34, Zeolite-13X, ZSM-5, etc.) (Jusoh et al. 2017, Xu et al. 2007), metal-organic framework (MOF) (e.g., MOF-5, Cu-BPY-HFS, $Zn(pyrz)_2(SiF_6)$, MIL-53, etc.) (Rowsell and Yaghi 2004, Kılıç et al. 2015, Gong et al. 2015, Kertik et al. 2016), zeolitic imidazolate frameworks (ZIFs, a sub-family of MOFs, ZIF-7, ZIF-8) (Wang et al. 2011, Askari and Chung 2013), carbon molecular sieves (CMS) (Bakhtiari et al. 2011, Ward and Koros 2011), mesoporous and nanoporous silica (MCM-41, SBA-12 and 15, fumed silica, etc.) (De Angelis et al. 2013, Kudasheva et al. 2015, Zornoza et al. 2015, Castro-Muñoz et al. 2017), carbon nanotubes (Bakhtiari et al. 2011), lamellar materials (MFI, MCM-22) (Choi and Tsapatsis 2009, Roth et al. 2014) and crystalline structures such as TiO_2 (Sun et al. 2014). The effects of different fillers on PI membranes can be diverse (Yampolskii et al. 2014, Castro-Muñoz et al. 2017).

Zeolite is preferable for gas separation due to its three-dimensional zeolite network that offers less restricted diffusion path (Tul Muntha et al. 2016). It is the first molecular sieve used as inorganic filler in the synthesis of MMMs (Khan et al. 2010). In Chen et al. (2012a)'s work, incorporation of amine-grafted FAU/EMT zeolite in 6FDA-ODA polyimide showed excellent CO_2/CH_4 separation, with the highest CO_2 permeability at 40.9 Barrer and ideal selectivity at 80.2. On the other hand, Musselman et al. (2009) reported the increment of CO_2/CH_4 selectivity from 34.7 to 57.4 by adding 10% of Y-type zeolite in Matrimid, as compared to the neat membrane. MMMs of PI and zeolite 13X, ZSM-5 and 4A were prepared by Chaidou et al. (2012) at various loadings ranging from 0 to 30 wt.%. Permeability of all gases (He, H_2, CO_2, and N_2) for PI-zeolite membranes improved with an increase in zeolite loading. However, the selectivity declined at high loading of zeolites 13X and ZSM-5 for H_2/N_2 separation.

Metal-Organic Framework (MOF) has been explored due to the difficulties in controlling the pores and flux properties of zeolites (Li et al. 2012). MOF consists of metal ions or oxide clusters coordinated to organic ligands to form a crystalline network that is less stiff and brittle compared to zeolites (Tan and Cheetham 2011). In addition, MOF has better compatibility with polymer due to the existence of organic linkers in MOF. This allows higher loading of filler in MMM to further increase the adsorption capacities for gas separation (Goh et al. 2011, Zhang et al. 2013). Notably, two unique features of MOF are: (1) tunable structure and composition of MOF (2) high flexibility in surface modification through the organic linker in MOF to improve the compatibility between inorganic phase and polymer (Jeazet et al. 2012, Sunarso et al. 2017). In a work carried out by Perez et al. (2009), the fabricated 30% MOF-5/Matrimid MMM demonstrated up to 120% increment in permeability but almost constant CO_2/CH_4 selectivity was observed from 41.7 (neat Matrimid membrane) to 44.7. High permeability was achieved due to the high surface area of the MOF-5 (3000 m^2/g), but at the same time, the selectivity remained unchanged because the permeabilities of all gases increased proportionally. Recently, Friebe et al. (2017) prepared a multilayer composite by coating a thin polyimide (Matrimid) on top of the 3D MOF structure UiO-66 [$Zr_6O_4(OH)_4(bdc)_6$] supported with α-Al_2O_3 (Fig. 7.7). Figure 7.7 showed a good contact between MOF/Matrimid interface without any detectable interfacial void. The composite membranes with MOF triangular pores of approximately 6 Å were evaluate in the separation of H_2 from different binary mixtures at room temperature. The Matrimid/UiO-66 multilayer membrane showed improved molecular sieving in H_2/CO_2 and H_2/CH_4 separation, in which the selectivity increased from 3 to 5 and from 16.5 to 80, respectively, as compared to neat Matrimid membrane.

Zeolitic Imidazolate Framework (ZIF), a sub-family of MOF have been extensively explored in recent years (Wang et al. 2011, Askari and Chung 2013). Similar to zeolite and MOF, ZIF is defined by its high micro-porosity, high thermal and chemical stability, and high surface area. ZIFs with tunable nano-pore size are produced by linking transition metal ions such as Fe, Zn, Co or Cu through N atom by imidazolate linkers (Schejn et al. 2014, Jusoh et al. 2016). 6FDA-DAM-ZIF-11 MMMs loaded with various ZIF-11 percentages were analyzed for H_2/CH_4 and CO_2/CH_4 separations (Boroglu and Yumru 2017). The MMM integrated with 20 wt.% ZIF-11 disclosed remarkable enhancement of CO_2 permeability (257.50 Barrer) from 20.60 Barrer of pristine 6FDA-DAM membrane. However, gas selectivity of

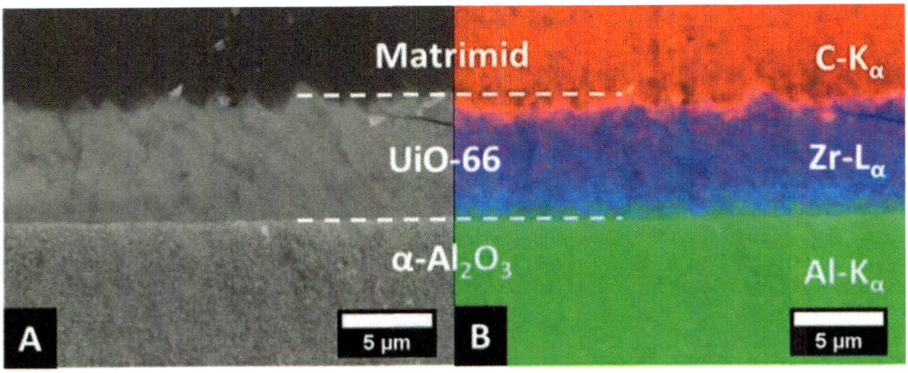

Reprinted

Figure 7.7. (A) SEM image and (B) EDXS mapping of the Matrimid/UiO-66 multilayer membrane. Reprinted with permission from Friebe et al. (2017). Copyright 2017 American Chemical Society.

MMMs remained almost constant or slightly decreased from pristine 6FDA-DAM membrane (H_2/CH_4 = 33.96 and CO_2/CH_4 = 32.69). When the dispersed inorganic filler was introduced into the polymer matrix, it will enhance the intersegmental mobility between the polymer chains to facilitate the diffusion of gas penetrants. Hence, permeabilities of all gases increased without affecting the selectivity (Yilmaz and Keskin 2012, Boroglu and Yumru 2017). The idea of double-layered ZIF-8/ZIF-9-P84 and ZIF-67/ZIF-9-P84 polyimide MMM proposed by Cacho-Bailo et al. (2017) had shed light on the possibility of H_2/CO_2 separation. The different metallic character of the ZIF materials (Zn in ZIF-8 and Co in ZIF-67 and ZIF-9) facilitated the molecular sieving of this membrane. In this work, ZIF-67 and ZIF-8 coatings with lower CO_2 adsorptive methylimidazolate were used to restrict the CO_2 concentration on the surface of the ZIF-9 layer. Meanwhile, ZIF-9 that functioned as the molecular sieve would further enhance the H_2 purification. This unique double-layered ZIF strategy led to a H_2/CO_2 separation selectivity of 9.6 at 150°C. Several researchers reported the enhanced gas separation performances of MMMs subjected to annealing (Hibshman et al. 2003, Ozturk and Demirciyeva 2013, Shahid and Nijmeijer 2014). Mahdi and Tan (2016) who focused on the mechanical properties and the viscoelastic response of ZIF-8/Matrimids MMM had identified the adverse effects of post annealing (> 180°C), i.e., mechanical degradation in terms of ductility and toughness that caused PI MMMs to be less suitable for practical use. Results showed that an annealed 10 wt% ZIF-8/Matrimid MMM was most practical for separation applications. In the report, it was concluded that an optimal combination of mechanical resilience and gas permselectivity was the prerequisites for the development of MMMs (Mahdi and Tan 2016). Examples of other inorganic fillers, such as carbon molecular sieves, silica, carbon nanotubes, graphene oxide, etc., embedded in PI MMMs are summarized in Table 7.2.

Improper material selection creates several technical problems in MMM fabrication including the issues on interfacial adhesion, blockage of inorganic phase and stress formation within the membrane matrix. Figure 7.8 demonstrates three common defects found in MMM, i.e., sieve-in-a-cage or leaky interface, plugged

Table 7.2. Gas separation performances of PI MMMs embedded with different inorganic fillers.

Polyimide as organic phase	Inorganic phase	Operating Condition	Separation performances	References
P84	[a]CMS	50-50% CO_2/CH_4 feed gas, Temperature = 35°C Pressure = 20 atm	CO_2 permeability = 500 Barrer CO_2/CH_4 selectivity = 89	Tin et al. (2004)
Matrimid	[a] CMS	10–90% CO_2/CH_4 and 20-80% CO_2/N_2 feed gas, Temperature = 25°C Pressure = 1 bar	CO_2 permeability = 12 Barrer CO_2/CH_4 selectivity = 33 CO_2/N_2 selectivity = 15	Fuertes et al. (1999)
Kapton	[a] CMS	Temperature = 25°C Pressure = 1 bar	CO_2 permeability = 92 Barrer CO_2/CH_4 selectivity = 16 CO_2/N_2 selectivity = 9	Fuertes et al. (1999)
[b] 6FDA-6FpDA	Silanized γ-Al_2O_3 (2 wt.%)	Temperature = 30°C Pressure = 3 bar	CO_2 permeability in CO_2/CH_4 = 51 Barrer CO_2/CH_4 selectivity = 52 CO_2 permeability in CO_2/N_2 = 51 Barrer CO_2/N_2 selectivity = 21	Tena et al. (2010)
Matrimid/ polysulfone	Silica (15.2 wt.%)	50-50% CO_2/CH_4 feed gas, Temperature = 25°C Pressure = 10 bar	CO_2 permeance = 90 GPU CO_2/CH_4 selectivity = 60	Rafiq et al. (2012)
[c] DMMDA-BTDA	[d] MWCNTs (3 wt.%)	Temperature = 15°C Pressure = 0.1 MPa	CO_2 permeability = 9.06 Barrer CO_2/CH_4 selectivity = 24.49 CO_2/N_2 selectivity = 37.74	Sun et al. (2017)
Polyimide	[e] MWCNT-GONRs (1 wt.%)	Temperature = 35°C Pressure = 1 bar	CO_2 permeability = 17 Barrer CO_2/CH_4 selectivity = 26 CO_2/N_2 selectivity = 25	Xue et al. (2017)
Matrimid 5218	Polyzwitterion coated CNT (5 wt.%)	Temperature = 30°C Pressure = 2 bar	CO_2 permeability = 103 Barrer CO_2/CH_4 selectivity = 36	Liu et al. (2014)
Matrimid	Ordered mesoporous COK-12 type silica (30 wt.%)	50-50% CO_2/CH_4 and CO_2/N_2 feed gas, Temperature = 25°C Pressure =10 bar	CO_2 permeability in CO_2/CH_4 = 14 Barrer CO_2/CH_4 selectivity = 32.5 CO_2 permeability in CO_2/N_2 = 13 Barrer CO_2/N_2 selectivity = 32.5	Khan et al. (2015)
6FDA-DAM	[f] Mg-MSSs	50-50% CO_2/CH_4 and CO_2/N_2 feed gas Temperature = 35°C Pressure =300 kPa	CO_2 permeability in CO_2/CH_4 = 1245 Barrer * CO_2/CH_4 selectivity = 315 CO_2 permeability in CO_2/N_2 = 1214 Barrer * CO_2/N_2 selectivity = 24.4	Zornoza et al. (2015)

Table 7.2 contd. ...

...Table 7.2 contd.

Polyimide as organic phase	Inorganic phase	Operating Condition	Separation performances	References
[g] BTDA–BAPF–DMMDA	TiO_2 (24 wt.%)	Temperature = 303K Pressure = 0.1 MPa	O_2 permeability = 4.5 Barrer * O_2/N_2 selectivity = 15.8	Sun et al. (2014)
Linear polyimide (LPI)	Neodymium magnetic powder MQP-14-12 (56-87 wt.% relative to LPI)	Synthetic air (21 vol.% O_2 and 79 vol.% N_2) as the feed gas	O_2 permeability = 52.5 Barrer O_2/N_2 selectivity = 6.22	Rybak et al. (2014)
Matrimid 9725	[h] SO_3H-MCM-41 (20 wt.%)	Temperature = 25°C Pressure = 10 bar	CO_2 permeability in CO_2/CH_4 = 8.2 Barrer CO_2/CH_4 selectivity = 38 CO_2 permeability in CO_2/N_2 = 7.2 Barrer CO_2/N_2 selectivity = 30.5	Khan et al. (2013)/ EndNote>
[i] A-PI	beta-cyclodextrin functionalized MWCNTs (6 wt.%)	Temperature = 25°C Pressure = 15 bar Gauge	CO_2 permeance = 2.2 GPU * CO_2/CH_4 selectivity = 62.86	Aroon et al. (2013)
[j] PMDA-ODA	SiO_2 (13.7 wt.%)	Room temperature Pressure = 4 bar	He permeability = 16.59 Barrer CO_2 permeability = 1.47 Barrer N_2 permeability = 0.12 Barrer * He/N_2 selectivity = 138.25 * CO_2/N_2 selectivity = 12.25	Lua and Shen (2013)
[k] 6FDA-4MPD / 6FDA-DABA	MCM-41 (8wt.%) / JDF-L1 (4 wt.%)	50-50% H_2/CH_4 feed gas, Temperature = 35°C Pressure = 340 kPa	H_2 permeability in H_2/CH_4 = 440 Barrer CO_2/CH_4 selectivity = 32 O_2 permeability in O_2/N_2 = 126 Barrer O_2/N_2 selectivity = 3.6	Galve et al. (2013)

* ideal selectivity

[a] carbon molecular sieve (CMS)

[b] 4,4'-(hexafluoroisopropylidene) diphthalic anhydride (6FDA) and 4,4'-(hexafluoroisopropylidene) dianiline (6FpDA)

[c] 3,3',4,4'-benzophenonetetracarboxylic dianhydride (BTDA) and 3,3'-dimethyl-4,4'-diaminodiphenylmethane (DMMDA)

[d] Multi-walled carbon nanotube (MWCNTs)

[e] Multi-walled carbon nanotube (MWCNTs)-Graphene oxide nanoribbons (GONRs)

[f] Grignard surface functionalized ordered mesoporous silica MCM-41 spheres (Mg-MSSs)

[g] 3,3',4,4'-benzophenonetetracarboxylic dianhydride (BTDA) 9,9'-bis (4-aminophenyl) fluorine (BAPF) and 4,4'-biamino-3,3'-dimethyldiphenyl-methane (DMMDA)

[h] sulfonic acid functionalized ordered mesoporous silica spheres (SO_3H-MCM-41)

[i] 1,3-Isobenzofurandione, 5,5'-carbonylbis-, polymer with 1 (or 3)-(4-aminophenyl)-2,3-dihydro-1,3,3 or (1,1,3)-trimethyl-1H-inden-5-amine polyimide (A-PI)

[j] pyromellitic dianhydride (PMDA, 97% purity) and 4,4'-oxydianiline (ODA, 98% purity)

[k] 6FDA (4,4'-hexafluoroisopropylidene diphthalic anhydride), 4MPD (2,3,5,6-tetramethyl-1,4-phenylene diamine) and DABA (3,5-diaminobenzoic acid)

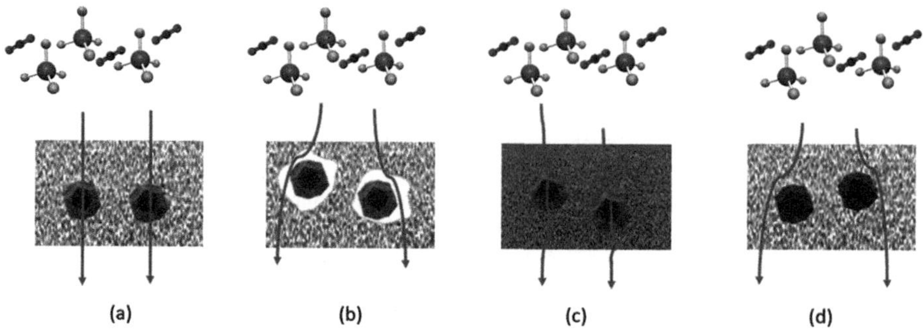

Figure 7.8. Ideal (a) and Non-ideal gas transport properties of MMMs: (b) interfacial voids, (c) matrix rigidification and (d) plugged sieves. Figures adapted with permission from Rezakazemi et al. (2014). Copyright 2014 Elsevier.

sieve and matrix rigidification. Poor interfacial adhesion between polymer-fillers interface creates undesirable interfacial voids (Fig. 7.8b) (Rezakazemi et al. 2014, Ismail and Kusworo 2014). These voids provide a diffusion path with the least resistance. Hence, gas molecules tend to pass through these voids instead of the discriminative pores of the inorganic fillers. MMMs with such voids usually show enhancement in permeability but lower selectivity than that of the pure polymer if the void size is larger than the gas molecules. Indeed, surface modification of fillers is an effective strategy to reduce the interfacial void and also to increase the solubility of gas penetrants in PI. In particular, a modified filler carrying partial charges tends to attract polar gases such as CO_2 (Ismail et al. 2011, Lin et al. 2015). Lin et al. (2015) produced 6FDA-durene MMMs by incorporating CNT-MOF fillers as the inorganic phase. MOF NH_2-MIL-101(Al) was first grown on the surface of CNTs prior to the MMM synthesis to introduce MIL-101 active sites for better sorption of CO_2. Interfacial voids between CNT-MOF fillers and polymer phase were undetectable under SEM, even at higher loading of 15%. The 10% CNT-MOF/PI MMM not only demonstrated higher permeability (1037 Barrer) but also good CO_2/CH_4 selectivity of 25.4 attributed to the larger free volume created by the incorporation of CNTs (high permeability) and $-NH_2$ groups in MIL-101(Al) that improved CO_2 adsorption selectivity (Ge et al. 2013, Lin et al. 2015).

The immobilization of polymer chains at polymer-filler interface will cause the formation of rigidified polymer layer or known as "matrix rigidification" (Fig. 7.8c) (Moore and Koros 2005, Souza and Quadri 2013). This rigidified polymer will restrict the polymer chain mobility and hence reduced the penetration of gas molecules without significant increase in selectivity. In the work carried out by Chung et al. (2003), benzylamine-modified C_{60} served as an impenetrable element to induce matrix rigidification of Matrimid, thereby reduced N_2 permeability and slightly increased He/N_2 selectivity. Figure 7.8d demonstrates another type of non-ideality in gas transport properties of MMMs known as "plugged sieve". Pore blockage of dispersed fillers always decreased the gas permeability. However, its effect on the selectivity was different when different inorganic fillers were used (Boroglu and Gurkaynak 2011). For example, Boroglu and Gurkaynak (2011) found that 4A zeolite with pore size comparable to the molecular diameter of gas penetrant will decrease the gas selectivity in O_2/N_2 and CO_2/CH_4 separations. On the contrary, zeolite 5A or beta zeolites with much larger original pore size than the gas penetrant showed the increase of MMMs selectivity.

Despite the successful tuning of membrane structure through the modification of polymer backbone or the integration of inorganic fillers to increase permeability and selectivity, the swelling of polymer network during high pressure separation processes could not be ignored. Glassy polymer such as polyimide tends to swell during sorption of carbon dioxide, which in turn leads to reduction in the membrane permselectivity. Such membrane swelling phenomenon is known as membrane plasticization, which will be discussed in the following section.

Improvement of Polyimide Membrane Plasticization Resistance

Plasticization

Apart from the permeability-selectivity trade-off, plasticization is also another limiting issue that captures plenty of attention for the application of polyimide membrane in CO_2 removal. With a high critical temperature (304.2 K), CO_2 is shown to be a highly condensing and sorbing gas penetrant (Scholes et al. 2010, Tan et al. 2016). Hence, CO_2 is anticipated to interact strongly with the polyimide matrix and causes polyimide chains swelling at high CO_2 concentration, which increases the segmental mobility of polyimide chains (Wind et al. 2003, Velioğlu et al. 2012). This leads to faster diffusion of all the gas penetrants across the polyimide matrix (Taubert et al. 2003) thus severely affecting the membrane selectivity. The impact of plasticization is better demonstrated in Fig. 7.9, in which the permeability increases at the expense of selectivity particularly at high CO_2 concentration (Wind et al. 2002, Kapantaidakis et al. 2003). The loss of membrane selectivity is unfavorable for industrial gas separation as the separation efficiency reduces greatly with increasing operating pressure. Therefore, plasticization is problematic especially in the CO_2 removal from

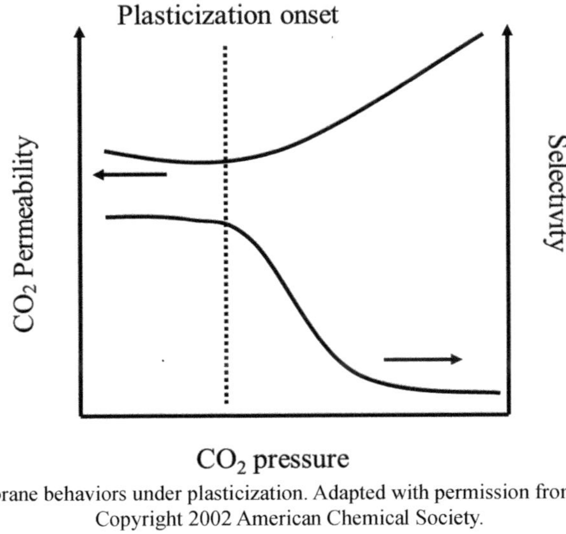

Figure 7.9. Membrane behaviors under plasticization. Adapted with permission from Wind et al. (2002). Copyright 2002 American Chemical Society.

either natural gas or flue gas stream which contains high CO_2 concentration. In view of this, there is a strong need to improve the plasticization resistance of a polyimide membrane in order to maintain a desired membrane selectivity even at high CO_2 pressure.

Crosslinking

One of the key points in improving the plasticization resistance of a polyimide membrane is to suppress the polyimide chains flexibility (Bos et al. 1998). As aforementioned, plasticization is characterized by the polyimide chains swelling that causes increase in segmental chain mobility. Hence, it is of interest to develop a polyimide membrane with restricted chain mobility in order to minimize the swelling effect caused by the gas penetrant. In this regard, crosslinking has been proposed to stabilize the polyimide membrane under high CO_2 concentration (Bos et al. 1998, Liu et al. 2001, Qiu et al. 2011). In the crosslinking strategy, an interwoven polyimide network is produced by linking the polyimides chains together, as demonstrated in Fig. 7.10, by using either ultraviolet (UV) irradiation, chemical or thermal induced

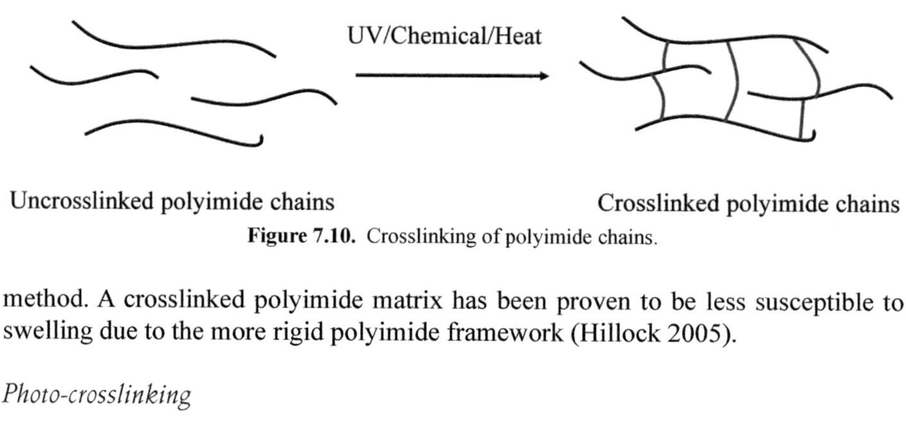

Figure 7.10. Crosslinking of polyimide chains.

method. A crosslinked polyimide matrix has been proven to be less susceptible to swelling due to the more rigid polyimide framework (Hillock 2005).

Photo-crosslinking

Generally, photo-crosslinking is a simple technique that is achieved by exposing a polyimide membrane to UV irradiation. However, this approach is only workable for the polyimide containing benzophenone moiety. Kita et al. (1994) had reported on the photo-crosslinking of BTDA-DAM and BTDA/6FDA:DAM polyimides, which were made possible by the presence of benzophenone group in BTDA dianhydride. During the photo-crosslinking reaction, the benzophenone group of BTDA reacted with the methyl group of DAM, creating two radicals through hydrogen abstraction. Crosslinking between two polyimide chains was then formed through the subsequent radicals coupling.

Kita et al. (1994) found that when both crosslinked polyimide (BTDA:DAM) and copolyimide (BTDA/6FDA:DAM) produced from 10 minutes UV irradiation were dissolved in N-methyl-2-pyrrolidone (NMP), a gel fraction of more than 90% was obtained. This reflected that the crosslinking occurred uniformly in both the surface and interior layer of the polyimide films. The critical role of benzophenone group in photocrosslinking was further reinforced by the higher gel fraction of BTDA-DAM polyimide film compared to BTDA/6FDA:DAM polyimide film. Although Kita et al. (1994) did not examine the plasticization resistance of their crosslinked BTDA-based polyimide membranes, it is believed that the polyimide membranes produced were sufficiently robust to suppress membrane swelling as the polyimide membranes hardly dissolved even in NMP, which is a well-known strong solvent.

With due consideration of the high permeability and selectivity offered by the 6FDA moiety (Neyertz et al. 2010), a continuously increasing number of researchers are prompted to develop 6FDA-based polyimide. Unfortunately, it is not feasible to carry out photo-crosslinking on this polyimide owing to the absence of benzophenone group in the structure of 6FDA. In order to address this shortcoming, Matsui et al. (1998) proposed to introduce a low molecular weight photosensitizer, such as benzophenone (BP), into the 6FDA-DAM polyimide matrix to initiate the crosslinking reaction. The successful polyimides crosslinking was proven from the gelation phenomenon observed in the membrane containing BP after it was dissolved in the solvent (Kita et al. 1994, Matsui et al. 1998).

In most of the reported literatures, photo-crosslinking was performed on membrane in solid state except the work demonstrated by Kang et al. (2000). Kang et al. (2000) attempted to crosslink the Matrimid polyimide membrane (containing benzophenone group in the structure) in liquid form by exposing the polymer cast solution to UV irradiation. It is believed that the imposition of UV irradiation on the polymer cast solution will greatly improve the efficiency of crosslinking. Unfortunately, this work did not disclose the plasticization resistance of the synthesized polyimide membrane.

Later, the photochemical stability of the BTDA-based polyimide under UV light of 300 nm wavelength was investigated by Rosu et al. (2011). Based on the Fourier transform infrared spectroscopy (FTIR) analysis, 9% of imide bonds and 20% of benzophenone groups in BTDA were destructed after 200 hours of UV irradiation. Rosu et al. (2011) hence proposed a mechanism for the photochemical degradation of polyimide film according to the changes in FTIR spectra during UV irradiation. In view of the study above, although UV irradiation can induce the photo-crosslinking in polyimide membrane, the irradiation time should be controlled to avoid any undesirable changes to the polyimide structure.

Regrettably, the application of photo-crosslinking is still limited by its non-reproducible crosslinking degree (Wind et al. 2002). Basically, the crosslinking degree depends strongly on the experimental conditions, for instance the source of UV irradiation and the irradiation time (Staudt-Bickel and Koros 1999). Besides, the uniformity of crosslinking within a polyimide film is not guaranteed (Liu et al. 2001). As the benzophenone groups in the polyimide absorb the UV light, the UV irradiation intensity reduces across the film thickness (Kang et al. 2000). This leads to a non-uniform distribution of radicals within the polyimide film, where more radicals are found at the film top surface. Hence, the crosslinking density at the top surface is usually higher compared to the internal film layer (Kang et al. 2000). Furthermore, undesirable photo-fries rearrangement reaction is possible under UV irradiation (Wright and Paul 1997). In view of these disadvantages, photo-crosslinking is strongly not favorable commercially. Therefore, more attention is averted towards chemical or thermal crosslinking in improving the plasticization resistance of a polyimide membrane.

Chemical crosslinking

Chemical crosslinking refers to the use of a chemical crosslinker to link two polyimide chains together. This approach has been abundantly reported particularly for DABA-containing polyimides. As DABA possesses carboxylic acid group, which defines as the crosslinking site, it is possible to crosslink the polyimide chains with diol via a two-step reaction scheme as described in Fig. 7.11 (Wind et al. 2002, 2003). Firstly, the carboxylic acid group of DABA reacts with the hydroxyl group of the diol to form a monoester polymer. This is then followed by vacuum heating of the monoester polymers at elevated temperature to initiate the polyimide chains crosslinking with the removal of a diol molecule during transesterification reaction.

Staudt-Bickel and Koros (1999) had explored the potential of diol crosslinking in improving the anti-plasticization characteristic of polyimide. By taking 6FDA-*m*PD polyimide as the reference material (plasticization at 5 atm pure CO_2 feed pressure), plasticization was surprisingly subdued up to 14 atm pure CO_2 feed pressure with just the incorporation of DABA moiety. This was presumably due to the hydrogen bonding between the polar carboxylic acid groups. Further chemical crosslinking of the 6FDA-*m*PD:DABA (9:1) polyimide with Ethylene Glycol (EG) managed to suppress the plasticization even more effectively, in which no plasticization was noticed up to 35 atm. In fact, the crosslinked polyimide membrane did not experience

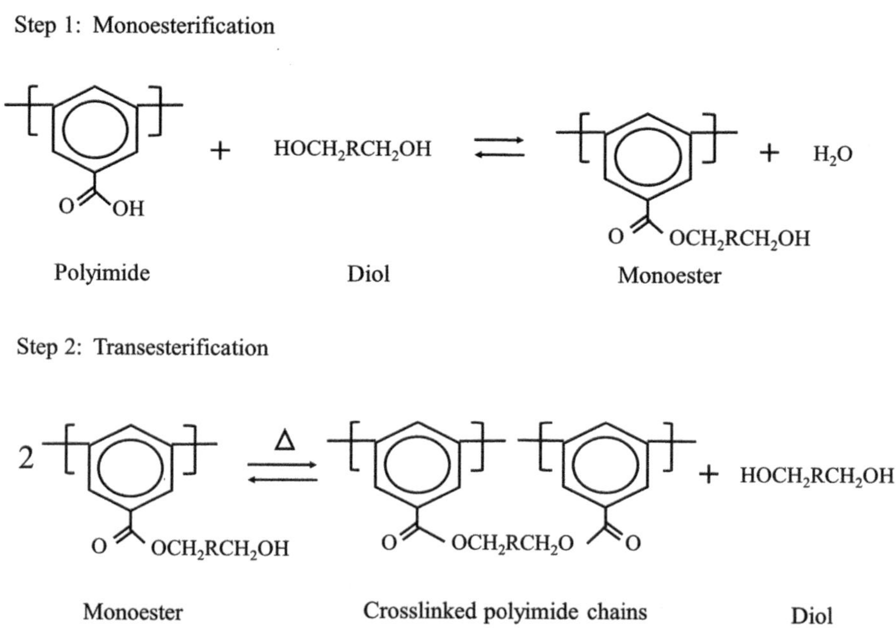

Figure 7.11. Two-step diol crosslinking reaction scheme for DABA-containing polyimide. Adapted with permission from Wind et al. (2003). Copyright 2003 American Chemical Society.

significant reduction in CO_2 permeability in relative to the uncrosslinked reference polyimide as the increased polyimide framework rigidity was compensated by the additional free volume introduced by the EG crosslinker in the polyimide matrix.

Wind et al. (2003) analyzed the impact of different diol crosslinking agents on the gas transport properties of 6FDA-based polyimide. In comparison to EG, Butylene Glycol (BG) and 1,4-cyclohexanedimethanol (CHDM) were the more effective crosslinker with the ability of suppressing plasticization up to 40 atm CO_2 feed pressure. Besides, Wind et al. (2003) also reported that annealing temperature during transesterification altered the polyimide chains rigidity and free volume distribution. In general, higher annealing temperature increased the crosslinking density. Therefore, higher free volume was obtained within the polyimide matrix due to the removal of more diol groups during transesterification. This led to higher gas permeability at higher annealing temperature.

The role of monoesterification catalyst in influencing the membrane plasticization resistance had also been highlighted by the work of Hess and Staudt (2007). In their study, only a slight improvement in plasticization resistance was achieved for the EG crosslinked polyimide membranes when no catalyst was employed in the monoesterification reaction (plasticization at 20 bar CO_2 feed pressure as compared to 15 bar CO_2 feed pressure for an uncrosslinked membrane). This indicated that the crosslinking degree was still insufficient to overcome the plasticization effectively. On the contrary, plasticization was repressed successfully up to 30 bar CO_2 feed pressure by applying *p*-toluenesulfonic acid (*p*-TSA) as the catalyst. However, this resulted in a low CO_2 permeability, presumably due to the reduced free volume caused by the presence of bulky *p*-TSA in the polyimide matrix. Therefore, Hess and Staudt (2007) attempted to use Methane Sulfonic Acid (MSA) with a smaller molecular size as the catalyst. In this case, a much higher CO_2 permeability was obtained without compromising the membrane stability up to 30 bar.

Besides covalent crosslinking with the use of diol as crosslinker, DABA-containing polyimide can also be crosslinked ionically by using inorganic crosslinker such as metal complexes. In the work carried out by Chua et al. (2014), iron (III) acetylacetonate (FeAc) was chosen as the crosslinker to establish an ionic crosslinking network in 6FDA-durene:DABA (9:1) polyimide by coordinating the multivalent cation (Fe^{3+}) with the carboxylate anion ($RCOO^-$). The organic acetylacetonate serves to improve the compatibility between the metal complex and the polymer matrix. Under pure gas permeation, the ionic crosslinking effectively increased the onset of CO_2 plasticization from 7 atm to 30 atm with the addition of 6 wt% FeAc into the polyimide matrix.

Comparisons between the effectiveness of covalent crosslinking and ionic crosslinking in suppressing the CO_2 plasticization of DABA-containing polyimide had been done by Wind et al. (2002). They concluded that the covalent crosslinking (EG as crosslinker) stabilized the polyimide membrane more effectively than the ionic crosslinking (aluminium acetylacetonate as crosslinker). Specifically, the covalently crosslinked sample was not plasticized up to CO_2 partial pressure of 30 bar during 50:50 CO_2/CH_4 mixed gas permeation test. Meanwhile, the ionically crosslinked sample could only withstand high pressure operation up to 17 bar of CO_2 partial pressure without suffering from plasticization.

Apart from the aforementioned crosslinking methods, Liu et al. (2001) introduced another simple room temperature crosslinking technique for the polyimide without carboxylic acid group by employing diamine as the crosslinker. By using 6FDA-durene as an example, the polyimide chains were crosslinked by just immersing the dense film in a *p*-xylenediamine methanol solution for certain time interval. The crosslinking was established through the reaction between the amino groups of *p*-xylenediamine and the imide groups of polyimide, in which the imide groups in the polyimide backbone were eventually converted to the amide groups, as shown in Fig. 7.12. For this crosslinking approach, polyimide film swelling by the methanol solution is the prerequisite in order to allow the penetration of *p*-xylenediamine into the polyimide film. Zhou et al. (2003) adopted the same crosslinking procedure and they found that the diamine crosslinked ultra-thin polyimide films exhibited suppressed CO_2 plasticization up to 10 bar.

Later, Powell et al. (2007) added the diamine crosslinker in a different way as compared to the study of Liu et al. (2001), in which the crosslinker was directly dissolved in the casting solution to promote a more complete crosslinking reaction. Similarly, plasticization resistance of the polyimide membrane was enhanced. Powell et al. (2007) also reported that the reaction between the diamine crosslinker and the polyimide was reversible under vacuum heating.

The effectiveness of both addition methods of diamine crosslinker into the polyimide, either through direct mixing of crosslinker in the casting solution or immersing the polyimide film in a diamine containing methanol solution, was

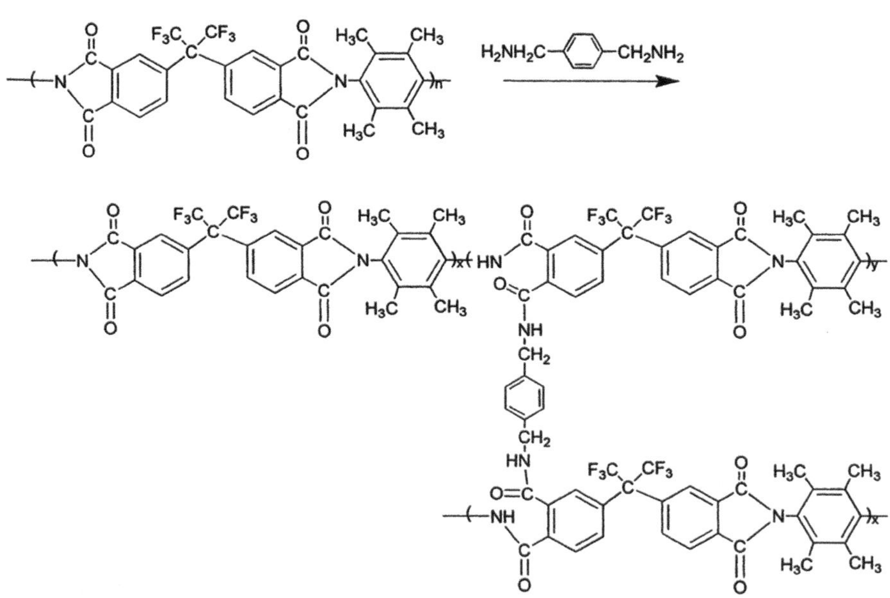

Figure 7.12. Mechanisms of diamine crosslinking by *p*-xylenediamine. Reprinted with permission from Liu et al. (2001). Copyright 2001 Elsevier.

investigated and compared by Chen et al. (2012b). Surprisingly, both methods were equally capable in producing a membrane that can withstand high pressure up to 45 atm without experiencing CO_2 plasticization.

Thermal crosslinking

Although diol crosslinking has been proven to be effective in stabilizing the polyimide membrane against plasticization, the ester linkages formed are easily hydrolyzed under the condition of aggressive acid gas feed stream (Kratochvil and Koros 2008, Qiu et al. 2011, Cui et al. 2011). This will diminish the effect of crosslinking in enhancing the plasticization resistance of membrane and undermine the membrane separation efficiency (Cui et al. 2011). In view of this, Kratochvil and Koros (2008) introduced a new decarboxylation-induced thermal crosslinking approach for the carboxylic acid-containing polyimide. The membrane was thermally crosslinked by heating it to a temperature 15°C higher than its T_g. In this technique, no vulnerable ester bonds are formed as in the diol crosslinking. It was found that the thermally crosslinked membrane was resistant to plasticization even under high CO_2 pressure of 48 bar. Kratochvil and Koros (2008) also excluded charge transfer complexing, polymer decomposition, oligomer crosslinking and dianhydride formation as the possible reasons of the improved plasticization resistance. Instead, it was attributed to the decarboxylation of the acid pendant group at high annealing temperature.

The detailed mechanism of the decarboxylation-induced thermal crosslinking in 6FDA-DAM:DABA (3:2) polyimide was described by Qiu et al. (2011), as depicted in Fig. 7.13. Firstly, the two adjacent carboxylic acid groups of DABA react at high

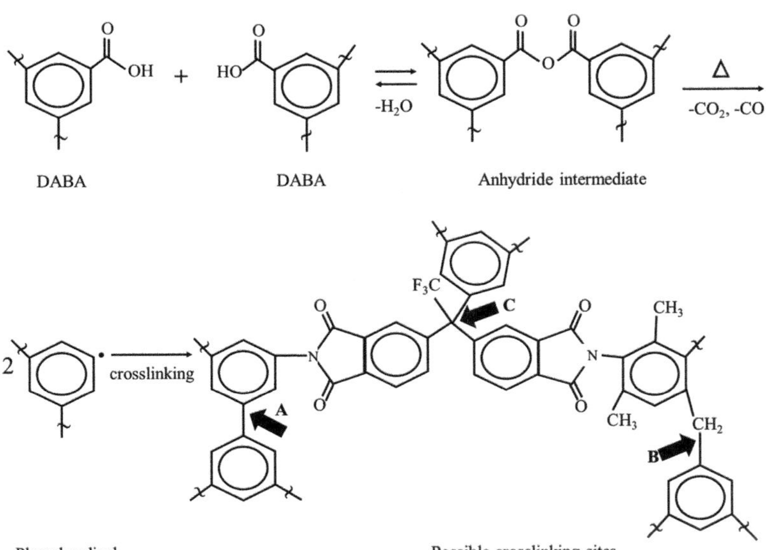

Figure 7.13. Mechanism for the decarboxylation-induced thermal crosslinking. Adapted with permission from Qiu et al. (2011). Copyright 2011 American Chemical Society.

temperature to form an anhydride intermediate. The anhydride intermediate is then decarboxylated to create two phenyl free radicals with the release of CO_2 and CO molecule. The phenyl radicals are most likely to link together to yield a biphenyl crosslinking (crosslinking site A in Fig. 7.13) and to abstract a hydrogen from the methyl group of DAM to form a methyl radical (crosslinking site B in Fig. 7.13). Lastly, the linkage of $–CF_3$ groups in 6FDA may cleave under high temperature to form a free radical that functions as another crosslinking site (crosslinking site C in Fig. 7.13).

In contrast to Kratochvil and Koros (2008), Qiu et al. (2011) attempted to carry out the decarboxylation-induced thermal crosslinking at a temperature below the T_g of polyimide membrane. They claimed that the high temperature (15°C above T_g) employed in Kratochvil's work will cause structural collapse in the asymmetric hollow fiber, which deteriorates the separation performance of polyimide membrane. Surprisingly, Qiu et al. (2011) noticed that the 6FDA-DAM:DABA (3:2) polyimide membrane could be crosslinked sufficiently by sub-T_g thermal crosslinking at a temperature as low as 330°C. The thermally crosslinked membrane demonstrated excellent plasticization resistance up to 48 bar pure CO_2 feed pressure and 34 bar partial pressure of CO_2 for 50:50 CO_2/CH_4 mixed gas feed.

In all the previous works, DABA was incorporated into the polyimide backbone to provide carboxylic acid group for the thermal crosslinking. However, the biphenyl linkage (C-C single bond) formed from the decarboxylation (primary crosslinking site A in Fig. 7.13) has a bond length of 1.55 Å only, which resulted in a membrane with low fractional free volume and gas permeability (Zhang et al. 2017). Hence, Zhang et al. (2017) synthesized new diamines with a rigid phenyl group between the carboxylic acid and the main backbone chain, named CADA1 and CADA2 as shown in Fig. 7.14. After decarboxylation, the interchain distance was expected to increase to 5.57 Å due to the existence of phenyl group between the polyimide chains. Indeed, this predicted value was very close to the *d*-spacing of the fully crosslinked polyimide determined experimentally. As compared to the uncrosslinked polyimide membrane, the CO_2 permeability of the thermally crosslinked polyimide membrane increased 100 times without selectivity loss. Besides, no plasticization was observed up to CO_2 pressure of 30 atm for the polyimide membrane thermally treated at a temperature beyond 425°C.

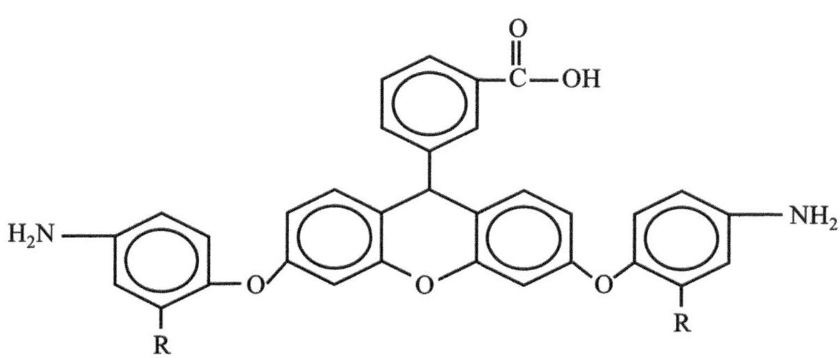

Figure 7.14. Structure of diamine synthesized by Zhang et al. (2017). CADA1, R= CF_3; CADA2, R=H. Adapted with permission from Zhang et al. (2017). Copyright 2017 Elsevier.

Molecular Simulation of Polyimide Membrane for Gas Separation

As aforementioned, polyimides with different chemical structures are available attributed to the diversity of monomer combinations. However, it is impractical to evaluate the gas separation performances of all polyimide structures experimentally as the process involves high chemical usage and is time consuming. In view of this, Molecular Dynamics (MD) simulation serves as an alternative technique to simulate the gas transport properties of polyimide matrix such as the gas diffusivity and solubility. Besides, fundamental insight into the molecular mobility of gas penetrant and polyimide chain as well as the interaction between them is also made possible by the simulation study.

A polyimide model is first required in order to determine the gas diffusivity and solubility. In general, polyimide model is constructed atomistically by packing the polyimide chains into a cubic simulation box under periodic boundary condition. The polyimide model is typically packed at low initial packing density with small spacer molecules being introduced as obstacles to avoid packing artefacts such as ring catenation and spearing (Zhang et al. 2010, Hölck et al. 2013, Park et al. 2014). After a series of equilibration procedures comprising energy minimization and annealing, the polyimide model is ready for the estimation of gas transport properties.

In MD simulation, the targeted gas penetrant molecules are inserted into the equilibrated polyimide model to determine their mean square displacement ($s(t) = \langle|r(t)-r(0)|^2\rangle$, where $r(t)$ is the trajectory of gas penetrant), which is then used to calculate the diffusivity according to Einstein equation ($D = s(t)/6t$) (Hofmann et al. 2000). Einstein equation is only valid under normal diffusion regime, in which the gas penetrants follow random walk, which is characterized by a slope near to one for log ($s(t)$) – log (t) plot. However, it is very difficult to achieve random walk condition within the time scale available for MD simulation at low temperature simulation (the temperature of experimental permeation test usually ranges from 25–35°C). Hence, high temperature extrapolation approach was suggested by Neyertz et al. (2010), since the random walk condition is only readily to be achieved within the achievable time scale of MD simulation under high temperature condition. In this approach, a series of high temperature simulations were first performed to obtain the diffusivities at high temperature. The linear relationship between the logarithm of diffusivity (In D) and reciprocal temperature ($1/T$) was then used for extrapolation to estimate the diffusivity at lower permeation temperature. The predicted diffusivity obtained using this approach showed good agreement with the experimental data (Neyertz et al. 2010).

Meanwhile, the gas sorption isotherm of a polyimide membrane can be analyzed using the Grand Canonical Monte Carlo (GCMC) simulation. This approach allows the estimation of the amount of gas molecules that sorbed into the polyimide matrix by considering the phase equilibrium between the polymer matrix and the ambient environment at specific temperature and pressure. However, GCMC assumes a rigid polymer framework in the calculation (no dilation), in which the significant structural relaxation and swelling during sorption process are neglected (Hölck et al. 2013). Hence, the simulated sorption isotherm differed from the experimental result when pure polyimide model (non-swollen state) was employed in the sorption simulation. In view of this, Hölck et al. (2013) had proposed to simulate the sorption isotherm of a polyimide model in the swollen state as well. A polyimide model in the swollen state is referred to a model that containing both the polyimide chain and CO_2 molecules at the initial state; followed by CO_2 removal after the equilibration steps. Thus, the membrane fractional free volume in the swollen polyimide model is expected to be higher. The linear combination of both sorption isotherms attained in non-swollen and swollen state allowed better prediction of the actual gas sorption isotherm for the polyimide membrane.

Besides gas transport performances, CO_2-induced membrane plasticization has captured a lot of attention lately. Generally, plasticization originates from the interaction between gas penetrant and polymer. Zhang et al. (2010) simulated the interaction between CO_2 and 6FDA-ODA polyimide. The radial distribution function, which is based on the probability of finding an atom from the reference particle in a function of distance, is used to analyze the favorable interaction sites of polyimide chain with CO_2. In their study (Zhang et al. 2010), it was found that the imide group was the most preferable sorption site for CO_2; followed by ether and CF_3 group. However, the simulation results also revealed that ether group had more significant effect on plasticization under high CO_2 loading. Therefore, a polyimide structure without ether group was highly preferable to suppress membrane plasticization.

Feasibility in Industrial Application

In reality, a lot of minor components exist in the untreated flue gas stream of industry such as water vapor are detrimental to the membrane gas separation performances (Ansaloni et al. 2014). In view of this, more considerable attention must be devoted to the feasibility of polyimide membrane in industrial application as compared to the lab-scale testing that usually adopts pure gas separation or binary gas separation at the desired composition. In this context, the gas separation performances of polyimide membrane under real industrial process conditions and its economic practicability need to be assessed thoroughly.

Ansaloni et al. (2014) evaluated the effect of relative humidity on the gas transport behaviors of Matrimid® membrane on lab-scale by introducing water vapor as the third component in the feed gas stream. The increasing concentration of water vapor in the membrane was found to be detrimental to the gas permeability. In particular, the permeability reduced to 50% of the dry gas permeability, with CO_2 permeability of 5.1 Barrer and CO_2/N_2 selectivity of 36 at 75% relative humidity and temperature of 35°C. Although the separation performances of Matrimid® membrane deteriorated with relative humidity, its gas separation efficiency was still comparable with other commercial membrane materials.

Li et al. (2015) also tested the separation performances of Matrimid® membrane containing 4 wt.% Silwet® additive using the real flue gas stream from Vales Point Power Station, Australia. The flue gas composition is summarized in Table 7.3. The pilot scale membrane unit consists of three parallel hollow fiber modules aimed to treat a feed with flow rate of 6 L/min. Each module has 100 hollow fibers with 500 mm length and 0.8 mm outer diameter. With the existence of minor components such as water vapor, SO_x and NO_x in the flue gas stream, a decrease in both CO_2 permeance (up to 15 GPU) and CO_2/N_2 selectivity (up to 15) was noticed as

Table 7.3. Wet flue gas composition from Vales Pont Power Station (Li et al. 2015).

Component	Unit	Value
N_2	%	75–90
CO_2	%	10–15
H_2O	%	N/A*
O_2	%	7
NO_x	ppm	100
SO_x	ppm	200

* The exact water content cannot be measured due to flooding.

compared to the lab-scale pure gas separation. However, there was only marginal loss of performance over three operation days, which suggested that the Matrimid® membrane exhibits good integrity against real industrial conditions (Li et al. 2015). As compared to the performance of PEBAX 2533 membrane in pilot scale syngas separation (CO_2 permeance = 2.3 GPU and CO_2/N_2 selectivity = 3) (Scholes et al. 2012), the separation efficiency of Matrimid® membrane was considerably better. However, there is still much room for improvement in obtaining a more feasible polyimide membrane with even higher productivity and separation efficiency.

Besides gas separation performances, the economic feasibility of polyimide membrane in industrial application is also of great interest. Kazama et al. (2004) performed a simple economic analysis on a cardo polyimide hollow fiber membrane for industrial CO_2 separation. The calculation was based on capture of 180 ton-CO_2/hr. The total cost for CO_2 separation from the flue gas of steel work (CO_2 concentration = 26.8%) was estimated at US$43 /ton-$CO_2$. On the contrary, the conventional amine absorption technique required the separation cost of US$ 47 /ton-CO_2 for similar steel work. Clearly, CO_2 removal using polyimide membrane is a more cost effective process and hence is highly feasible to be implemented in industrial scale.

Conclusion

Intensive industrial development has necessitated more attention towards CO_2 removal in minimizing the impact of global warming. In this regard, polyimide membrane with superior gas separation performances is proposed as one of the promising separation tools. However, the permeability-selectivity trade-off and plasticization occurrence in the polyimide membrane have restricted its widespread industrial applications. Hence, much research interest is directed towards overcoming these two limitations in order to obtain a polyimide membrane with separation performances beyond the Robeson's 2008 upper bound plot. The structural flexibility of polyimide allows the adjustment of monomer combination in producing a polyimide membrane with improved gas permeability and selectivity. Despite exhibiting encouraging gas separation performances, most of the structurally modified pristine polyimide membranes still demonstrate separation performances below the Robeson's 2008 trade-off plot. Therefore, a more advanced approach such as polyimide mixed matrix membrane is introduced to push the membrane gas separation efficiency above the trade-off plot. In solving the membrane plasticization problem, crosslinking via UV irradiation, chemical or thermal induced methods is proven to be effective in imparting the polyimide membrane with antiplasticization characteristic. The separation performances of polyimide membrane implemented in pilot plant show that there is still much room for improvement in order to achieve a more feasible polyimide membrane for industrial applications. Meanwhile, economic analysis suggests that polyimide membrane is more cost effective compared to the conventional amine absorption in CO_2 separation from flue gas stream.

Acknowledgements

The authors wish to thank MOHE Fundamental Research Grant Scheme FRGS/1/2017/TK02/USM/02/3 (203.PJKIMIA.6071367) for the financial support. P.C. Tan is financially assisted by the Human Life Advancement Foundation (HLAF) scholarship.

References

Ansaloni, L., Minelli, M., Baschetti, M.G. and Sarti, G.C. 2014. Effect of relative humidity and temperature on gas transport in Matrimid®: Experimental study and modeling. J. Membr. Sci. 471: 392–401.

Aroon, M., Ismail, A. and Matsuura, T. 2013. Beta-cyclodextrin functionalized MWCNT: A potential nano-membrane material for mixed matrix gas separation membranes development. Sep. Purif. Technol. 115: 39–50.

Askari, M. and Chung, T.S. 2013. Natural gas purification and olefin/paraffin separation using thermal cross-linkable co-polyimide/ZIF-8 mixed matrix membranes. J. Membr. Sci. 444: 173–183.

Ayala, D., Lozano, A.E., de Abajo, J., Garcıa-Perez, C., de la Campa, J.G., Peinemann, K.V. et al. 2003. Gas separation properties of aromatic polyimides. J. Membr. Sci. 215: 61–73.

Bakhtiari, O., Mosleh, S., Khosravi, T. and Mohammadi, T. 2011. Synthesis and characterization of polyimide mixed matrix membranes. Sep. Sci. Technol. 46: 2138–2147.

Boroglu, M.S. and Gurkaynak, M.A. 2011. Fabrication and characterization of silica modified polyimide–zeolite mixed matrix membranes for gas separation properties. Polym. Bull. 66: 463–478.

Boroglu, M.S. and Yumru, A.B. 2017. Gas separation performance of 6FDA-DAM-ZIF-11 mixed-matrix membranes for H_2/CH_4 and CO_2/CH_4 separation. Sep. Purif. Technol. 173: 269–279.

Bos, A., Pünt, I.G.M., Wessling, M. and Strathmann, H. 1998. Plasticization-resistant glassy polyimide membranes for CO_2/CO_4 separations. Sep. Purif. Technol. 14: 27–39.

Cacho-Bailo, F., Matito-Martos, I., Perez-Carbajo, J., Etxeberría-Benavides, M., Karvan, O., Sebastián, V. et al. 2017. On the molecular mechanisms for the H_2/CO_2 separation performance of zeolite imidazolate framework two-layered membranes. Chem. Sci. 8: 325–333.

Calle, M., García, C., Lozano, A.E., de la Campa, J.G., de Abajo, J. and Álvarez, C. 2013. Local chain mobility dependence on molecular structure in polyimides with bulky side groups: Correlation with gas separation properties. J. Membr. Sci. 434: 121–129.

Castro-Muñoz, R., Fíla, V. and Dung, C.T. 2017. Mixed matrix membranes based on PIMs for gas permeation: Principles, synthesis, and current status. Chem. Eng. Commun. 204: 295–309.

Chaidou, C.I., Pantoleontos, G., Koutsonikolas, D.E., Kaldis, S.P. and Sakellaropoulos, G.P. 2012. Gas separation properties of polyimide-zeolite mixed matrix membranes. Sep. Sci. Technol. 47: 950–962.

Chen, X.Y., Nik, O.G., Rodrigue, D. and Kaliaguine, S. 2012a. Mixed matrix membranes of aminosilanes grafted FAU/EMT zeolite and cross-linked polyimide for CO_2/CH_4 separation. Polymer 53: 3269–3280.

Chen, X.Y., Rodrigue, D. and Kaliaguine, S. 2012b. Diamino-organosilicone APTMDS: A new cross-linking agent for polyimides membranes. Sep. Purif. Technol. 86: 221–233.

Chen, X.Y., Vinh-Thang, H., Ramirez, A.A., Rodrigue, D. and Kaliaguine, S. 2015. Membrane gas separation technologies for biogas upgrading. RSC Adv. 5: 24399–24448.

Choi, J. and Tsapatsis, M. 2009. MCM-22/silica selective flake nanocomposite membranes for hydrogen separations. J. Am. Chem. Soc. 132: 448–449.

Chua, M.L., Xiao, Y. and Chung, T.S. 2014. Using iron (III) acetylacetonate as both a cross-linker and micropore former to develop polyimide membranes with enhanced gas separation performance. Sep. Purif. Technol. 133: 120–128.

Chung, T.S., Chan, S.S., Wang, R., Lu, Z. and He, C. 2003. Characterization of permeability and sorption in Matrimid/C_{60} mixed matrix membranes. J. Membr. Sci. 211: 91–99.

Coleman, M.R. and Koros, W.J. 1990. Isomeric polyimides based on fluorinated dianhydrides and diamines for gas separation applications. J. Membr. Sci. 50: 285–297.

Cong, H., Radosz, M., Towler, B.F. and Shen, Y. 2007. Polymer–inorganic nanocomposite membranes for gas separation. Sep. Purif. Technol. 55: 281–291.

Cui, L., Qiu, W., Paul, D.R. and Koros, W.J. 2011. Responses of 6FDA-based polyimide thin membranes to CO_2 exposure and physical aging as monitored by gas permeability. Polymer 52: 5528–5537.

De Angelis, M.G., Gaddoni, R. and Sarti, G.C. 2013. Gas solubility, diffusivity, permeability, and selectivity in mixed matrix membranes based on PIM-1 and fumed silica. Ind. Eng. Chem. Res. 52: 10506–10520.

Friebe, S., Geppert, B., Steinbach, F. and Caro, J.r. 2017. Metal–organic framework UiO-66 layer: A highly oriented membrane with good selectivity and hydrogen permeance. ACS Appl. Mater. Interfaces 9: 12878–12885.

Fu, S., Sanders, E.S., Kulkarni, S.S. and Koros, W.J. 2015. Carbon molecular sieve membrane structure–property relationships for four novel 6FDA based polyimide precursors. J. Membr. Sci. 487: 60–73.

Fuertes, A.B., Nevskaia, D.M. and Centeno, T.A. 1999. Carbon composite membranes from Matrimid® and Kapton® polyimides for gas separation. Microporous Mesoporous Mater. 33: 115–125.

Galve, A., Sieffert, D., Staudt, C., Ferrando, M., Güell, C., Tellez, C. et al. 2013. Combination of ordered mesoporous silica MCM-41 and layered titanosilicate JDF-L1 fillers for 6FDA-based copolyimide mixed matrix membranes. J. Membr. Sci. 431: 163–170.

Ge, L., Wang, L., Rudolph, V. and Zhu, Z. 2013. Hierarchically structured metal–organic framework/vertically-aligned carbon nanotubes hybrids for CO_2 capture. RSC Adv. 3: 25360–25366.

Goh, P.S., Ismail, A.F., Sanip, S.M., Ng, B.C. and Aziz, M. 2011. Recent advances of inorganic fillers in mixed matrix membrane for gas separation. Sep. Purif. Technol. 81: 243–264.

Gong, H., Nguyen, T.H., Wang, R. and Bae, T.H. 2015. Separations of binary mixtures of CO_2/CH_4 and CO_2/N_2 with mixed-matrix membranes containing $Zn(pyrz)_2(SiF_6)$ metal-organic framework. J. Membr. Sci. 495: 169–175.

Heck, R., Qahtani, M.S., Yahaya, G.O., Tanis, I., Brown, D., Bahamdan, A.A. et al. 2017. Block copolyimide membranes for pure-and mixed-gas separation. Sep. Purif. Technol. 173: 183–192.

Hess, S. and Staudt, C. 2007. Variation of esterfication conditions to optimize solid-state crosslinking reaction of DABA-containing copolyimide membranes for gas separations. Desalination 217: 8–16.

Hibshman, C., Cornelius, C.J. and Marand, E. 2003. The gas separation effects of annealing polyimide–organosilicate hybrid membranes. J. Membr. Sci. 211: 25–40.

Hillock, A.M.W. 2005. Crosslinkable Polyimide Mixed Matrix Membranes for Natural Gas Purification. Ph.D. Thesis, Georgia Institute of Technology, Georgia, US.

Hillock, A.M.W. and Koros, W.J. 2007. Cross-linkable polyimide membrane for natural gas purification and carbon dioxide plasticization reduction. Macromolecules 40: 583–587.

Hillock, A.M.W., Miller, S.J. and Koros, W.J. 2008. Crosslinked mixed matrix membranes for the purification of natural gas: Effects of sieve surface modification. J. Membr. Sci. 314: 193–199.

Hirayama, Y., Yoshinaga, T., Kusuki, Y., Ninomiya, K., Sakakibara, T. and Tamari, T. 1996. Relation of gas permeability with structure of aromatic polyimides I. J. Membr. Sci. 111: 169–182.

Hofmann, D., Fritz, L., Ulbrich, J., Schepers, C. and Böhning, M. 2000. Detailed-atomistic molecular modeling of small molecule diffusion and solution processes in polymeric membrane materials. Macromol. Theory Simul. 9:293–327.

Hölck, O., Böhning, M., Heuchel, M., Siegert, M.R. and Hofmann, D. 2013. Gas sorption isotherms in swelling glassy polymers—detailed atomistic simulations. J. Membr. Sci. 428: 523–532.

Ismail, A.F., Rahim, N.H., Mustafa, A., Matsuura, T., Ng, B.C., Abdullah, S. et al. 2011. Gas separation performance of polyethersulfone/multi-walled carbon nanotubes mixed matrix membranes. Sep. Purif. Technol. 80: 20–31.

Ismail, A.F. and Kusworo, T.D. 2014. Studies on as separation behaviour of polymer blending PI/PES hybrid mixed membrane: Effect of polymer concentration and zeolite loading. Int. J. Sci. Eng. 6: 144–148.

Japip, S., Wang, H., Xiao, Y. and Chung, T.S. 2014. Highly permeable zeolitic imidazolate framework (ZIF)-71 nano-particles enhanced polyimide membranes for gas separation. J. Membr. Sci. 467: 162–174.

Jeazet, H.B.T., Staudt, C. and Janiak, C. 2012. Metal–organic frameworks in mixed-matrix membranes for gas separation. Dalton Trans. 41: 14003–14027.

Jung, C.H., Lee, J.E., Han, S.H., Park, H.B. and Lee, Y.M. 2010. Highly permeable and selective poly(benzoxazole-co-imide) membranes for gas separation. J. Membr. Sci. 350: 301–309.

Jusoh, N., Yeong, Y.F., Chew, T.L., Lau, K.K. and Shariff, A.M. 2016. Current development and challenges of mixed matrix membranes for CO_2/CH_4 separation. Sep. Purif. Rev. 45: 321–344.

Jusoh, N., Yeong, Y.F., Lau, K.K. and Shariff, A.M. 2017. Enhanced gas separation performance using mixed matrix membranes containing zeolite T and 6FDA-durene polyimide. J. Membr. Sci. 525: 175–186.

Kammakakam, I., Yoon, H.W., Nam, S.Y., Park, H.B. and Kim, T.H. 2015. Novel piperazinium-mediated crosslinked polyimide membranes for high performance CO_2 separation. J. Membr. Sci. 487: 90–98.

Kanehashi, S., Kishida, M., Kidesaki, T., Shindo, R., Sato, S., Miyakoshi, T. et al. 2013. CO_2 separation properties of a glassy aromatic polyimide composite membranes containing high-content 1-butyl-3-methylimidazolium bis(trifluoromethylsulfonyl)imide ionic liquid. J. Membr. Sci. 430: 211–222.

Kang, J.S., Won, J., Park, H.C., Kim, U.Y., Kang, Y.S. and Lee, Y.M. 2000. Morphology control of asymmetric membranes by UV irradiation on polyimide dope solution. J. Membr. Sci. 169: 229–235.

Kapantaidakis, G.C., Koops, G.H., Wessling, M., Kaldis, S.P. and Sakellaropoulos, G.P. 2003. CO_2 plasticization of polyethersulfone/polyimide gas-separation membranes. AIChE J. 49: 1702–1711.

Kazama, S., Morimoto, S., Tanaka, S., Mano, H., Yashima, T., Yamada, K. et al. 2004. Cardo polyimide membranes for CO_2 capture from flue gases. Greenhouse Gas Control Technologies 1: 75–82.

Kertik, A., Khan, A.L. and Vankelecom, I.F.J. 2016. Mixed matrix membranes prepared from non-dried MOFs for CO_2/CH_4 separations. RSC Adv. 6: 114505–114512.

Khan, A.L., Cano-Odena, A., Gutiérrez, B., Minguillón, C. and Vankelecom, I.F.J. 2010. Hydrogen separation and purification using polysulfone acrylate–zeolite mixed matrix membranes. J. Membr. Sci. 350: 340–346.

Khan, A.L., Klaysom, C., Gahlaut, A., Khan, A.U. and Vankelecom, I.F.J. 2013. Mixed matrix membranes comprising of Matrimid and $-SO_3H$ functionalized mesoporous MCM-41 for gas separation. J. Membr. Sci. 447: 73–79.

Khan, A.L., Sree, S.P., Martens, J.A., Raza, M.T. and Vankelecom, I.F.J. 2015. Mixed matrix membranes comprising of matrimid and mesoporous COK-12: Preparation and gas separation properties. J. Membr. Sci. 495: 471–478.

Kılıç, A., Atalay-Oral, Ç., Sirkecioğlu, A., Tantekin-Ersolmaz, Ş.B. and Ahunbay, M.G. 2015. *Sod*-ZMOF/Matrimid® mixed matrix membranes for CO_2 separation. J. Membr. Sci. 489: 81–89.

Kim, J.H., Lee, S.B. and Kim, S.Y. 2000. Incorporation effects of fluorinated side groups into polyimide membranes on their physical and gas permeation properties. J. Appl. Polym. Sci. 77: 2756–2767.

Kim, J.H., Koros, W.J. and Paul, D.R. 2006a. Effects of CO_2 exposure and physical aging on the gas permeability of thin 6FDA-based polyimide membranes: Part 2. with crosslinking. J. Membr. Sci. 282: 32–43.

Kim, S., Pechar, T.W. and Marand, E. 2006b. Poly (imide siloxane) and carbon nanotube mixed matrix membranes for gas separation. Desalination 192: 330–339.

Kita, H., Inada, T., Tanaka, K. and Okamoto, K.I. 1994. Effect of photocrosslinking on permeability and permselectivity of gases through benzophenone-containing polyimide. J. Membr. Sci. 87: 139–147.

Kraftschik, B., Koros, W.J., Johnson, J.R. and Karvan, O. 2013. Dense film polyimide membranes for aggressive sour gas feed separations. J. Membr. Sci. 428: 608–619.

Kratochvil, A.M. and Koros, W.J. 2008. Decarboxylation-induced cross-linking of a polyimide for enhanced CO_2 plasticization resistance. Macromolecules 41: 7920–7927.

Kudasheva, A., Sorribas, S., Zornoza, B., Téllez, C. and Coronas, J. 2015. Pervaporation of water/ethanol mixtures through polyimide based mixed matrix membranes containing ZIF-8, ordered mesoporous silica and ZIF-8-silica core-shell spheres. J. Chem. Technol. Biotechnol. 90: 669–677.

Li, H., Chen, V., Dong, G. and Hou, J. 2015. Evaluation of CO_2 capture with high performance hollow fibre membranes from flue gas: Final report. ANLEC report. In Cooperative Research Centre for Greenhouse Gas Technologie. Canberra, Australia.

Li, J.R., Sculley, J. and Zhou, H.C. 2012. Metal–organic frameworks for separations. Chem. Rev. 112: 869–932.

Lin, R., Ge, L., Liu, S., Rudolph, V. and Zhu, Z. 2015. Mixed-matrix membranes with metal-organic framework-decorated CNT fillers for efficient CO_2 separation. ACS Appl. Mater. Interfaces 7: 14750–14757.

Liu, Y., Wang, R. and Chung, T.S. 2001. Chemical cross-linking modification of polyimide membranes for gas separation. J. Membr. Sci. 189: 231–239.

Liu, Y., Peng, D., He, G., Wang, S., Li, Y., Wu, H. et al. 2014. Enhanced CO_2 permeability of membranes by incorporating polyzwitterion@ CNT composite particles into polyimide matrix. ACS Appl. Mater. Interfaces 6: 13051–13060.

Lua, A.C. and Shen, Y. 2013. Preparation and characterization of polyimide–silica composite membranes and their derived carbon–silica composite membranes for gas separation. Chem. Eng. J. 220: 441–451.

Mahdi, E.M. and Tan, J.C. 2016. Mixed-matrix membranes of zeolitic imidazolate framework (ZIF-8)/Matrimid nanocomposite: Thermo-mechanical stability and viscoelasticity underpinning membrane separation performance. J. Membr. Sci. 498: 276–290.

Matsui, S., Sato, H. and Nakagawa, T. 1998. Effects of low molecular weight photosensitizer and UV irradiation on gas permeability and selectivity of polyimide membrane. J. Membr. Sci. 141: 31–43.

Maya, E.M., Tena, A., de Abajo, J., de La Campa, J.G. and Lozano, A.E. 2010. Partially pyrolyzed membranes (PPMs) derived from copolyimides having carboxylic acid groups. Preparation and gas transport properties. J. Membr. Sci. 349: 385–392.

Moore, T.T. and Koros, W.J. 2005. Non-ideal effects in organic–inorganic materials for gas separation membranes. J. Mol. Struct. 739: 87–98.

Musselman, I., Balkus, K., Jr. and Ferraris, J. 2009. Mixed-matric membranes for CO_2 and H_2 gas separations using metal-organic framework and mesoporus hybrid silicas. University of Texas at Dallas, US.

Neyertz, S., Brown, D., Pandiyan, S. and van der Vegt, N.F.A. 2010. Carbon dioxide diffusion and plasticization in fluorinated polyimides. Macromolecules 43: 7813–7827.

Nik, O.G., Chen, X.Y. and Kaliaguine, S. 2011. Amine-functionalized zeolite FAU/EMT-polyimide mixed matrix membranes for CO_2/CH_4 separation. J. Membr. Sci. 379: 468–478.

Ohya, H., Kudryavsev, V.V. and Semenova, S.I. 1996. Polyimide Membranes: Applications, Fabrications and Properties. Kodansha Ltd., Tokyo.

Ozturk, B. and Demirciyeva, F. 2013. Comparison of biogas upgrading performances of different mixed matrix membranes. Chem. Eng. J. 222: 209–217.

Park, C.H., Tocci, E., Kim, S., Kumar, A., Lee, Y.M. and Drioli, E. 2014. A simulation study on OH-containing polyimide (HPI) and thermally rearranged polybenzoxazoles (TR-PBO): relationship between gas transport properties and free volume morphology. J. Phys. Chem. B 118: 2746–2757.

Park, S.H., Kim, K.J., So, W.W., Moon, S.J. and Lee, S.B. 2003. Gas separation properties of 6FDA-based polyimide membranes with a polar group. Macromol. Res. 11: 157–162.

Perez, E.V., Balkus, K.J. Jr., Ferraris, J.P. and Musselman, I.H. 2009. Mixed-matrix membranes containing MOF-5 for gas separations. J. Membr. Sci. 328: 165–173.

Powell, C.E. and Qiao, G.G. 2006. Polymeric CO_2/N_2 gas separation membranes for the capture of carbon dioxide from power plant flue gases. J. Membr. Sci. 279: 1–49.

Powell, C.E., Duthie, X.J., Kentish, S.E., Qiao, G.G. and Stevens, G.W. 2007. Reversible diamine cross-linking of polyimide membranes. J. Membr. Sci. 291: 199–209.

Qiu, W., Chen, C.C., Xu, L., Cui, L., Paul, D.R. and Koros, W.J. 2011. Sub-T_g cross-linking of a polyimide membrane for enhanced CO_2 plasticization resistance for natural gas separation. Macromolecules 44: 6046–6056.

Qiu, W., Xu, L., Chen, C.C., Paul, D.R. and Koros, W.J. 2013. Gas separation performance of 6FDA-based polyimides with different chemical structures. Polymer 54: 6226–6235.

Rafiq, S., Man, Z., Maulud, A., Muhammad, N. and Maitra, S. 2012. Separation of CO_2 from CH_4 using polysulfone/polyimide silica nanocomposite membranes. Sep. Purif. Technol. 90: 162–172.

Ramírez-Santos, Á.A., Castel, C. and Favre, E. 2017. Utilization of blast furnace flue gas: Opportunities and challenges for polymeric membrane gas separation processes. J. Membr. Sci. 526: 191–204.

Rezakazemi, M., Amooghin, A.E., Montazer-Rahmati, M.M., Ismail, A.F. and Matsuura, T. 2014. State-of-the-art membrane based CO_2 separation using mixed matrix membranes (MMMs): An overview on current status and future directions. Prog. Polym. Sci. 39: 817–861.

Robeson, L.M. 1991. Correlation of separation factor versus permeability for polymeric membranes. J. Membr. Sci. 62: 165–185.

Robeson, L.M. 2008. The upper bound revisited. J. Membr. Sci. 320: 390–400.

Rosu, L., Sava, I. and Rosu, D. 2011. Modification of the surface properties of a polyimide film during irradiation with polychromic light. Appl. Surf. Sci. 257: 6996–7002.

Roth, W.J., Nachtigall, P., Morris, R.E. and Cejka, J. 2014. Two-dimensional zeolites: current status and perspectives. Chem. Rev. 114: 4807–4837.

Rowsell, J.L.C. and Yaghi, O.M. 2004. Metal–organic frameworks: a new class of porous materials. Microporous Mesoporous Mater. 73: 3–14.

Rybak, A., Dudek, G., Krasowska, M., Strzelewicz, A., Grzywna, Z. and Sysel, P. 2014. Magnetic mixed matrix membranes in air separation. Chem. Pap. 68: 1332–1340.

Schejn, A., Balan, L., Falk, V., Aranda, L., Medjahdi, G. and Schneider, R. 2014. Controlling ZIF-8 nano-and microcrystal formation and reactivity through zinc salt variations. CrystEngComm 16: 4493–4500.

Scholes, C.A., Tao, W.X., Stevens, G.W. and Kentish, S.E. 2010. Sorption of methane, nitrogen, carbon dioxide, and water in Matrimid 5218. J. Appl. Polym. Sci. 117: 2284–2289.

Scholes, C.A., Bacus, J., Chen, G.Q., Tao, W.X., Li, G., Qader, A. et al. 2012. Pilot plant performance of rubbery polymeric membranes for carbon dioxide separation from syngas. J. Membr. Sci. 389: 470–477.

Shahid, S. and Nijmeijer, K. 2014. Performance and plasticization behavior of polymer–MOF membranes for gas separation at elevated pressures. J. Membr. Sci. 470: 166–177.

Shao, L., Chung, T.S., Goh, S.H. and Pramoda, K.P. 2005. The effects of 1, 3-cyclohexanebis(methylamine) modification on gas transport and plasticization resistance of polyimide membranes. J. Membr. Sci. 267: 78–89.

Shao, L., Liu, L., Cheng, S.X., Huang, Y.D. and Ma, J. 2008. Comparison of diamino cross-linking in different polyimide solutions and membranes by precipitation observation and gas transport. J. Membr. Sci. 312: 174–185.

Souza, V.C. and Quadri, M.G.N. 2013. Organic-inorganic hybrid membranes in separation processes: a 10-year review. Braz. J. Chem. Eng. 30: 683–700.

Staudt-Bickel, C. and Koros, W.J. 1999. Improvement of CO_2/CH_4 separation characteristics of polyimides by chemical crosslinking. J. Membr. Sci. 155: 145–154.

Sun, H., Ma, C., Yuan, B., Wang, T., Xu, Y., Xue, Q. et al. 2014. Cardo polyimides/TiO_2 mixed matrix membranes: Synthesis, characterization, and gas separation property improvement. Sep. Purif. Technol. 122: 367–375.

Sun, H., Wang, T., Xu, Y., Gao, W., Li, P. and Niu, Q.J. 2017. Fabrication of polyimide and functionalized multi-walled carbon nanotubes mixed matrix membranes by in-situ polymerization for CO_2 separation. Sep. Purif. Technol. 177: 327–336.

Sunarso, J., Hashim, S.S., Lin, Y.S. and Liu, S.M. 2017. Membranes for helium recovery: An overview on the context, materials and future directions. Sep. Purif. Technol. 176: 335–383.

Tan, J.C. and Cheetham, A.K. 2011. Mechanical properties of hybrid inorganic–organic framework materials: establishing fundamental structure–property relationships. Chem. Soc. Rev. 40: 1059–1080.

Tan, P.C., Jawad, Z.A., Ooi, B.S., Ahmad, A.L. and Low, S.C. 2016. Correlation between polymer packing and gas transport properties for CO_2/N_2 separation in glassy fluorinated polyimide membrane. J. Eng. Sci. Technol. 11: 935–946.

Tanaka, K., Kita, H., Okano, M. and Okamoto, K.I. 1992a. Permeability and permselectivity of gases in fluorinated and non-fluorinated polyimides. Polymer 33: 585–592.

Tanaka, K., Okano, M., Toshino, H., Kita, H. and Okamoto, K.I. 1992b. Effect of methyl substituents on permeability and permselectivity of gases in polyimides prepared from methyl-substituted phenylenediamines. J. Polym. Sci. Part B: Polym. Phys. 30: 907–914.

Taubert, A., Wind, J.D., Paul, D.R., Koros, W.J. and Winey, K.I. 2003. Novel polyimide ionomers: CO_2 plasticization, morphology, and ion distribution. Polymer 44: 1881–1892.

Tena, A., Fernández, L., Sánchez, M., Palacio, L., Lozano, A.E., Hernández, A.E. et al. 2010. Mixed matrix membranes of 6FDA-6FpDA with surface functionalized γ-alumina particles. An analysis of the improvement of permselectivity for several gas pairs. Chem. Eng. Sci. 65: 2227–2235.

Tin, P.S., Chung, T.S., Liu, Y. and Wang, R. 2004. Separation of CO_2/CH_4 through carbon molecular sieve membranes derived from P84 polyimide. Carbon 42: 3123–3131.

Tong, H., Hu, C., Yang, S., Ma, Y., Guo, H. and Fan, L. 2015. Preparation of fluorinated polyimides with bulky structure and their gas separation performance correlated with microstructure. Polymer 69: 138–147.

Tul Muntha, S., Kausar, A. and Siddiq, M. 2016. A review on zeolite-reinforced polymeric membranes: Salient features and applications. Polym. Plast. Technol. Eng. 55: 1971–1987.

Vanherck, K., Koeckelberghs, G. and Vankelecom, I.F.J. 2013. Crosslinking polyimides for membrane applications: a review. Prog. Polym. Sci. 38: 874–896.

Velioğlu, S., Ahunbay, M.G. and Tantekin-Ersolmaz, S.B. 2012. Investigation of CO_2-induced plasticization in fluorinated polyimide membranes via molecular simulation. J. Membr. Sci. 417: 217–227.

Wang, F., Liu, Z.S., Yang, H., Tan, Y.X. and Zhang, J. 2011. Hybrid zeolitic imidazolate frameworks with catalytically active TO_4 building blocks. Angew. Chem. Int. Ed. 50: 450–453.

Wang, S., Tian, Z., Feng, J., Wu, H., Li, Y., Liu, Y. et al. 2015. Enhanced CO_2 separation properties by incorporating poly(ethylene glycol)-containing polymeric submicrospheres into polyimide membrane. J. Membr. Sci. 473: 310–317.

Wang, Y.C., Huang, S.H., Hu, C.C., Li, C.L., Lee, K.R., Liaw, D.J. et al. 2005. Sorption and transport properties of gases in aromatic polyimide membranes. J. Membr. Sci. 248: 15–25.

Ward, J.K. and Koros, W.J. 2011. Crosslinkable mixed matrix membranes with surface modified molecular sieves for natural gas purification: I. Preparation and experimental results. J. Membr. Sci. 377: 75–81.

Wind, J.D., Staudt-Bickel, C., Paul, D.R. and Koros, W.J. 2002. The effects of crosslinking chemistry on CO_2 plasticization of polyimide gas separation membranes. Ind. Eng. Chem. Res. 41: 6139–6148.

Wind, J.D., Staudt-Bickel, C., Paul, D.R. and Koros, W.J. 2003. Solid-state covalent cross-linking of polyimide membranes for carbon dioxide plasticization reduction. Macromolecules 36: 1882–1888.

Wright, C.T. and Paul, D.R. 1997. Gas sorption and transport in UV-irradiated polyarylate copolymers based on tetramethyl bisphenol-A and dihydroxybenzophenone. J. Membr. Sci. 124: 161–174.

Xiao, Y., Low, B.T., Hosseini, S.S., Chung, T.S. and Paul, D.R. 2009. The strategies of molecular architecture and modification of polyimide-based membranes for CO_2 removal from natural gas-A review. Prog. Polym. Sci. 34: 561–580.

Xu, R., Gao, Z., Chen, J. and Yan, W. 2007. From Zeolite to Porous MOF Materials–the 40th Anniversary of International Zeolite Conference. Elsevier, UK.

Xue, Q., Pan, X., Li, X., Zhang J. and Guo, Q. 2017. Effective enhancement of gas separation performance in mixed matrix membranes using core/shell structured multi-walled carbon nanotube/graphene oxide nanoribbons. Nanotechnology 28: 065702.

Xuesong, G. and Fengcai, L. 1995. Gas transport properties of polyimides and polypyrrolone containing ester linkage. Polymer 36: 1035–1038.

Yampolskii, Y.P., Starannikova, L.E. and Belov, N.A. 2014. Hybrid gas separation polymeric membranes containing nanoparticles. Pet. Chem. 54: 637–651.

Yilmaz, G. and Keskin, S. 2012. Predicting the performance of zeolite imidazolate framework/polymer mixed matrix membranes for CO_2, CH_4, and H_2 separations using molecular simulations. Ind. Eng. Chem. Res. 51: 14218–14228.

Yong, H.H., Park, H.C., Kang, Y.S., Won, J. and Kim, W.N. 2001. Zeolite-filled polyimide membrane containing 2, 4, 6-triaminopyrimidine. J. Membr. Sci. 188: 151–163.

Yoshioka, T., Kojima, K., Shindo, R. and Nagai, K. 2017. Gas-separation properties of amine-crosslinked polyimide membranes modified by amine vapor. J. Appl. Polym. Sci. 134.

Zhang, C., Li, P. and Cao, B. 2017. Decarboxylation crosslinking of polyimides with high CO_2/CH_4 separation performance and plasticization resistance. J. Membr. Sci. 528: 206–216.

Zhang, L., Xiao, Y., Chung, T.-S. and Jiang, J. 2010. Mechanistic understanding of CO_2-induced plasticization of a polyimide membrane: a combination of experiment and simulation study. Polymer 51: 4439–4447.

Zhang, Y., Sunarso, J., Liu, S. and Wang, R. 2013. Current status and development of membranes for CO_2/CH_4 separation: A review. Int. J. Greenh. Gas Con. 12: 84–107.

Zhou, C., Chung, T.S., Wang, R., Liu, Y. and Goh, S.H. 2003. The accelerated CO_2 plasticization of ultra-thin polyimide films and the effect of surface chemical cross-linking on plasticization and physical aging. J. Membr. Sci. 225: 125–134.

Zornoza, B., Téllez, C., Coronas, J., Esekhile, O. and Koros, W.J. 2015. Mixed matrix membranes based on 6FDA polyimide with silica and zeolite microsphere dispersed phases. AIChE J. 61: 4481–4490.

8

Zeolite Membranes for CO_2 Permeation and Separation

Thiam Leng Chew,[1,2,*] *Tiffany Yit Siew Ng*[1,2] and *Yin Fong Yeong*[1,2]

INTRODUCTION OF ZEOLITE FRAMEWORK

Zeolites are crystalline aluminosilicate or silica polymorph corner-shared by TO_4 (T = Si, Al or P) tetrahedral building blocks with tridimensional framework structure that form uniformly well-defined sized pores of molecular dimensions (Burton 2003). The captivating properties of zeolite are usually characterized by the uniform dimensions of pore opening and unique structure (Feng et al. 2015). Therefore, every zeolite is different from each other based on its properties, crystal framework and chemical composition. Furthermore, the particular specific properties of zeolite can be deduced by the nature of species within the channels, chemical composition of the crystal structure and the methods of post-synthesis modification (McCusker and Baerlocher 2007).

Generally, zeolites include small, medium and large pore structures. Small pore structures have pore apertures comprising of six, eight or nine tetrahedral (6-, 8- and 9-membered rings) with pore size below 4.5 Å, medium pore zeolites constitute of 10-membered rings with pore aperture of 4.5–6 Å, while large pore framework composed of 12-membered rings with pore diameter of 6–8 Å (Tavolaro and Drioli 1999, Flanigen et al. 2010). Table 8.1 illustrates the categories of zeolites.

[1] Department of Chemical Engineering, Faculty of Engineering, University Teknologi PETRONAS, 32610 Bandar Seri Iskandar, Perak, Malaysia.

[2] CO_2 Research Centre (CO_2RES), Institute of Contaminant Management, University Teknologi PETRONAS, 32610 Bandar Seri Iskandar, Perak, Malaysia.
Email: tiffany_16001691@utp.edu.my; yinfong.yeong@utp.edu.my

* Corresponding author: thiamleng.chew@utp.edu.my

Table 8.1. Categories of Zeolites with its Examples according to the International Zeolite Association (IZA) Database.

Category	Examples of Zeolites	Structural Type	Dimension	Pore Size (Å)
Small Pore	AlPO-18	AEI	3D	3.8 × 3.8
	SAPO-34	CHA	3D	3.8 × 3.8
	ZSM-58	DDR	2D	3.6 × 4.4
	BeAsO-RHO	RHO	3D	3.6 × 3.6
Medium pore	ZSM-11	MEL	3D	5.3 × 5.4
	ZSM-5	MFI	2D	5.1 × 5.5 and 5.3 × 5.6
	Stilbite	STI	2D	4.7 × 5.0 and 2.7 × 5.6
Large pore	Beta	BEA	3D	6.6 × 6.7
	Na-Y	FAU	3D	7.4 × 7.4
	Zeolite L	LTL	1D	7.1 × 7.1

Zeolites having an aluminosilicate framework are displayed as a negatively charged molecule due to the presence of alumina but it is neutral in the presence of aluminophosphate. This enable zeolites to trap positive ions such as Na^+, K^+, Ca^{2+}, Mg^{2+} and other ions to exchange with the exact cations number effortlessly to counterbalance the negative electric charge in the framework, producing a strong electrostatic field on the zeolite's surface (Rhodes 2010). Subsequently, the properties of the zeolites, including pore size, can be tailored via an exchange process. For example, a smaller sodium ion in zeolite which has opening pore aperture of 4 Å can decrease its size to 3 Å by exchanging with a larger potassium ion (Feng at al. 2015). Larger cations can help in the reduction of pore opening size and influence the properties of zeolites.

Synthesis of Zeolite Membranes

Generally, zeolite membranes are synthesized by subjecting precursor solution (which consists of aluminium and silica sources) to hydrothermal synthesis at certain temperature and autogenous pressure (Barrer 1982). During the hydrothermal synthesis, nucleation and followed by zeolite crystal growth occur. Nonetheless, in the crystallization duration, overabundance of chemical reactions, equilibria and variations of solubility that happen in the synthesis solution will result in complicated mechanism of zeolite formation (Byrappa and Yoshimura 2013). There are commonly few methods established in the preparation of zeolite membranes such as conventional direct *in situ* hydrothermal synthesis, secondary (seeded) growth, microwave-assisted and ultrasonic-assisted synthesis.

Direct in situ hydrothermal synthesis

Conventionally, synthesis of zeolite membranes is performed via direct *in situ* hydrothermal synthesis method. Numerous research works have been reported on

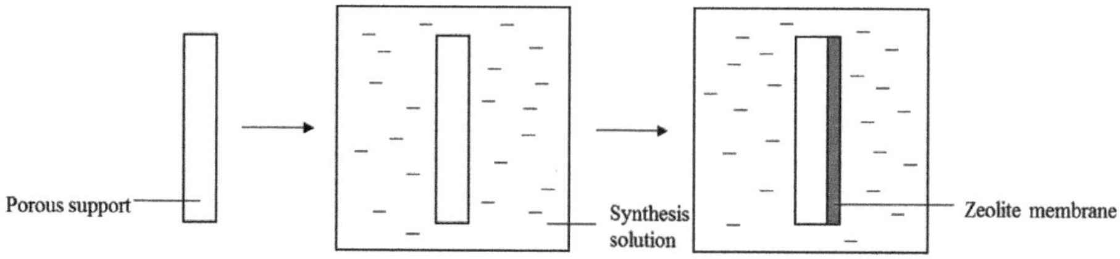

Figure 8.1. Direct *In situ* hydrothermal synthesis of zeolite membranes.

the direct *in situ* hydrothermal synthesis for zeolites including RHO (Liu et al. 2013), ZSM-5 (Yaripour et al. 2015), SAPO-34 (Najafi et al. 2014) and LTA (Anbia et al. 2017).

Figure 8.1 illustrates the direct *in situ* hydrothermal synthesis for the formation of zeolite membranes. Generally, the synthesis solution can be prepared by mixing amorphous silica or alumina sources, sodium hydroxide, distilled or deionized water and an organic Structure-Directing Agent (SDA) or also known as template (Bowen et al. 2004). In this method, the synthesis solution is placed in contact with a porous support in an autoclave under hydrothermal conditions. The reaction happens in an aqueous solution under required temperature and autogeneous pressure in a closed system (Caro and Noack 2010). Composition of precursor solution and time for crystallization basically depends on the type of zeolite membranes to be synthesized (Gorgojo et al. 2008). The porous supports made of alumina or stainless steel can be used for the deposition of the zeolite membrane layer, where the nucleation of zeolite crystals and followed by the formation of continuous zeolite membrane layer occur on the support (Lin et al. 2000, Gorgojo et al. 2008).

Figure 8.2 illustrates the mechanism for the growth of zeolite membrane. Mintova and his research team members proposed that the mechanism starts from primary particles which are initially formed in amorphous aluminosilicate suspension (Mintova et al. 1999). The primary particles with least stable colloidal will aggregate or agglomerate into amorphous secondary particles with certain density and composition and nucleate into amorphous zeolite crystals upon heating (Hould et al. 2009). The aggregative growth mechanism dominates the early stages of crystal growth (Davis et al. 2006). The high supersaturation level in the particles coupled with reorganization at the solid-liquid interface of the suspension becomes the driving force for zeolite nucleation (Valtchev and Bozhilov 2005). The crystals grow due to solution mass transfer that supplies the precursor material (Mintova et al. 1999).

The number and distribution of the zeolite nuclei on the support is highly dependent on the surface properties of the support which are hard to be control (Daramola et al. 2012). This is because the conventional hydrothermal synthesis can be affected easily by various factors such as the conditions of the synthesis process, support materials, composition of the synthesis solution, contact between the zeolite precursor solution and membrane support (O'Brien 2009). Direct *in situ* hydrothermal synthesis brings about drawbacks including the long induction period which typically requires few days in synthesis (Julbe 2007).

In recent years, template free synthesis of zeolite membranes is gaining popularity for the cost limitation process. In addition, the template removal step

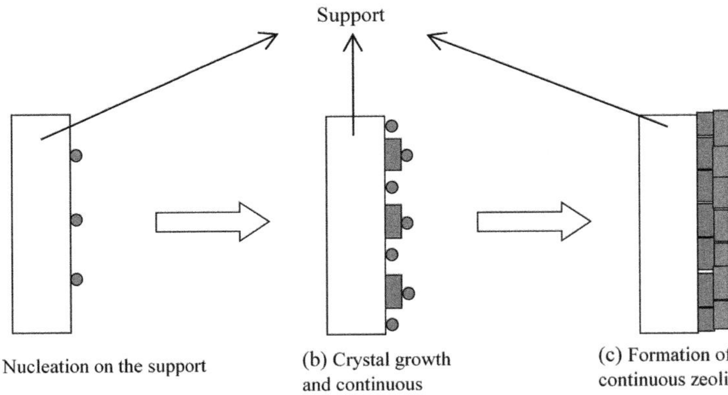

Figure 8.2. Mechanism of Growth of Zeolite Membrane (Vilaseca et al. 2004).

could be excluded if the zeolite membranes are formed via template free synthesis method. The removal of the template or SDA by calcination at high temperature will cause formation of post synthesis defect and eventually affects the final performance of the final zeolite membrane (Caro and Noack 2010). There have been research works reported on template free synthesis of zeolite membranes including NaA (Morigami et al. 2001), ZSM-5 (Pan and Lin 2001), Na-A (Pera-Titus et al. 2006), silicalite (Kanezashi et al. 2006), RHO (Mousavi et al. 2013) and BEA (Tang et al. 2014). In the research work by Gopalakrishnan et al. (2006), membranes without organic template displayed lower permeance and different selectivity compared to those of with template. The membranes performed differently due to the absence of protruding hydroxyls which block the zeolite pores and alter the surface polarity and adsorption capacity (Gopalakrishnan et al. 2006).

Secondary (Seeded) growth

An alternative approach that has been proposed to prepare supported zeolite membranes is the seeded secondary growth by using externally prepared seeds. This technique was first suggested by Boudreau et al. (1999) to grow zeolite LTA supported membranes, where it was found that it was difficult to synthesize zeolite LTA membrane directly from a sol in contact with porous support due to the absence of structure-directing agent (SDA) (Boudreau et al. 1999). Then, Tsapatsis et al. (1994) reported successful synthesis of MFI membranes by secondary growth from seeds in 1994 (Tsapatsis et al. 1994). Secondary growth can be assumed as the most extensible and appealing method for the formation of thin membranes such as FAU, SOD, LTA and high oriented MFI films (Julbe 2005).

Figure 8.3 shows the secondary growth method for the development of zeolite membranes. In secondary growth, a closely packed layer of zeolite seeds is attached on the support surface before the hydrothermal synthesis. The pre-attached zeolite seeds serve as nuclei for the growth of crystal during hydrothermal synthesis with the aim to decrease any imperfections in the crystalline structure (Hedlund et al. 1999). The zeolite seeds can grow in low concentration solutions without undergoing the secondary nucleation (Julbe 2007). By precisely controlling the deposition of zeolite

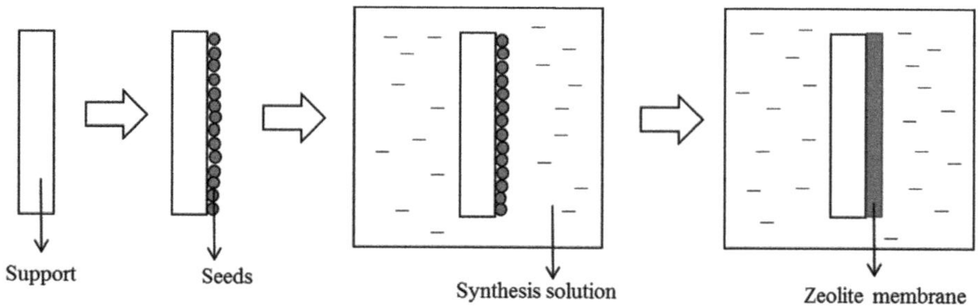

Figure 8.3. Secondary Growth in Zeolite Membranes.

seeds on the support, bulk crystallization can be avoided and favors the formation of continuous crystal zeolite film (Caro and Noack 2010).

Microwave-assisted hydrothermal synthesis

In recent decades, heating and driving chemical reactions by microwave energy has been providing a fast, simple and energy efficient tool in the science industry, especially in the preparation of zeolite materials. The initial research on microwave-assisted synthesis of zeolite was detected in 1988 where researchers from Mobil Oil Corporation filed a patent on new crystallization method utilizing microwave radiation on microporous zeolites such as zeolite MFI (ZSM-5) and LTA (NaA) (Chu et al. 1988). It was found that microwave irradiation was able to accelerate the zeolite crystallization significantly by exposing the synthesis solution to microwave radiation. Thenceforth, microwave-assisted synthesis has been implemented for various zeolites including Na-Y (FAU), ZSM-5 (MFI) (Arafat et al. 1993), $AlPO_4$-5 (AFI) (Girnus et al. 1995), Beta (BEA) (Panzarella et al. 2007), Na-A (LTA), silicalite-1 (Hu et al. 2009) and SAPO-34 (CHA) (Jun et al. 2011). The number of publications in microwave radiation synthesis of zeolites has been growing steadily year by year. In 1998, Cundy and Zhao (1998) were able to review comprehensively on the microwave synthesis and modification of zeolites while Tompsett et al. (2006) produced an excellent summary on the microwave synthesis of nanoporous materials and outlined the zeolites, mixed oxides and mesoporous molecular sieves preparation by applying microwave radiation (Tompsett et al. 2006).

Microwave is an electromagnetic wave with wavelength in the range of 0.01–1 m and frequencies between 0.3–300 GHz, located in between the infrared and radio waves in the electromagnetic spectrum. Most of the reported microwave synthesis experiments were conducted at 2450 MHz due to maximum microwave energy absorption by water (Li and Yang 2008). Figure 8.4 shows the comparison between microwave-assisted hydrothermal synthesis and conventional direct *in situ* hydrothermal synthesis for zeolite membranes. The introduction of microwave heating provides instantaneous and volumetric heating to the synthesis solution where the solution becomes high supersaturation rapidly. Unlike conventional direct *in situ* hydrothermal synthesis, the volumetric microwave heating without heat diffusion effect enables better control in zeolite crystal size distribution. Hence, microwave-assisted hydrothermal synthesis is able to form thinner zeolite membrane with

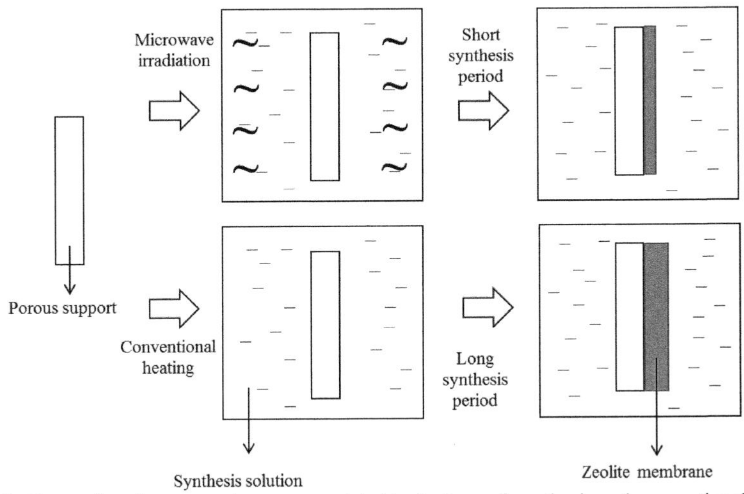

Figure 8.4. Comparison between microwave-assisted hydrothermal synthesis and conventional direct *in situ* hydrothermal synthesis for zeolite membranes.

narrower crystal size distribution (Chew et al. 2011a, Li and Yang 2008, Tompsett et al. 2006).

Microwave-assisted hydrothermal synthesis has been demonstrated to be beneficial in synthesizing good quality zeolite membranes within a very short period. Successful synthesis of Na-Y membranes was reported within 100 minutes by employing microwave-assisted hydrothermal synthesis instead of 12 hours via conventional heating (Panzarella et al. 2007). SAPO-5 membranes were obtained after 1–2 hours of microwave irradiation compared to the 4 days of conventional heating at 190°C (Utchariyajit and Wongkasemjit 2010). Li and Yang (2008) reported that the 3 hours of conventional heating time for zeolite LTA was able to be reduced to within 20 minutes via microwave-assisted hydrothermal synthesis. Besides, the microwave-assisted hydrothermal synthesis was reported to be able to decrease the membrane thickness and enhance the flux (Li and Yang 2008). In the research work by Chen et al. (2005), NaA membranes were obtained after 20 minutes of microwave irradiation and SOD phase was avoided to form.

Ultrasonic-assisted hydrothermal synthesis

In recent years, successful attempts of applying the sonochemical method to synthesize nanostructured materials has been proven as a potential technique to produce various inorganic compounds (Bang and Suslick 2010). The utilization of ultrasonic-assisted hydrothermal synthesis offers a facile and versatile synthesis for nano and microstructure compounds, compared to long crystallization and low rate of mixing in conventional heating.

Sonochemistry via ultrasonic irradiation is the chemical effects of ultrasound acquired from acoustic cavitation (Chen et al. 2003). Ultrasound is higher frequency (20 kHz–10 MHz) sound wave compare to upper audible limit of human hearing (Gandhi and Kumar 1994). When ultrasound is transmitted through a liquid-solid system, a phenomenon known as acoustic cavitation happens. Sound waves oscillation occurs in the direction of wave and forms longitudinal waves which are series of compressions and expansions (rarefactions) area (Shah et al. 2012). During compression cycle, the liquid layer located nearest to the source of ultrasound will be replaced, promoting the neighboring layers reposition in a repetitive manner so that the layers can be in compressed. While during rarefaction cycle, some layers are expanded when others are compressed (Askari et al. 2009). These alternating expansive and compressive sound waves produce bubbles or cavities when the pressure in the system decreases below its vapor pressure (Shah et al. 2012).

Ultrasonic irradiation is beneficial to hydrothermal synthesis by shortening the crystallization time and producing zeolite crystals with smaller crystal size. The cavities accumulate the ultrasonic energy while growing to certain size and subsequently collapse, liberating the energy stored in the bubble within very short time (Suslick 1989). Due to the collapse of bubbles, extremely high temperature, pressure and cooling rate are produced (Askari and Halladj 2012). The energy provided during collapse of cavitation bubbles is sufficient for crystallization. The acoustic stream, produced by ultrasonic irradiation, passes through the liquid solid system and creates a stirring effect which reduces the thickness of diffusion layer and eventually enhances the mass transfer of the system. Subsequently, the crystal growth rate increases and it results in increase in nucleation rate (Askari et al. 2013). The advantages of ultrasonic irradiation in shortening the zeolite crystallization work was reported in the literature. In the research work reported by Bose et al. (2014), the synthesis duration of the DD3R seeds was able to be significantly reduced from 25 days to 2 days via ultrasonic-assisted hydrothermal synthesis. In addition, the occurrence of high temperature decrease rate during rapid collapse of the bubbles prevents the agglomeration of particles and hence results in formation of zeolite crystals with smaller size (Askari et al. 2013). The smaller zeolite crystal in essential for formation of more compact zeolite membrane with thinner layer.

Concluding remarks

Direct *in situ* hydrothermal synthesis is the widely applied technique for the synthesis of different zeolite membranes. In direct *in situ* hydrothermal synthesis, the surface of the porous support is in contact directly with zeolite precursor solution under controlled environment for nucleation and growth of a continuous zeolite membrane on the surface of the support. Secondary growth is the extensible and appealing method for the formation of high quality and thin continuous zeolite membranes. Through secondary growth, any imperfections or defects in the zeolite crystalline structure can be reduced by controlling the deposition of zeolite seeds on the support, resulting in the formation of zeolite membrane with improved quality. Besides that, microwave- and ultrasonic-assisted hydrothermal synthesis appear to be alternative approaches for facile synthesis of zeolite membranes. The application of microwave and ultrasound can help to shorten the crystallization time of zeolite and produce zeolites with narrower and more uniform crystal size distribution.

CO_2 Permeation and Separation Using Zeolite Membranes

The mechanism of gas permeation through porous inorganic membranes

The mechanisms of permeation through porous inorganic membranes, include zeolite membranes, are generally viscous flow, Knudsen diffusion, surface diffusion and molecular diffusion, as shown in Fig. 8.5. The gas permeation mechanisms can be influenced by a few aspects, including the gas molecules properties such as kinetic diameter and molecular weight, as well as the properties of the membranes such as pore size and surface properties.

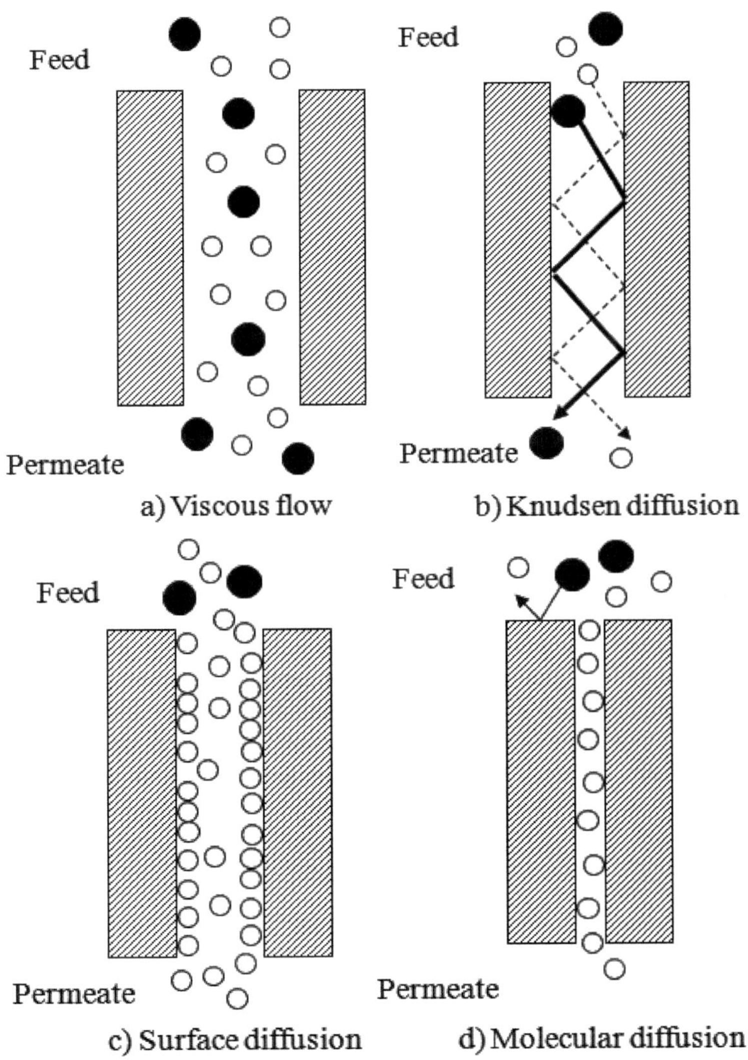

Figure 8.5. The general mechanisms of gas permeation in zeolite membranes.

Viscous flow

Generally, the diffusion is dominated by viscous flow in the porous inorganic membranes having pore diameter, d_p bigger than 50 nm. In this case, the gas molecules flow as a whole under pressure gradient. Viscous flow usually results in no separation among the gas molecules (Lito et al. 2009).

Knudsen diffusion

Knudsen diffusion happens through the pores in the porous inorganic membrane having diameter, d_p, smaller than the mean free path, λ, of the gas molecules in the gas phase (Park 2013). When the Knudsen number (λ/d) is greater than one, the collision frequency between the molecules inside the narrow pore channels with the wall is much higher than the collision frequency between gas molecules themselves. Under the scenario where Knudsen diffusion is predominant, the diffusion rate of different gas molecules is inversely proportional to the square root of the molecular weights of the gas molecules. In another way, lower molecular weight gas molecules can diffuse faster than those with higher molecular weight. If the Knudsen number is relatively small, both collisions between gas molecules themselves, and between gas molecules and pore walls will be rate determined (Shekhawat et al. 2003).

Surface diffusion

Surface diffusion is when the specific gas molecules are selectively adsorbed on the pore walls and hence transport through the pore framework in the direction of decreasing surface concentration (Shekhawat et al. 2003). The separation of gas molecules is achieved when the selectively adsorbed and diffusing specific gas molecules hinder the diffusion of the other gas molecule species through the pore framework. Surface diffusion usually occurs at low temperature when the gas molecules could not escape from the surface potential field due to stronger interaction between the inner surface pore wall and gas molecules compared to their kinetic energy (Oyama et al. 2011). Increase in temperature will result in decrease in surface diffusion. Chew and Ahmad (2016) reported higher surface diffusion of CO_2, compared to CH_4, through Ba-SAPO-34 zeolite membrane owing to stronger adsorption strength of CO_2 in Ba-SAPO-34 pore framework. However, the surface coverage of CO_2 on Ba-SAPO-34 pore framework decreased, causing decrease of CO_2 gas flux from 19.5 to 9.8 × 10^{-3} mol/m^2, when temperature was elevated from 30 to 180°C.

Molecular sieving/molecular diffusion

Molecular sieving or molecular diffusion is based on size exclusion of the gas molecules, which highly relies on the pore aperture of the membranes. As the pore size is approximately same with the kinetic diameter of the smaller gas molecules, smaller molecules in the gas mixtures are able to pass through the membrane pores while the gas molecules with larger kinetic diameter are excluded from diffusing through the pore framework, leading to selective preferential separation of smaller gas molecules species (Shekhawat et al. 2003). High permeability and selectivity of smaller gas molecules in the gas mixture can be achieved through the membranes based on molecular sieving. Hence, pore sizes of the membrane must be tailored accordingly in order to obtain high selectivities of the gas molecules (Rao and Sircar 1993).

Concluding remarks

For porous inorganic membranes include zeolite membranes, the few factors influencing the gas permeation mechanisms include the properties of gas molecules such as kinetic diameter and molecular weight, as well as the properties of the membranes such as pore size and surface properties. Despite several common gas permeation mechanisms through the zeolite membrane, usually the gas separation performance of the zeolite membranes is attributed by surface diffusion and molecular diffusion. Surface diffusion occurs due to strong interaction between the inner surface zeolite pore wall and specific gas molecules. Molecular sieving occurs when the gas molecules (with kinetic diameter smaller than the zeolite pore size) are able to pass through the pores of the membrane while the gas molecules with larger kinetic diameter are excluded from diffusing through the pore framework.

Permeation and separation of CO_2 from different systems using zeolite membranes

Removal of CO_2 from various gas systems, including CO_2/CH_4, CO_2/N_2 and CO_2/H_2, is important for gas processing industries. There are extensive research works done on investigating different zeolite membranes for CO_2 separation from different gas systems of CO_2/CH_4, CO_2/N_2 and CO_2/H_2, as will be analyzed in the next segment.

CO_2/CH_4 system

Zeolite membranes have been investigated extensively to remove CO_2 from CH_4. Table 8.2 shows the CO_2/CH_4 gas permeation and separation performance reported for assorted types of zeolite membranes. It can be observed from Table 8.2 that FAU and MFI zeolite membrane do not display high selectivity or separation factor for the separation of CO_2/CH_4 system owing to their larger pore aperture (> 0.5 nm) which is too far reaching to remove CO_2 from CH_4. The separation of CO_2/CH_4 using FAU and MFI zeolite membrane is usually dependent on competitive adsorption mechanism which did not supply high selectivity or separation factor.

On the contrary, small-pore zeolite membrane such as SAPO-34, T-type and DDR membrane, with comparable pore aperture to the kinetic diameter of CH_4 shows greater selectivity or separation factor for CO_2/CH_4 system, due to a combination effect of differences in diffusivity and competitive adsorption. Due to the significance of the gas permeance in industrial gas separation, 10–20 μm thick SAPO-34 membrane fabricated by Tian et al. (2009) was able to acquire high CO_2 permeance of 2.5×10^{-6} mol/m^2.s.Pa but low CO_2/CH_4 separation selectivity of 9 for equimolar gas mixture permeation and separation at room temperature and

Table 8.2. CO_2/CH_4 permeation and separation using zeolite membranes reported in the literature.

Membrane	T, °C	CO_2/CH_4 Selectivity or separation factor	CO_2 Permeance, $\times 10^8$mol/m^2.s.Pa	Reference
AlPO-18	25	120	20	Wu et al. (2014)
	22	52–60	6.6	Carreon et al. (2012)
SAPO-34	24	67	16	Li et al. (2004)
	22	133–170	38	Carreon et al. (2008)
	20	~ 28	~ 0.8	Tian et al. (2009)
	Room temp.	9	250	
	22	130–290	23–71	Li et al. (2010)
	22	283	240	Zhang et al. (2010)
Ba-SAPO-34	30	103	37.6	Chew et al. (2011b)
T-type	120	70.8	10^{-9}–10^{-6}	Mirfendereski et al. (2008)
	35	400	4.6	Cui et al. (2003)
	200	52	1.5	
DDR	25	5.22	24.6	Mubashir et al. (2016)
	25	200	30	Himeno et al. (2007)
	28	220	~ 6.3	Tomita et al. (2004)
	100	100	~ 3.3	
	–48	5000	10	van den Bergh et al. (2008)
	30	400	6.5	
FAU (NaX)	35	28	~ 27	Hasegawa et al. (2002)
FAU (K-Y)	35	40	70	Hasegawa et al. (2002)
MFI	30	2.5	-	van den Broeke et al. (1999)
	25	6.0	~ 30	Fan et al. (2013)

feed pressure of 120 kPa. The flow distance of gas molecules was reduced by thin SAPO-34 membrane attached on a macroporous stainless steel net, resulting in the increment of gas permeance (Tian et al. 2009). Li et al. (2010) was able to obtain CO_2/CH_4 selectivity as high as 290 using SAPO-34 zeolite membrane at 22°C. Cui et al. (2003) obtained CO_2/CH_4 selectivity of 400 and a CO_2 permeance of 4.6×10^{-8} mol/m^2.s.Pa at 35°C using T-type zeolite membrane at feed pressure of 0.1 MPa and vacuum on the permeate side. The highest CO_2/CH_4 separation factor of more than 5000 was reported by van den Bergh et al. (2008) using DDR zeolite membrane at –48°C where CH_4 molecules were unable to permeate through the membrane due to the inhibition of higher mobility and adsorption affinity of CO_2 in the zeolite pores (van den Bergh et al. 2008).

CO_2/N_2 system

Selective CO_2/N_2 separation can be obtained due to the stronger electrostatic quadrupole of CO_2 with the zeolite compared to N_2. Several zeolite membranes were

Table 8.3. CO_2/N_2 permeation and separation using zeolite membranes reported in the literature.

Membrane	T, °C	CO_2/N_2 Selectivity or separation factor	CO_2 Permeance, $\times 10^8$mol/m^2.s.Pa	References
SAPO-34	27	16	9.5	Poshusta et al. (2000)
	22	32	120	Li and Fan (2010)
Ba-SAPO-34	30	29.6	~ 9.4–19.5	Chew and Ahmad (2016)
T-type	35	107	3.8	Cui et al. (2004)
FAU (NaX)	35	78	70	Hasegawa et al. (2002)
FAU (NaY)	30	5	-	Clet et al. (2001)
	Room temp.	31.2	2.1	Gu et al. (2005)
Na-ZSM-5	25	54.3	~ 3.6	Shin et al. (2005)
	100	14.9	~ 1.0	
	27	13.7	260	Bernal et al. (2004)
B-ZSM-5	27	12.6	19	Bernal et al. (2004)
MFI	25	9.2	~ 30	Fan et al. (2013)
ETS-4	35	~ 0.3	-	Guan et al. (2001)
ETS-10	Room temp.	7-20	-	Tiscornia et al. (2010)

reported with CO_2 permeation dominant in surface diffusion while N_2 permeation via micropore diffusion (Bernal et al. 2004, Shin et al. 2005). High separation factor of CO_2/N_2 can be caused by the preferential adsorption of CO_2 inhibits the N_2 permeation. Table 8.3 illustrates the performance of different types of zeolite membrane in CO_2/N_2 separation system. Owing to its low Al content, MFI-type zeolite membrane is able to attract researchers' attention and these membranes are synthesized with good reproducibility and chemical stability. For the separation and permeation of CO_2/N_2 using MFI zeolite membrane, in Table 3.2, the best performance is reported by Shin et al. (2005) with selectivity of 54.3 and CO_2 permeance of ~ 3.6×10^{-8} mol/m^2.s.Pa at 25°C using Na-ZSM-5 zeolite membrane. ETS-4 zeolite membrane is another type of zeolite membrane used to remove N_2 from CO_2 in which it favorably adsorbs N_2 over CO_2 (Guan et al. 2001). Hasegawa et al. (2002) synthesized NaX zeolite membrane and was able to obtain CO_2/N_2 selectivities as high as 78. Research works have been carried out on investigating the CO_2/N_2 permeation and separation performance using small pore zeolite membrane such as SAPO-34 and T-type zeolite membranes. In the research works reported by Cui et al. (2004), the synthesized T-type zeolite membrane displayed high CO_2/N_2 selectivity of 107 with CO_2 permeance of 3.8×10^{-8} mol/m^2.s.Pa at 35°C.

CO_2/H_2 system

Table 8.4 shows CO_2/H_2 gas permeation and separation studies analyzed for various types of zeolite membranes. It can be seen that the CO_2/H_2 selectivities reported for several zeolite membranes, such as MFI, NaY, DDR, AM-3 and, etc., were much smaller than CO_2/N_2 and CO_2/CH_4 selectivities. This can be explained by

Table 8.4. CO_2/H_2permeation and separation using zeolite membranes reported in the literature.

Membrane	T, °C	CO_2/H_2 selectivity or separation factor	CO_2 Permeance, $\times 10^8$ mol/m².s.Pa	Reference
SAPO-34	–20	~ 136.00	~ 2.70	Hong et al. (2008)
	35	~ 16.00	~ 3.70	
	200	~ 0.5	-	
HZSM-5	–38–37	17–210	620–780	Korelskiy et al. (2015)
Na-LTA	30	~ 1.20	~ 1.40	Varela-Gandiaet al. (2011)
	150	~ 0.50	~ 0.60	
MFI	25	~ 4.50	~ 18.00	Yuan et al. (2011)
	450	~ 0.10	~ 5.00	
	22	~ 0.64-1.09	-	Lindmark and Hedlund (2010)
NaY	65	~ 0.01	0.55	Dey et al. (2013)
P/NaX	16	6.40	1.80	Yin et al. (2007)
	75	4.10	15.80	
DDR	30	~ 1.00	-	van den Bergh et al. (2010)
	400	~ 2.00	-	
AM-3	25	~ 0.40	~ 60.00	Li et al. (2011)
ETS-10	Room temp.	9	-	Tiscornia et al. (2010)

the extremely small kinetic diameter of H_2 molecule burdens the gas separation. Although most of the reported MFI zeolite membranes were not effective for the CO_2/H_2 separation, Korelskiy et al. (2015) reported CO_2/H_2 selectivities of 17–210 with CO_2 permeance of 620–780 × 10^{-8} mol/m².s.Pa using prepared HZSM-5 zeolite membrane for temperature range from –38 to 37°C. This was due to much stronger CO_2 adsorption on the membrane compared to H_2 (Korelskiy et al. 2015). The CO_2 adsorption on the membrane decreased with increasing temperature, resulted in decrease in CO_2/H_2 selectivity (Korelskiy et al. 2015).

SAPO-34 emerges as a promising choice in the removal of H_2 from CO_2 at low temperature. Hong et al. (2008) obtained CO_2/H_2 selectivity of ~ 136 for SAPO-34 membrane at –20°C with CO_2 permeance of 2.7 × 10^{-8} mol/m².s.Pa as the permeation of H_2 through SAPO-34 zeolite membrane was hindered by strong CO_2 adsorption at low temperature. Yet, SAPO-34 zeolite membrane became H_2-selective at high temperature where the H_2/CO_2 separation factor of only 2 was acquired for SAPO-34 at 200°C (Hong et al. 2008).

Concluding remarks

Different zeolite membranes have been extensively investigated for the separation system of CO_2/CH_4, CO_2/N_2 and CO_2/H_2. CO_2/H_2 selectivities for several zeolite membranes were reported to be smaller than CO_2/N_2 and CO_2/CH_4 selectivities. This can be explained by the extremely small kinetic diameter of H_2 molecule that causes the separation a difficult task. Generally, small pore zeolite membranes are able to display higher gas separation performance compared to medium pore and large pore zeolite membranes.

Effect of water vapor on gas permeation and separation performance of zeolite membranes

The presence of water vapor in the gas streams might affect the gas processing operation or the separation performance of the zeolite membranes. Water vapor in natural gas stream has always been a source of trouble especially in the transportation of natural gas. According to US pipelines specifications, water vapor tends to form hydrates along the pipelines when the system reaches –15°C dew point temperature (Sanders 2004, Hammerschmidt 1934). There has been research works reported on the influence of water vapor on the separation performance of the zeolite membranes.

Moisture or water vapor was reported to post suppression effect on gas permeation in MFI zeolite membranes at low temperature (< 300°C) (Chau et al. 2000). When water molecules adsorb on the surface of zeolite through dipole-field interactions, it also forms hydrogen bonding with the remaining hydroxyl groups (Kawai and Tsutsumi 1992).The amount of water molecules adsorb on the zeolite surface depends on the amount of aluminum atoms in the framework. Therefore, the affinity of zeolites with water molecules is influenced by the Si/Al ratio of the zeolite (Calero et al. 2014). Takeuchi et al. (2007) studied the adsorption of water molecules on various ZSM-5 zeolites with different Si/Al ratios. When the composition of Al_2O_3 increased, the Si/Al ratio decreased resulting in increased amount of water adsorbed on the surface. This was because water molecules adsorb on the adsorption sites of Al^{3+} ions. Electrostatic interactions enable strong adsorption of water molecules on the cationic sites of oxide surfaces while hydrogen bonding lead to weak adsorption of water molecules on the surface of hydroxyl groups (Takeuchi et al. 2007).

Sublet et al. (2012) investigated the effect of moisture on the performance of MFI zeolite membrane for CO_2/N_2 permeation and separation. In the presence of 10% water moisture at 100°C, the gas permeance through the MFI membrane reduced approximately 40%. However, the CO_2/N_2 separation factor of the MFI membranes improved when water moisture was incorporated, which might be due to defective pore blockage and formation of bicarbonate stable intermediates (Sublet et al. 2012).

Bernal et al. (2004) studied the separation performance of B-ZSM-5 membranes for CO_2/N_2 saturated with water vapor at room temperature. The CO_2 flux dropped by 54% at 27°C and retentate pressure of 170 kPa. The capillary condensation of water blocked small defects on the membranes and reduced the non-selective transport through the membrane at room temperature. Anyway, the presence of moisture posted small effect on the separation factor, which might be due to the limited hydrophilicity of ZSM-5 membranes (Bernal et al. 2004).

SAPO-34 was reported with low stability in humid conditions due to its hydrophilicity (Lutz et al. 2010). It was reported by Poshusta et al. (2001) that the water moisture has significant effect on the gas permeation through SAPO-34 membrane, where months of exposure to the laboratory atmosphere degraded the

SAPO-34 membrane. The rate of degradation was found to be accelerating once the membrane degradation began. The non-SAPO pores, which were larger than the SAPO-34 pores, were created during membrane degradation and then influenced significantly the gas separation performance of SAPO-34 (Poshusta et al. 2001).

The effects of humidity on CO_2/CH_4 separation of DDR zeolite membrane were studied by Himeno et al. (2007). The CO_2 permeance of the DDR membrane dropped from 2.5×10^{-7} to 1.1×10^{-7} mol/m^2.s.Pa and the CO_2/CH_4 selectivity decreased to half that of the dry CO_2/CH_4 mixture, in the presence of water vapor. Anyway, the CO_2 permeance and CO_2/CH_4 selectivity were restored after introducing feed gas mixture without water into the system (Himeno et al. 2007).

Gu et al. (2005) synthesized NaY zeolite membranes and studied the separation of CO_2/N_2 at dry and moist conditions at atmospheric pressure. Gu et al. (2005) obtained CO_2/N_2 selectivity of about 31.2 with CO_2 permeance of 2.1×10^{-8} mol/m^2.s.Pa under dry condition at room temperature. Introducing water vapor into the feed stream resulted drop in CO_2 and N_2 permeance for temperature range from 23 to 200°C. The CO_2/N_2 selectivity decreased for temperature below 80°C, but increased for temperature between 110–200°C, in the presence of water vapor (Gu et al. 2005).

Conclusions

Extensive research works have been reported for synthesis and gas performance testing of zeolite membranes. Newer technologies such as microwave-assisted hydrothermal synthesis and ultrasonic-assisted hydrothermal synthesis have recently gained popularity due to their advantages in shortening the synthesis time and enhancing the properties of the zeolite membrane. The gas permeation and separation performance of the zeolite membranes is dependent on few factors including the properties of gas molecules such as kinetic diameter and molecular weight, as well as the properties of the membranes such as pore size and surface properties. Small pore zeolite membranes have been reported to be able to display high selectivities or separation factors for different gas systems of CO_2/N_2, CO_2/CH_4 and CO_2/H_2, depending on the conditions such as pressure and temperature. The presence of water vapor in the gas streams posted effect on the gas permeation and separation performance of several zeolite membranes. The extent of the effect by the water vapor depends on factors such as type of zeolite membrane, duration of exposure to water vapor, temperature when water vapor is introduced and etc. Therefore, necessary steps can be carried out including the use of water adsorber to reduce the moisture content in the gas stream before feeding the gas stream to zeolite membrane separation unit. Selection of the correct type of zeolite membrane is also crucial for the separation of the gas system under designated condition.

Acknowledgements

The authors acknowledge the support provided by Universiti Teknologi PETRONAS, Institute of Contaminant Management UTP and CO_2 Research Centre (CO2RES).

References

Anbia, M., Koohsaryan, E. and Borhani, A. 2017. Novel hydrothermal synthesis of hierarchically-structured zeolite LTA microspheres. Mater. Chem. Phys. 193: 380–390.

Arafat, A., Jansen, J.C., Ebaid, A.R. and van Bekkum, H. 1993. Microwave preparation of zeolite Y and ZSM-5. Zeolites 13(3): 162–165.

Askari, S. and Halladj, R. 2012. Ultrasonic pretreatment for hydrothermal synthesis of SAPO-34 nanocrystals. Ultrason. Sonochem 19(3): 554–559.

Askari, S., Halladj, R. and Nasernejad, B. 2009. Characterization and preparation of sonochemically synthesized silver–silica nanocomposites. Mater. Sci. Pol. 27(2): 397–405.

Askari, S., Alipour, S.M., Halladj, R. and Farahani, M.H.D.A. 2013. Effects of ultrasound on the synthesis of zeolites: A Review. J. Porous Mater. 20(1): 285–302.

Bang, J.H. and Suslick, K.S. 2010. Applications of ultrasound to the synthesis of nanostructured materials. Adv. Mater. 22(10): 1039–1059.

Barrer, R.M. 1982. Hydrothermal chemistry of zeolites. Academic Press, London.

Bernal, M. P., Coronas, J., Menendez, M. and Santamaria, J. 2004. Separation of CO_2/N_2 mixtures using MFI-type zeolite membranes. AIChE Journal. 50(1): 127–135.

Boudreau, L.C., Kuck, J.A. and Tsapatsis, M. 1999. Deposition of oriented zeolite A films: in situ and secondary growth. J. Membr. Sci. 152(1): 41–59.

Bose, A., Sen, M., Das, J.K., and Das, N. 2014. Sonication mediated hydrothermal process–an efficient method for the rapid synthesis of DDR zeolite membranes. RSC Adv. 4(36): 19043–19052.

Bowen, T.C., Noble, R.D. and Falconer, J.L. 2004. Fundamentals and applications of pervaporation through zeolite membranes. J. Membr. Sci. 245(1): 1–33.

Burton, A. 2003. Zeolites: porous architectures. Nat. Mater. 2(7): 438–440.

Byrappa, K. and Yoshimura, M. 2012. Handbook of hydrothermal technology. William Andrew, London.

Calero, S. and Gómez-Álvarez, P. 2014. Hydrogen bonding of water confined in zeolites and their zeolitic imidazolate framework counterparts. RSC Advances. 4(56): 29571–29580.

Caro, J. and Noack, M. 2010. Zeolite membranes–status and prospective. Adv. Nanoporous Mater. 1: 1–96.

Carreon, M.A., Li, S., Falconer, J.L. and Noble, R.D. 2008. SAPO-34 Seeds and Membranes Prepared Using Multiple Structure Directing Agents. Adv. Mater. 20(4): 729–732.

Carreon, M.L., Li, S. and Carreon, M.A. 2012. AlPO-18 membranes for CO_2/CH_4 separation. Chem. Comm. 48(17): 2310–2312.

Chau, J.L.H., Tellez, C., Yeung, K.L. and Ho, K. 2000. The role of surface chemistry in zeolite membrane formation. J. Membr. Sci. 164(1): 257–275.

Chen, W., Zhang, J. and Cai, W. 2003. Sonochemical preparation of Au, Ag, Pd/SiO_2 mesoporous nanocomposites. Scr. Mater. 48(8): 1061-1066.

Chen, X., Yang, W., Liu, J. and Lin, L. 2005. Synthesis of zeolite NaA membranes with high permeance under microwave radiation on mesoporous-layer-modified macroporous substrates for gas separation. J. Membr. Sci. 255(1): 201–211.

Chew, T.L., Ahmad, A.L. and Bhatia, S. 2011a. Rapid synthesis of thin SAPO-34 membranes using microwave heating. J. Porous Mater. 18: 355–360.

Chew, T.L., Ahmad, A.L. and Bhatia, S. 2011b. Ba-SAPO-34 membrane synthesized from microwave heating and its performance for CO_2/CH_4 gas separation. Chem. Eng. J. 171(3): 1053–1059.

Chew, T.L. and Ahmad, A.L. 2016. Gas permeation properties of modified SAPO-34 zeolite membranes. Procedia Eng. 148: 1225–1231.

Chu, P., Dwyer, F.G. and Vartuli, J.C. 1988. Crystallization method employing microwave radiation. U.S. Patent # 4,778,666.

Clet, G., Gora, L., Nishiyama, N., Jansen, J.C., van Bekkum, H. and Maschmeyer, T. 2001. An alternative synthesis method for zeolite Y membranes. Chem. Comm. (1): 41–42.

Cui, Y., Kita, H. and Okamoto, K.I. 2003. Preparation and gas separation properties of zeolite T membrane. Chem. Comm. (17): 2154–2155.

Cui, Y., Kita, H. and Okamoto, K.I. 2004. Preparation and gas separation performance of zeolite T membrane. J. Mater. Chem. 14(5): 924–932.

Cundy, C. and Zhao, J. 1998. Remarkable synergy between microwave heating and the addition of seed crystals in zeolite synthesis—a suggestion verified. Chem. Comm. (14): 1465–1466.

Daramola, M.O., Aransiola, E.F. and Ojumu, T.V. 2012. Potential applications of zeolite membranes in reaction coupling separation processes. Materials, 5(11): 2101–2136.

Davis, T.M., Drews, T.O., Ramanan, H., He, C., Dong, J., Schnablegger, H. et al. 2006. Mechanistic principles of nanoparticle evolution to zeolite crystals. Nat. Mater. 5(5): 400.

Dey, K.P., Kundu, D., Chatterjee, M. and Naskar, M.K. 2013. Preparation of NaA zeolite membranes using poly (ethyleneimine) as buffer layer, and study of their permeation behavior. J. Am. Ceram. Soc. 96(1): 68–72.

Fan, S., Liu, J., Zhang, F., Zhou, S. and Sun, F. 2013. Fabrication of zeolite MFI membranes supported by α-Al_2O_3 hollow ceramic fibers for CO_2 separation. J. Mater. Res. 28(13): 1870–1876.

Feng, C., Khulbe, K.C., Matsuura, T., Farnood, R. and Ismail, A.F. 2015. Recent progress in zeolite/zeotype membranes. J. Membr. Sci. Res. 1(2): 49–72.

Flanigen, E., Broach, R. and Wilson, S. 2010. Introduction. pp. 2. *In*: Kulprathipanja, S. (ed.). Zeolites in Industrial Separation and Catalysis. John Wiley and Sons, New Jersey.

Gabriel, C., Gabriel, S., Grant, E.H., Halstead, B.S.J. and Mingos, D.M.P. 1998. Dielectric parameters relevant to microwave dielectric heating. Chem. Soc. Rev. 27(3): 213–224.

Gandhi, K. and Kumar, R. 1994. Sonochemical reaction engineering. Sadhana Bangalore, 19: 1055–1076.

Girnus, I., Pohl, M.M., Richter-Mendau, J., Schneider, M., Noack, M., Venzke, D. et al. 1995. Synthesis of AIPO4-5 aluminumphosphate molecular sieve crystals for membrane applications by microwave heating. Adv. Mater. 7(8): 711–714.

Gopalakrishnan, S., Yamaguchi, T. and Nakao, S.I. 2006. Permeation properties of templated and template-free ZSM-5 membranes. J. Membr. Sci. 274(1): 102–107.

Gorgojo, P., de la Iglesia, O. and Coronas, J. 2008. Preparation and characterization of zeolite membranes. pp. 135–175. *In*: Mallada, R. and Menéndez, M. (eds.). Membrane Science and Technology. 13. Elsevier Science, Amsterdam.

Gu, X., Dong, J. and Nenoff, T.M. 2005. Synthesis of defect-free FAU-type zeolite membranes and separation for dry and moist CO_2/N_2 mixtures. Ind. Eng. Chem. Res. 44(4): 937–944.

Guan, G., Kusakabe, K. and Morooka, S. 2001. Synthesis and permeation properties of ion-exchanged ETS-4 tubular membranes. Microporous Mesoporous Mater. 50(2): 109–120.

Hammerschmidt, E.G. 1934. Formation of gas hydrates in natural gas transmission lines. Ind. Eng. Chem. 26(8): 851–855.

Hasegawa, Y., Tanaka, T., Watanabe, K., Jeong, B.H., Kusakabe, K. and Morooka, S. 2002. Separation of CO_2-CH_4 and CO_2-N_2 systems using ion-exchanged FAU-type zeolite membranes with different Si/Al ratios. Korean J. Chem. Eng. 19(2): 309–313.

Hedlund, J., Noack, M., Kölsch, P., Creaser, D., Caro, J. and Sterte, J. 1999. ZSM-5 membranes synthesized without organic templates using a seeding technique. J. Membr. Sci. 159(1): 263–273.

Himeno, S., Tomita, T., Suzuki, K., Nakayama, K., Yajima, K. and Yoshida, S. 2007. Synthesis and permeation properties of a DDR-type zeolite membrane for separation of CO_2/CH_4 gaseous mixtures. Ind. Eng. Chem. Res. 46(21): 6989–6997.

Hong, M., Li, S., Falconer, J.L. and Noble, R.D. 2008. Hydrogen purification using a SAPO-34 membrane. J. Membr. Sci. 307(2): 277–283.

Hould, N.D., Kumar, S., Tsapatsis, M., Nikolakis, V. and Lobo, R.I.F. 2009. Structure and colloidal stability of nanosized zeolite beta precursors. Langmuir. 26(2): 1260–1270.

Hu, Y., Liu, C., Zhang, Y., Ren, N. and Tang, Y. 2009. Microwave-assisted hydrothermal synthesis of nanozeolites with controllable size. Microporous Mesoporous Mater. 119(1): 306–314.

Julbe, A. 2005. Zeolite membranes—a short overview. pp. 135. *In*: Čejka, J. and van Bekkum, H. (eds.). Zeolites and Ordered Mesoporous Materials: Progress and Prospects. 157. Elsevier Science, Amsterdam.

Julbe, A. 2007. Zeolite Membranes—Synthesis, Characterization and Application. pp. 181–219. *In*: Cejka, J., van Bekkum, H., Corma, A. and Schueth, F. (eds.). Introduction to Zeolite Science and Practice. 168. Elsevier Science, Amsterdam.

Jun, J.W., Lee, J.S., Seok, H.Y., Chang, J.S., Hwang, J.S. and Jhung, S.H. 2011. A facile synthesis of SAPO-34 molecular sieves with microwave irradiation in wide reaction conditions. Bull. Korean. Chem. Soc. 32(6): 1957–1964.

Kanezashi, M., O'Brien, J. and Lin, Y.S. 2006. Template-free synthesis of MFI-type zeolite membranes: permeation characteristics and thermal stability improvement of membrane structure. J. Membr. Sci. 286(1): 213–222.

Kawai, T. and Tsutsumi, K. 1992. Evaluation of hydrophilic-hydrophobic character of zeolites by measurements of their immersional heats in water. Colloid Polym. Sci. 270(7): 711–715.

Korelskiy, D., Ye, P., Fouladvand, S., Karimi, S., Sjöberg, E. and Hedlund, J. 2015. Efficient ceramic zeolite membranes for CO_2/H_2 separation. J. Mater. Chem. A. 3(23): 12500–12506.

Li, S. and Fan, C.Q. 2010. High-flux SAPO-34 membrane for CO_2/N_2 separation. Ind. Eng. Chem. Res. 49(9): 4399–4404.

Li, S., Carreon, M.A., Zhang, Y., Funke, H.H., Noble, R.D. and Falconer, J.L. 2010. Scale-up of SAPO-34 membranes for CO_2/CH_4 separation. J. Membr. Sci. 352(1): 7–13.

Li, S., Falconer, J.L. and Noble, R.D. 2004. SAPO-34 membranes for CO_2/CH_4 separation. J. Membr. Sci. 241(1): 121–135.

Li, S., Martinek, J.G., Falconer, J.L., Noble, R.D. and Gardner, T.Q. 2005. High-pressure CO_2/CH_4 separation using SAPO-34 membranes. Ind. Eng. Chem. Res. 44(9): 3220–3228.

Li, X., Zhou, C., Lin, Z., Rocha, J., Lito, P.F., Santiago, A.S. et al. 2011. Titanosilicate AM-3 membrane: A new potential candidate for H_2 separation. Microporous Mesoporous Mater. 137(1): 43–48.

Li, Y. and Yang, W. 2008. Microwave synthesis of zeolite membranes: a review. J. Membr. Sci. 316(1): 3–17.

Lin, X., Kita, H. and Okamoto, K.I. 2000. A novel method for the synthesis of high performance silicalite membranes. Chem. Comm. (19): 1889–1890.

Lindmark, J. and Hedlund, J. 2010. Carbon dioxide removal from synthesis gas using MFI membranes. J. Membr. Sci. 360(1): 284–291.

Lito, P.F., Magalhães, A.L., Silva, C.M. and Fernandes, D.L. 2009. Permeation of Adsorbable and Non-Adsorbable Gases in Microporous Zeolite Membranes. J. Chem. Educ. 86(8): 976.

Liu, X., Zhang, X., Chen, Z. and Tan, X. 2013. Hydrothermal synthesis of zeolite Rho using methylcellulose as the space-confinement additive. Ceram. Int. 39(5): 5453–5458.

Lutz, W., Kurzhals, R., Sauerbeck, S., Toufar, H., Buhl, J.C., Gesing, T. et al. 2010. Hydrothermal stability of zeolite SAPO-11. Microporous Mesoporous Mater. 132(1): 31–36.

McCusker, L.B. and Baerlocher, C. 2007. Zeolite Structures. pp. 13. *In*: Cejka, J., van Bekkum, H., Corma, A. and Schueth, F. (eds.). Introduction to Zeolite Science and Practice. 168. Elsevier Science, Amsterdam.

Mintova, S., Olson, N.H., Valtchev, V. and Bein, T. 1999. Mechanism of zeolite A nanocrystal growth from colloids at room temperature. Science 283(5404): 958–960.

Mirfendereski, S.M., Mazaheri, T., Sadrzadeh, M. and Mohammadi, T. 2008. CO_2 and CH_4 permeation through T-type zeolite membranes: effect of synthesis parameters and feed pressure. Sep. Purif. Technol. 61(3): 317–323.

Morigami, Y., Kondo, M., Abe, J., Kita, H. and Okamoto, K. 2001. The first large-scale pervaporation plant using tubular-type module with zeolite NaA membrane. Sep. Purif. Technol. 25(1): 251–260.

Mousavi, S.F., Jafari, M., Kazemimoghadam, M. and Mohammadi, T. 2013. Template free crystallization of zeolite Rho via Hydrothermal synthesis: Effects of synthesis time, synthesis temperature, water content and alkalinity. Ceram. Int. 39(6): 7149–7158.

Mubashir, M., Yeong, Y.F. and Lau, K.K. 2016. Ultrasonic-assisted secondary growth of deca-dodecasil 3 rhombohedral (DD3R) membrane and its process optimization studies in CO_2/CH_4 separation using response surface methodology. J. Nat. Gas Sci. Eng. 30: 50–63.

Najafi, N., Askari, S. and Halladj, R. 2014. Hydrothermal synthesis of nanosized SAPO-34 molecular sieves by different combinations of multi templates. Powder Technol. 254: 324–330.

O' Brien, J. 2009. A Study of the Microstructure-Property Relationship for MFI-Type Zeolite Membranes for Xylene Separation. Ph.D. Thesis, Arizona State University, Arizona, US.

Oyama, S.T., Yamada, M., Sugawara, T., Takagaki, A. and Kikuchi, R. 2011. Review on Mechanisms of Gas Permeation through Inorganic Membranes. J. Jpn. Pet. Inst. 54(5): 298–309.

Pan, M. and Lin, Y.S. 2001. Template-free secondary growth synthesis of MFI type zeolite membranes. Microporous Mesoporous Mater. 43(3): 319–327.

Panzarella, B., Tompsett, G., Conner, W.C. and Jones, K. 2007. *In situ* SAXS/WAXS of zeolite microwave synthesis: NaY, NaA, and Beta zeolites. ChemPhysChem. 8(3): 357–369.

Park, H.B. 2013. Gas Separation Membranes. pp. 1–32. *In:* Hoek, E.M.V. and Tarabara, V.V. (eds.). Encyclopedia of Membrane Science and Technology. John Wiley and Sons, New Jersey.

Pera-Titus, M., Mallada, R., Llorens, J., Cunill, F. and Santamaría, J. 2006. Preparation of inner-side tubular zeolite NaA membranes in a semi-continuous synthesis system. J. Membr. Sci. 278(1): 401–409.

Poshusta, J.C., Tuan, V.A., Falconer, J.L. and Noble, R.D. 1998. Synthesis and permeation properties of SAPO-34 tubular membranes. Ind. Eng. Chem. Res. 37(10): 3924–3929.

Poshusta, J.C., Tuan, V.A., Pape, E.A., Noble, R.D. and Falconer, J.L. 2000. Separation of light gas mixtures using SAPO-34 membranes. AIChE Journal. 46(4): 779–789.

Poshusta, J.C., Noble, R.D. and Falconer, J.L. 2001. Characterization of SAPO-34 membranes by water adsorption. J. Membr. Sci. 186(1): 25–40.

Rao, M.B. and Sircar, S. 1993. Nanoporous carbon membranes for separation of gas mixtures by selective surface flow. J. Membr. Sci. 85(3): 253–264.

Rhodes, C.J. 2010. Properties and applications of zeolites. Sci. Prog. 93(3): 223–284.

Sanders, E. 2004. BCFD-scale membrane separation systems for CO_2 removal application in oil and gas production. 83rd GPA Annual Convention. New Orleans, LA.

Shah, Y.T., Pandit, A. and Moholkar, V. 2012. Cavitation reaction engineering. Springer Science & Business Media. New York.

Shekhawat, D., Luebke, D.R. and Pennline, H.W. 2003. A review of carbon dioxide selective membranes. US Department of Energy: 9–11.

Shin, D.W., Hyun, S.H., Cho, C.H. and Han, M.H. 2005. Synthesis and CO_2/N_2 gas permeation characteristics of ZSM-5 zeolite membranes. Microporous Mesoporous Mater. 85(3): 313–323.

Sublet, J., Pera-Titus, M., Guilhaume, N., Farrusseng, D., Schrive, L., Chanaud, P. et al. 2012. Technico-economical assessment of MFI-type zeolite membranes for CO_2 capture from postcombustion flue gases. AIChE Journal 58(10): 3183–3194.

Suslick, K.S. 1989. The chemical effects of ultrasound. Sci. Am. 260(2): 80–86.

Takeuchi, M., Kimura, T., Hidaka, M., Rakhmawaty, D. and Anpo, M. 2007. Photocatalytic oxidation of acetaldehyde with oxygen on TiO_2/ZSM-5 photocatalysts: effect of hydrophobicity of zeolites. J. Catal. 246(2): 235–240.

Tang, Y., Liu, X., Nai, S. and Zhang, B. 2014. Template-free synthesis of beta zeolite membranes on porous α-Al_2O_3 supports. Chem. Comm. 50(64): 8834–8837.

Tavolaro, A. and Drioli, E. 1999. Zeolite membranes. Adv. Mater. 11(12): 975–996.

Tian, Y., Fan, L., Wang, Z., Qiu, S. and Zhu, G. 2009. Synthesis of a SAPO-34 membrane on macroporous supports for high permeance separation of a CO_2/CH_4 mixture. J. Mater. Chem. 19(41): 7698–7703.

Tiscornia, I., Kumakiri, I., Bredesen, R., Téllez, C. and Coronas, J. 2010. Microporous titanosilicate ETS-10 membrane for high pressure CO_2 separation. Sep. Purif. Technol. 73(1): 8–12.

Tomita, T., Nakayama, K. and Sakai, H. 2004. Gas separation characteristics of DDR type zeolite membrane. Microporous Mesoporous Mater. 68(1): 71–75.

Tompsett, G.A., Conner, W.C. and Yngvesson, K.S. 2006. Microwave synthesis of nanoporous materials. ChemPhysChem. 7(2): 296–319.

Tsapatsis, M., Okubo, T., Lovallo, M. and Davis, M.E. 1994. Synthesis and structure of ultrafine zeolite KL (LTL) crystallites and their use for thin film zeolite processing. MRS Online Proceedings Library Archive: 371.

Utchariyajit, K. and Wongkasemjit, S. 2010. Effect of synthesis parameters on mesoporous SAPO-5 with AFI-type formation via microwave radiation using alumatrane and silatrane precursors. Microporous Mesoporous Mater. 135(1): 116–123.

Valtchev, V.P. and Bozhilov, K.N. 2005. Evidences for zeolite nucleation at the solid– liquid interface of gel cavities. J. Am. Chem. Soc. 127(46): 16171–16177.

Van den Bergh, J., Zhu, W., Kapteijn, F., Moulijn, J.A., Yajima, K., Nakayama, K. et al. 2008. Separation of CO_2 and CH_4 by a DDR membrane. Res. Chem. Intermed. 34(5): 467–474.

Van den Bergh, J., Tihaya, A. and Kapteijn, F. 2010. High temperature permeation and separation characteristics of an all-silica DDR zeolite membrane. Microporous Mesoporous Mater. 132(1): 137–147.

Van den Broeke, L.J.P., Kapteijn, F. and Moulijn, J.A. 1999. Transport and separation properties of a silicalite-1 membrane—II. Variable separation factor. Chem. Eng. Sci. 54(2): 259–269.

Varela-Gandía, F.J., Berenguer-Murcia, Á., Lozano-Castelló, D. and Cazorla-Amorós, D. 2011. Zeolite A/ carbon membranes for H_2 purification from a simulated gas reformer mixture. J. Membr. Sci. 378(1): 407–414.

Vilaseca, M., Mateo, E., Palacio, L., Prádanos, P., Hernández, A., Paniagua, A. et al. 2004. AFM characterization of the growth of MFI-type zeolite films on alumina substrates. Microporous Mesoporous Mater. 71(1): 33–37.

Wu, T., Wang, B., Lu, Z., Zhou, R. and Chen, X. 2014. Alumina-supported AlPO-18 membranes for CO_2/CH_4 separation. J. Membr. Sci. 471: 338–346.

Yaripour, F., Shariatinia, Z., Sahebdelfar, S. and Irandoukht, A. 2015. Conventional hydrothermal synthesis of nanostructured H-ZSM-5 catalysts using various templates for light olefins production from methanol. J. Nat. Gas Sci. Eng. 22: 260–269.

Yin, X., Zhu, G., Wang, Z., Yue, N. and Qiu, S. 2007. Zeolite P/NaX composite membrane for gas separation. Microporous Mesoporous Mater. 105(1): 156–162.

Yuan, W., Wang, D. and Li, L. 2011. MFI-type zeolite membrane on hollow fiber substrate for hydrogen separation. Chin. Sci. Bull. 56(23): 2416–2418.

Zhang, Y., Avila, A.M., Tokay, B., Funke, H.H., Falconer, J.L. and Noble, R.D. 2010. Blocking defects in SAPO-34 membranes with cyclodextrin. J. Membr. Sci. 358(1): 7–12.

9

Rheological Evaluation of the Fabrication Parameters of Cellulose Acetate Butyrate Membrane on CO_2/N_2 Separation Performance

R.J. Lee,[1] *Z.A. Jawad,*[1,*] *A.L. Ahmad,*[2] *H.B. Chua,*[1] *H.P. Ngang*[2] and *S.H.S. Zein*[3]

INTRODUCTION

There is a trend of rapid increase in world population, which is expected to hit 10 billion by 2050 (Lalia et al. 2013). In this regard, higher demand in energy will be required for the 21st century to meet the urgent needs. It is predicted that the energy demand will increase by 57% in 2030 (Conti et al. 2016). As a major contributor to the world energy supply, fossil fuel solely contributes around 40% of the total carbon dioxide (CO_2) emission into the environment, which is mainly attributed to the massive coal combustion activities (Carapellucci and Milazzo 2003). Global warming has become a genuine problem due to the excessive discharge of pollutants emitted from the combustion activities in the primary industries (Yang et al. 2008).

[1] School of Engineering and Science, Department of Chemical Engineering, Curtin University Malaysia, CDT 250, Miri 98009, Sarawak, Malaysia.
Email: jacklee93@postgrad.curtin.edu.my; chua.han.bing@curtin.edu.my

[2] School of Chemical Engineering, Engineering Campus, Universities Sains Malaysia, 14300 Nibong Tebal, Penang, Malaysia.
Email: chlatif@usm.my; hueyping0404@gmail.com

[3] Chemical Engineering, School of Engineering and Computer Science, University of Hull, Hull, HU6 7RX, United Kingdom.
Email: s.h.zein@hull.ac.uk

* Corresponding author: zeinab.aj@curtin.edu.my

In the past few decades through their efforts, researchers have contributed in combating this global issue to limit and minimize the impact of greenhouse gases (GHGs). They have outlined three feasible options. The first comprises of saving energy used intensively with methods that are more efficient. The second option is to minimize the usage of carbon-based material source or replace it with renewable energy, and the third is to improve the effectiveness of CO_2 sequestration with more advanced technology development (Yang et al. 2008). For the past few years, membrane separation technology has been utilized intensively for both water treatment and gas separation purpose (Yang et al. 2008, Kappel et al. 2014, Barnes et al. 2014, Zhu et al. 2014). The membrane's chemical and physical properties, and interaction between permeance and membrane are relatively crucial factors in determining the diffusion characteristics of the gas separation field (Shekhawat 2003). This is because the separation selectivity and permeance are two critical parameters that indicate membrane separation performance. In an ideal situation, high selectivity and permeance are preferred as both induce less capital costs and operating expenses for the industries (Paradise and Goswami 2007, Low et al. 2013). Hence, the selection of material plays an influential role, in determining the specific gas separation performance (Lalia et al. 2013, Zha et al. 2015, Feng et al. 2015).

The Cellulose Acetate Butyrate (CAB) possesses few interesting characteristics that include, film-forming properties, acetyl and butyryl functional groups, which can effectively improve and further expand the capacity of cellulose chain giving high sorption characteristic, as well as high impact, weather and chemical resistant (Feng et al. 2015, Basu et al. 2010, Kunthadong et al. 2015). The CAB was first investigated and studied by Sourirajan back in 1958, then followed by Manjikian and others in reverse osmosis (RO) separation (Wang et al. 1994). They reported that the CAB membrane owned high solute separation with tolerable membrane flux result, and also provided ease of fabrication as some pre-treatment was negligible (Ohya et al. 1980, Wang et al. 1994). However, limited studies have been conducted on the effects of the acetyl group content on CAB membranes in the CO_2/N_2 gas separation field. Further, no reports or systematic studies have been performed on the effects of membrane production procedure and fabrication parameters. This includes membrane-casting thickness, solvent exchange time for both isopropyl alcohol and n-hexane with different CAB molecular weights as well as the polymer matrix material structure and performance of CAB membranes. Therefore, the primary objective of this study is to investigate the effects of membrane production procedure and fabrication parameters. Discussions on how the mentioned parameters can affect the membrane in terms of morphology and gas separation performance are presented in this chapter. The separation performance of the synthesized CAB membrane was selected to evaluate the specified parameters towards CO_2/N_2.

Methodology

Materials

The cellulose acetate butyrate (CAB, $M_n \sim 12000, 65000, 70000$) in powder form was purchased from Sigma-Aldrich (Malaysia) for membrane preparation. Solutions required for membrane preparation, i.e., chloroform, isopropyl alcohol, and n-hexane were purchased from Merck (Malaysia). Distilled water was used for the phase-inversion steps, specifically for immersion precipitation for membrane formation.

CAB Polymer dope preparation

The CAB membrane was prepared using the wet-phase inversion method, followed by solvent exchange to dry the membrane. A dope solution consisting of 4 wt% CAB (Mn = 70000) powders and 96 wt% chloroform was prepared following the condition of each parameter. The solution was stirred for 24 hours, and then sonicated for 20 minutes to eliminate the gas bubbles in the solution (Ahmad et al. 2014, Feng et al. 2015). The solution was then poured into space within the casting bars with glass plate underneath. An automatic film applicator (Elcometer 4340, E.U.) was then used for the casting of the membrane. Referring to our previous work, 5 minutes of solvent evaporation time was allowed following each parameter's condition before immersing the membrane in distilled water (27°C) for a duration of 24 hours (Minhas 1992, Lee et al. 2017). The solvent exchange was performed on the as-spun membrane first with 60 minutes immersion period in isopropyl alcohol and then another 60 minutes immersion period in n-hexane. The resultant membrane was then dried at ambient temperature to eliminate the remaining volatile liquid in between two glass plates filled with filter paper for 24 hours before use (Minhas 1992, Jawad et al. 2015a)

Effect of casting thickness

The membranes were prepared at different casting thicknesses following the fabrication method as described next where the study range for this parameter was from 200 μm (CAB-200), 250 μm (CAB-250) to 300 μm (CAB-300).

Effect of exchange time with isopropyl alcohol

Meanwhile, for the effect of solvent exchange time, the membranes were prepared following the fabrication method as described earlier. The solvent exchange duration was studied for 15 minutes (CAB-15Iso), 30 minutes (CAB-30Iso), and 60 minutes (CAB-60Iso) for isopropyl alcohol, followed by 60 minutes of n-hexane.

Effect of exchange time with n-hexane

In addition, the solution of the solvent exchange time with n-hexane was prepared following the fabrication method as described earlier. The resultant membranes were first solvent exchanged with isopropyl alcohol for 30 minutes followed by solvent exchange times ranging from 15 minutes (CAB-15H), 30 minutes (CAB-30H) to 60 minutes (CAB-60H) for n-hexane.

Effect of CAB at different molecular weight (M_n)

The membranes were prepared with different CAB molecular weights (M_n) of 12000 (CAB-12000), 65000 (CAB-65000), and 70000 (CAB-70000) for the preparation of the dope solution. After that, following the fabrication method as mentioned earlier, the solvent exchange time for isopropyl alcohol and n-hexane were set for 30 minutes each.

Membrane permeability test

The procedure for gas permeation measurement was discussed in our previous published work (Lee et al. 2017).

Membrane characterization

Scanning Electron Microscopy (SEM)

The CAB membrane structures including surface and cross-sectional, were observed via SEM (Hitachi TM3000, Tokyo, Japan). Each membrane sample was cut into small pieces, and then kept on a plastic petri dish in the cryogenic freezer at a temperature of up to –80°C for 24 hours to give a consistent and clean-cut by freezing. The samples were coated with a platinum layer to prevent high-energy beam damage before the characterization works. Further, each sample's average membrane thickness was calculated based on the frequency count as measured by the Image-J software. Approximately, 100 measurements were taken to confirm the average membrane thickness.

X-ray Photoelectron Spectroscopy (XPS)

The CAB membranes fabricated at different molecular weights were characterized with the High Resolution Multi-Technique X-Ray Spectrometer (Axis Ultra DLD XPS, Kratos, Shimadzu Corporation, Japan). The analysis was carried out using a PHI 1600 spectrometer with hybrid lens mode, 150 W (Anode: Mono), 1000 meV step, and 5 sweeps for each membrane at room temperature.

Results and Discussion

Effect of casting thickness

The effect of casting thickness on the structure and performance of the CAB membrane was investigated, as depicted in Fig. 9.1. As observed from Fig. 9.1a, the structure of CAB-200 (200 μm) was porous. As the casting thickness of the membranes increased, a smooth surface was observed for CAB-250 (250 μm), as demonstrated in Fig. 9.1c. Alternatively, a rough surface was formed for CAB-300 (250 μm), as seen in Fig. 9.1e. The change in the structure was due to the different rates of demixing that occurred as the phase precipitation proceeded when high casting thickness was applied, causing the deposition speed of the membrane to reduce during the membrane formation phase. The slow deposition rate avoids rapid exchange of non-solvent and solvent within the membrane. As a result, the surface structure of the CAB membrane was built-up based on the sufficient phase precipitation period given (Ahmad et al. 2013, Thomas et al. 2014).

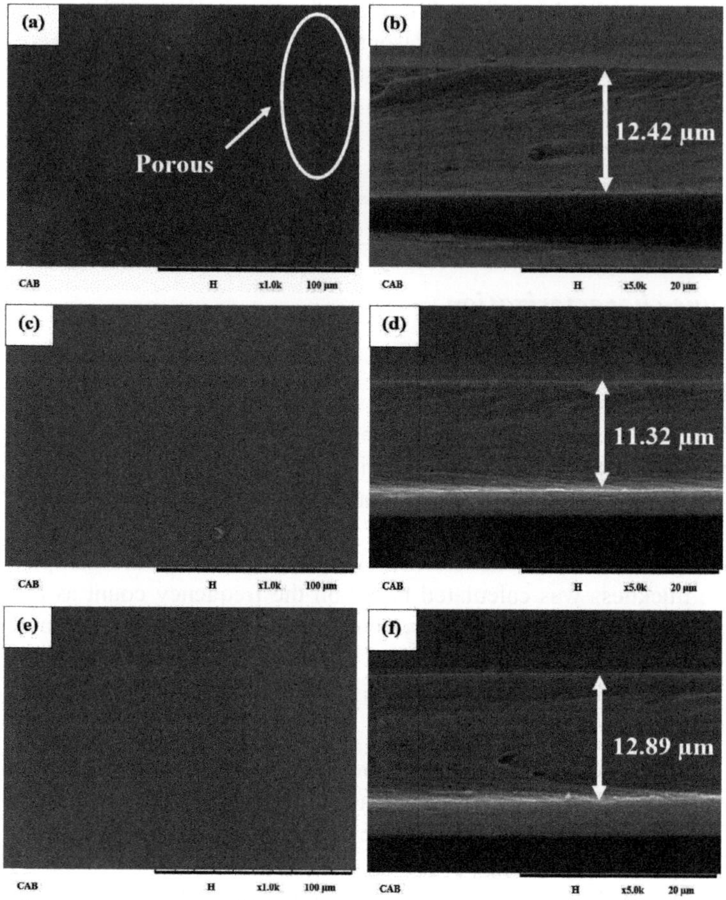

Figure 9.1. Top and cross-sectional SEM of CAB membrane at casting thickness (a-b) 200 μm (CAB-200), (c-d) 250 μm (CAB-250), and (e-f) 300 μm (CAB-300), with 4 wt% CAB polymer concentration and 5 minutes solvent evaporation time.

The cross-sectional micrographs of the fabricated CAB membrane at casting thickness of 200 μm (CAB-200), 250 μm (CAB-250), and 300 μm (CAB-300) were revealed in Figs. 9.1b, d, and f, respectively. From the micrographs, dense structures were depicted from all the cross-sectionals of the CAB membranes. The dense structure formation was due to the densification of the membrane during the immersion period, whereby the remaining solvent imbedded in the polymer matrix was replaced by distilled water. As the volatility of the solvent was generally higher than distilled water the membrane thickness changed from 12.42 ± 0.05 μm to 11.32 ± 0.06 μm and 12.89 ± 0.10 μm for CAB-200, CAB-250 and CAB-300, respectively. The reduction of membrane thickness from 12.42 ± 0.05 μm (CAB-200) to 11.32 ± 0.06 μm (CAB-250) was due to thicker casting thickness applied during membrane fabrication, which allows more solvent embedded in the polymer matrix to be

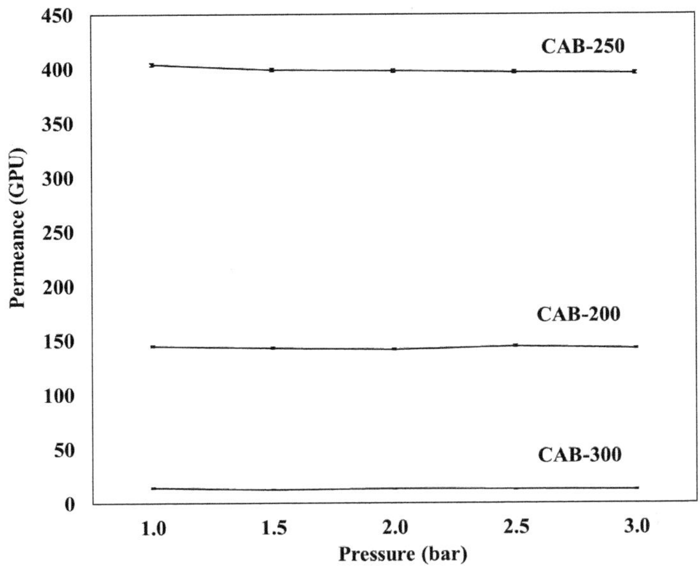

Figure 9.2. CO_2 permeance for membrane fabricated at 200 µm (CAB-200), 250 µm (CAB-250), and 300 µm (CAB-300), with 4 wt% CAB polymer concentration and 5 minutes solvent evaporation time.

replaced by non-solvent (H_20) during the immersion period, resulting in a denser and thinner membrane thickness for CAB-250 (Ahmad et al. 2013). In contrast, a thicker membrane was obtained when increasing the membrane thickness further to 300 µm for CAB-300 (12.89 ± 0.10 µm). This is correlated to the increase resistance of inward diffusion of non-solvent, due to higher casting thickness applied, causing a delay transition demixing in the film membrane (Tiraferri et al. 2011).

The CO_2 permeance of CAB-200, CAB-250, and CAB-300 are illustrated in Fig. 9.2. Notably, CAB-250 demonstrated a higher permeance result of 398.46 ± 1.43 GPU, as compared to CAB-200 (143.03 ± 0.62 GPU) and CAB-300 (12.93 ± 0.34 GPU). This was because of the reduction in its membrane thickness (11.32 µm, Fig. 9.1d) and its selective smooth surface structure, which allowed the solution diffusion mechanism to occur efficiently. Therefore, the CO_2 permeance of CAB-250 increased (Jawad et al. 2015a). Meanwhile, the CO_2 permeance of CAB-300 reduced to 12.93 ± 0.34 GPU, indicating that a higher casting thickness beyond 250 µm can exert extra resistance towards gas diffusion within the membrane, which in turn affects the efficiency of gas permeation due to the thick dense membrane synthesized (Fig. 9.1f).

On the other hand, the N_2 permeance for CAB-200, CAB-250 and CAB-300 were 112.83 ± 0.85, 121.55 ± 1.30, and 11.26 ± 0.31 GPU, respectively, as illustrated in Fig. 9.3. The CAB-250 exhibited higher N_2 permeance results. This was due to the initial casting thickness applied, resulting in a smooth membrane structure, which created less resistance towards the permeance of N_2 gas within the membrane (Freeman 1999). The low N_2 permeance result yield for CAB-300 (11.26 ± 0.31 GPU)

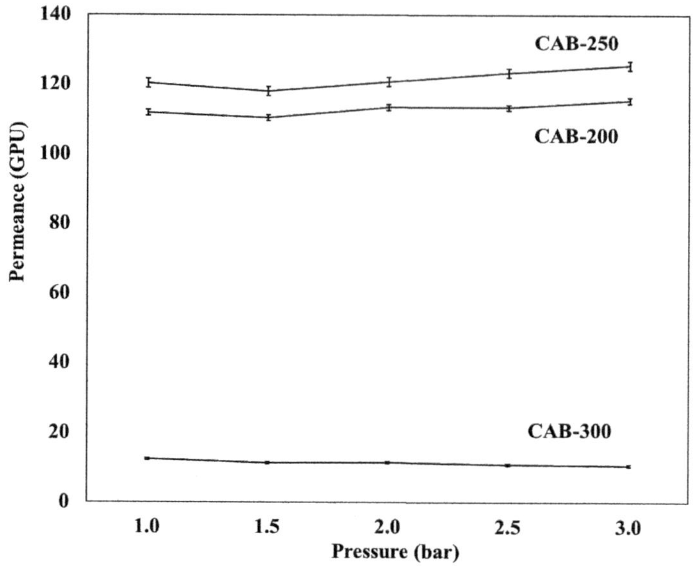

Figure 9.3. N_2 permeance for membrane fabricated at 200 μm (CAB-200), 250 μm (CAB-250), and 300 μm (CAB-300), with 4 wt% CAB polymer concentration and 5 minutes solvent evaporation time.

was due mainly to the thick dense membrane structure (12.89 ± 0.10 μm), which ultimately governed the solution diffusion rate of the membrane, as a thicker membrane usually induces more resistance to gas diffusion (Koros et al. 1988a).

The ideal selectivity of CO_2/N_2 separation performance for CAB-200, CAB-250, and CAB-300 are shown in Fig. 9.4. As observed from the results when increasing the casting thickness from 200 μm to 250 μm the selectivity increased from 1.27 ± 0.01 GPU (CAB-200) to 3.28 ± 0.04 GPU (CAB-250). The acceptable result obtained for CAB-250 was due to the membrane structure formation, which eventually increased the CO_2 permeance against the N_2 permeance attained. However, the selectivity reduced to 1.15 ± 0.01 GPU when the higher casting thickness (300 *μm)* was implemented for CAB-300. Even though the thickness of the membrane was essential for effective gas separation, however, excessive membrane thickness restricted the gas diffusion within the membrane.

Effect of solvent exchange with isopropyl alcohol

The solvent exchange was performed after the precipitation immersion process of the CAB membrane with the purpose of drying or removing any remaining volatile liquid in the membrane. As displayed in Figs. 9.5a and b, the CAB-15Iso (15 minutes) exhibited a porous surface and irregular dense cross-sectional structure with a membrane thickness of 13.87 ± 0.23 μm. This porous structure surface was caused by the rapid solvent exchange between the water molecules available within the CAB structure and the first solvent (isopropyl alcohol) (Lui et al. 1988). During the first step of the solvent exchange process, an enormous amount of water molecules embedded in the membrane were generally replaced by isopropyl alcohol.

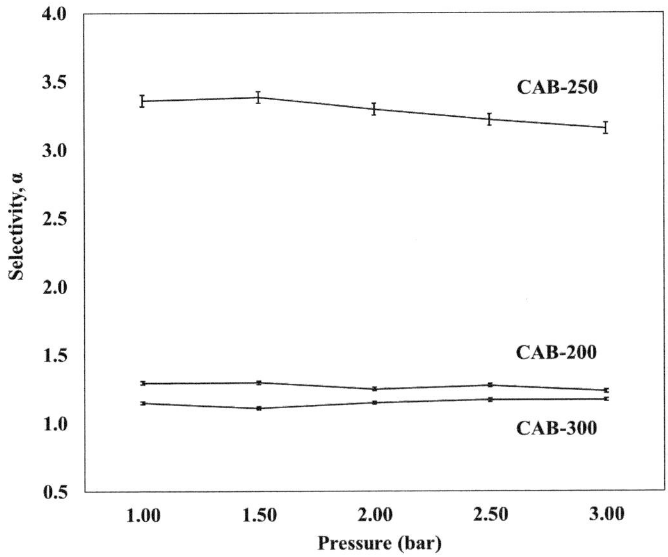

Figure 9.4. Ideal selectivity of CO_2/N_2 for membranes fabricated at different casting thickness 200 μm (CAB-200), 250 μm (CAB-250), and 300 μm (CAB-300), with 4 wt% CAB polymer concentration and 5 minutes solvent evaporation time

As a result, due to the short 15 minutes solvent exchange immersion period allocated, vigorous pore formation appeared throughout the film membrane of CAB-15Iso (Lui et al. 1988). Thus, CAB-15Iso demonstrated thick membrane thickness because of the short solvent exchange time applied, resulting in more water molecules being retained inside the membrane.

Meanwhile, when the isopropyl alcohol solvent exchange time was increased to 30 minutes (CAB-30Iso) and then subsequently to 60 minutes (CAB-60Iso), both revealed a smooth surface (Figs. 9.5c and e) with thin dense membrane thickness of 9.45 ± 0.06 μm and 9.30 ± 0.05 μm, respectively, as demonstrated in Figs. 9.5d and f. The formation of a smooth surface and thin membrane was because of the longer immersion period allocated. Therefore, this provided more relaxation time for the non-solvent (H_2O) imbedded in the film membrane to exchange with the isopropyl alcohol (Radjabian et al. 2014). This also allowed the formation of a thin dense membrane with homogeneous smooth surface structure as revealed from CAB-30Iso and CAB-60Iso.

As shown in Fig. 9.6, the CO_2 permeance rates increased from 65.53 ± 0.34 GPU (CAB-15Iso) to 262.29 ± 0.16 GPU (CAB-30Iso) and increased further to 398.82 ± 0.94 GPU (CAB-60Iso) by changing the solvent exchange duration of isopropyl from 15 minutes to 30 minutes and subsequently to 60 minutes, respectively. This resulted in extensive water content reduction within the membrane structure due to longer immersion period allocated. The steady exchange rate of water with isopropyl alcohol within the CAB polymer matrix caused less CO_2 molecules to interact with the water, therefore allowing more CO_2 gas to permeate through the membrane (Jawad et al. 2015b). In the meantime, the high CO_2 permeance rate for CAB-60Iso (60 minutes) contributed to the thin dense membrane structure,

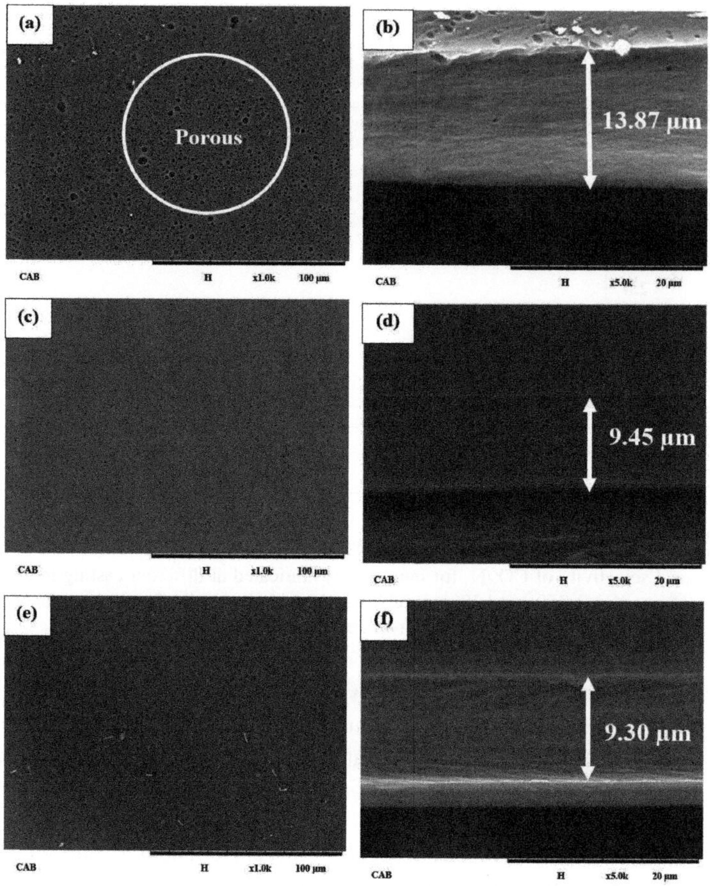

Figure 9.5. Surface and cross-sectional SEM of CAB membrane dried with isopropyl alcohol first for a solvent exchange duration of (a-b) 15 minutes (CAB-15Iso), (c-d) 30 minutes (CAB-30Iso), and (e-f) 60 minutes (CAB-60Iso); then subsequently solvent exchanged with 60 minutes of n-hexane as the final solvent, at casting thickness of 250 μm and 5 minutes solvent evaporation time

which allowed the CO_2 feed gas to pass through the membrane with least resistance pathway as compared to the thick dense membrane (Tiraferri et al. 2011). Thus, the CAB-60Iso (60 minutes) yielded the highest CO_2 permeance rate amongst the other membranes (CAB-15Iso and CAB-30Iso).

The N_2 permeance rates for CAB-15Iso, CAB-30Iso, and CAB-60Iso are depicted in Fig. 9.7. The results obtained for N_2 permeance were 64.59 ± 0.41, 70.49 ± 0.33, and 121.76 ± 0.83 GPU for CAB-15Iso, CAB-30Iso, and CAB-60Iso, respectively. The possible explanation for this trend was due to the reduction in the membrane thickness from 13.87 μm to 9.3 μm (Fig. 9.5). In addition, as isopropyl alcohol was mainly made up from non-polar molecules, the remaining molecules within the CAB structure can easily attract light gas molecules (Katayama and Nitta 1976). Thus, with longer solvent exchange duration, more isopropyl alcohol was retained within the polymer matrix, hence, attracting more N_2 gas molecules and

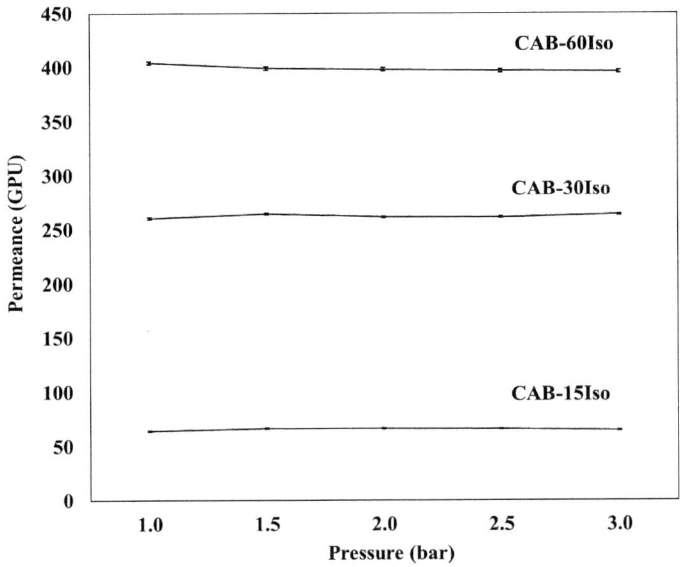

Figure 9.6. CO_2 permeance for membrane dried with 15 minutes (CAB-15Iso), 30 minutes (CAB-30Iso), and 60 minutes (CAB-60Iso) of isopropyl alcohol; then subsequently solvent exchanged with 60 minutes of n-hexane as the final solvent, at casting thickness of 250 µm and 5 minutes solvent evaporation time.

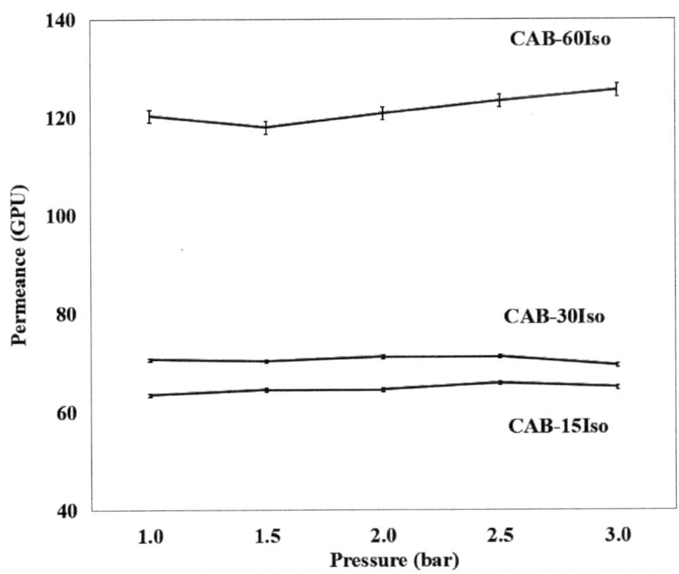

Figure 9.7. N_2 permeance for membrane dried with 15 minutes (CAB-15Iso), 30 minutes (CAB-30Iso), and 60 minutes (CAB-60Iso) of isopropyl alcohol; then subsequently solvent exchanged with 60 minutes of n-hexane as the final solvent, at casting thickness of 250 µm and 5 minutes solvent evaporation time.

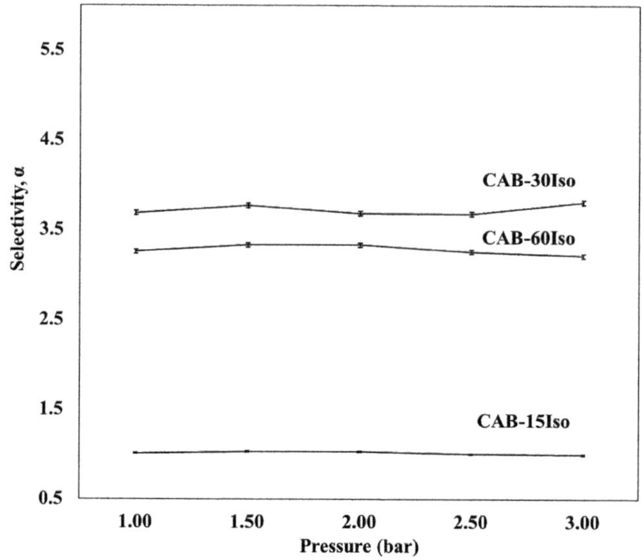

Figure 9.8. Ideal selectivity of CO_2/N_2 for membranes synthesised with solvent exchange duration of 15 minutes (CAB-15Iso), 30 minutes (CAB-30Iso), and 60 minutes (CAB-60Iso); then subsequently exchanged with 60 minutes of n-hexane as the final solvent, at casting thickness of 250 μm and 5 minutes solvent evaporation time.

resulting in the high N_2 permeance rate for CAB-60Iso (60 minutes). Eventually as the solvent exchange duration decreased, the N_2 permeance rate for CAB-15Iso and CAB-30Iso reduced as well.

As discussed previously, CAB-60Iso (60 minutes) showed a thin dense membrane formation with high CO_2 and N_2 permeance rates. However, based on Fig. 9.8, the CAB-30Iso (30 minutes) yielded the best selectivity performance. This was due to the smooth homogeneous surface and superior cross-sectional morphology, which selectively allowed a predetermined amount of CO_2 and N_2 to pass through the dense membrane. On the contrary, the CAB-15Iso (15 minutes) demonstrated low selectivity (Fig. 9.8). This was due to the presence of a thick irregular surface morphology (Figs. 9.5a and b), which imposed an undesirable effect on membrane permeance performance due to extra resistance pathway generated (Rahimpour et al. 2008, Yang and Wang 2006). Therefore, CAB-30Iso (30 minutes) was preferred as compared to CAB-15Iso (15 minutes) and CAB-60Iso (60 minutes) because of its excellent morphology and good selectivity performance.

Effect of exchange time with n-hexane

As discussed earlier, the best solvent exchange time for isopropyl alcohol was 30 minutes (CAB-30Iso). Subsequently, the CAB membrane was subjected to further optimization with the drying time of n-hexane. In this study, the CAB membranes were dried with solvent exchange times of 15 minutes (CAB-15H), 30 minutes (CAB-30H), and 60 minutes (CAB-60H) using n-hexane. As revealed from the SEM image in Fig. 9.9, the surface of CAB-15H (15 minutes) exhibited a porous structure, while CAB-30H (30 minutes) and CAB-60H (60 minutes) showed smooth surfaces. The main reason for the porous structure showed by CAB-15H was due to the rapid evaporation of the volatile solvent from the membrane structure itself and short duration of immersion period implemented (Chung and Kafchinski 1997). Gradually by increasing the solvent exchange immersion period, the membrane had sufficient time for the solvent exchange to occur between isopropyl alcohol and n-hexane at a consistent and steady rate. Hence, suppressing a vigorous solvent exchange process within the polymer, resulted in a smooth homogeneous surface as observed for CAB-30H (30 minutes) and CAB-60H (60 minutes) (Choi et al. 2006).

As presented in Fig. 9.9, the membrane thickness for CAB-15H (15 minutes), CAB-30H (30 minutes), and CAB-60H (60 minutes) were 11.79 ± 0.18, 9.50 ± 0.10, and 9.45 ± 0.06 μm, respectively. As seen from these results, the increased exchange time of n-hexane caused the CAB membrane to become more compact due to membrane densification as time passed (Sabde et al. 1997). In addition, the main reason for the reduction in the membrane thickness was due to the isopropyl alcohol imbedded within the membrane slowly being replaced by n-hexane with time. The replacement of isopropyl alcohol with n-hexane occurred when the molecular affinity of n-hexane was greater than isopropyl alcohol (Hansen 2007). Referring to the Hansen solubility chart, the solubility for isopropyl alcohol, n-hexane, and water are 23.6, 14.9, and 47.9 $MPa^{1/2}$, respectively (Egan and Dufresne 2008, Hansen 2007). Therefore, the molecular affinity is in the order of CAB-water > CAB-isopropyl alcohol > CAB-n-hexane. The order of the molecular affinity represents the attraction force between the polymer and the solvent and non-solvent used (Kim and Oh 2001).

Meanwhile, Fig. 9.11 illustrated a drastic increase of N_2 permeance from 10.03 ± 0.02 GPU to 37.28 ± 0.54 GPU when the solvent exchange time of n-hexane was increased from 15 minutes (CAB-15H) to 30 minutes (CAB-30H). The reason for this increment was mainly due to the thin dense membrane structure of CAB-30H (9.50 ± 0.10 μm), which allowed the feed of N_2 gas to pass through a least resistance pathway. However, the high N_2 permeance for CAB-60H (70.49 ± 0.33 GPU) was due to stress of surface tension caused by high capillary forces because of the evaporation of residual n-hexane within the membrane, which led to the collapse in the structure (Matsuyama et al. 2002).

As seen in Fig. 9.12, the CAB-30H membrane showed the highest gas selectivity, which was achieved at 6.12 ± 0.09. This result further proved that to have a high gas separation performance a smooth surface with regular thin dense membrane morphology was preferable (Figs. 9.9c and d) (Huang and Feng 1995, Jansen et al. 2005, Matsuyama et al. 2002, Lui et al. 1988). On the other hand, CAB-15H showed a lower separation performance of 2.15 ± 0.17. This was due to the collapse in the membrane structure caused by the short solvent immersion time, thereby generating an uneven porous surface and thick dense membrane structure, as presented in Figs. 9.9a and b. However, CAB-60H exhibited a smooth surface and thinner dense membrane morphology (9.45 ± 0.06 μm), as depicted in Figs. 9.9e and f. In addition, the low selectivity performance for CAB-60H (3.72 ± 0.03) was as a result of the

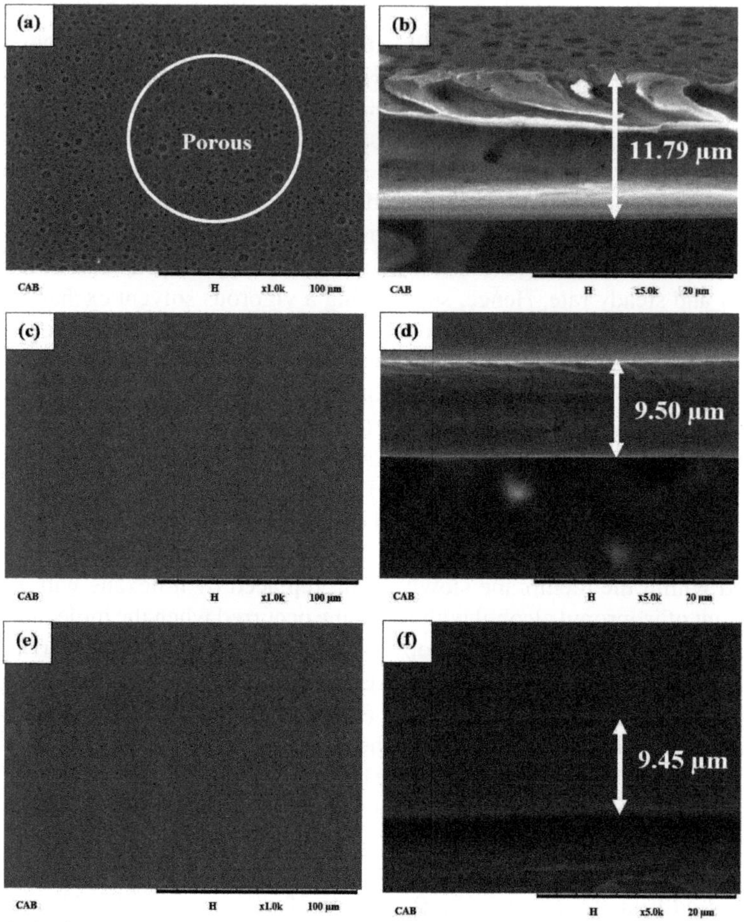

Figure 9.9. Surface and cross-sectional SEM of CAB membrane dried with 30 minutes of isopropyl alcohol first then followed by; (a-b) 15 minutes (CAB-15H), (c-d) 30 minutes (CAB-30H), and (e-f) 60 minutes (CAB-60H) of solvent exchange time using n-hexane, at casting thickness of 250 µm and 5 minutes solvent evaporation time.

excessive exchange time with n-hexane, which deformed the functionality of the membrane and therefore, generated moderate selectivity performance (Budd et al. 2005).

The effect of CAB polymer at different molecular weight (M_n)

According to Coltelli et al. (2008), the acetyl group has been deduced to have a prominent effect on the membrane gas separation performance, as excessive acetyl composition in the membrane could promote plasticisation within the membrane (Coltelli et al. 2008, Ismail and Lorna 2002). Thus, different CAB molecular weights with different acetyl, butyryl, and hydroxyl groups were investigated, as demonstrated in Fig. 9.13.

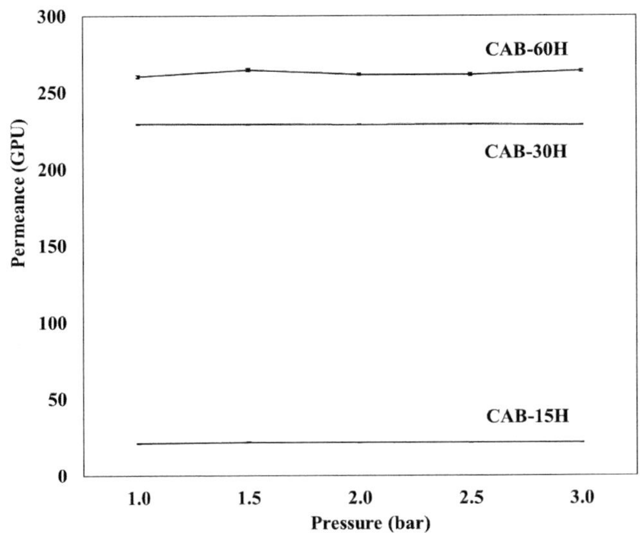

Figure 9.10. CO_2 permeance for membrane dried with 15 minutes (CAB-15H), 30 minutes (CAB-30H), and 60 minutes (CAB-60H) of n-hexane, at casting thickness of 250 μm and 5 minutes solvent evaporation time.

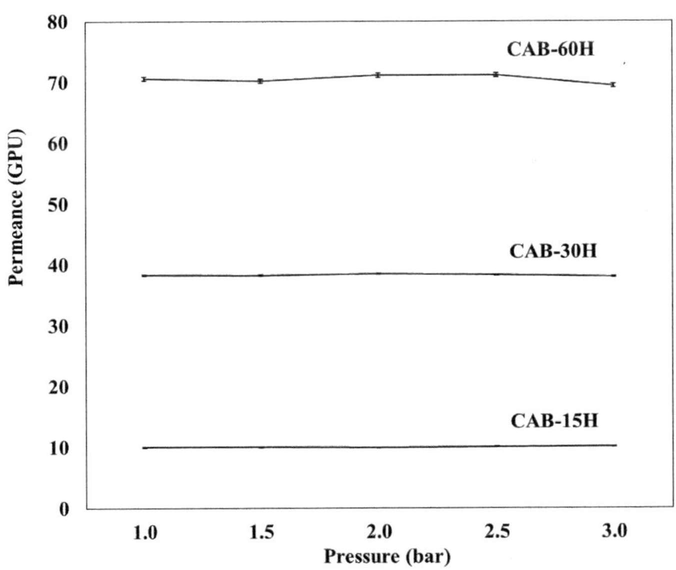

Figure 9.11. N_2 permeance for membrane dried with 15 minutes (CAB-15H), 30 minutes (CAB-30H), and 60 minutes (CAB-60H) of n-hexane, at casting thickness of 250 μm and 5 minutes solvent evaporation time.

As depicted in Figs. 9.13a and c, a porous structure was observed for both CAB-12000 (M_n = 12000) and CAB-65000 (M_n = 65000), while CAB-70000 (M_n = 70000) showed a smooth surface (Fig. 9.13e). The reason the membrane surface changed from porous to smooth was due to the high molecular weights of CAB, which caused

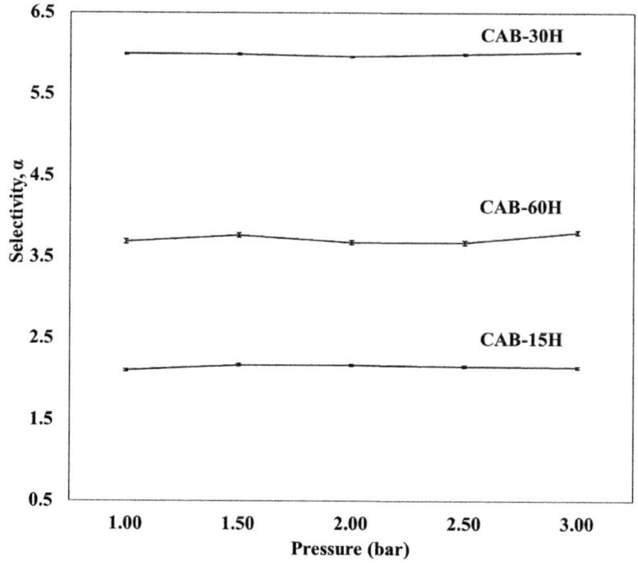

Figure 9.12. Ideal selectivity of CO_2/N_2 for CAB membrane dried with 30 minutes of isopropyl alcohol first then followed by; 15 minutes (CAB-15H), 30 minutes (CAB-30H), and 60 minutes (CAB-60H) of solvent exchange with n-hexane, at casting thickness of 250 μm and 5 minutes solvent evaporation time.

the increase in the number of entanglements between the macromolecular chains in the solution (Jansen et al. 2006). Therefore, the high molecular weights of CAB favoured the gelation of the polymer rich phase after the phase-inversion occurred and hence, suppressed the formation of the porous structure during the early stages (Jansen et al. 2005).

Based on Figs. 9.13b, d, and f, the thickness of CAB-12000, CAB-65000, and CAB-70000 were 10.96 ± 0.10, 16.05 ± 0.17, and 9.50 ± 0.10 μm, respectively. The increment in the CAB molecular weights further influenced the membrane thickness through the rheological properties of the casting solution (Jansen et al. 2005). This was due to the high molecular weights of the CAB polymer being utilized for membrane fabrication, which gave the rapid gelation (Jansen et al. 2005). After the rapid gelation, the porous structure was greatly suppressed and further evaporation of solvent and non-solvent from the polymer matrix resulted in gradual shrinkage of the structure (Jansen et al. 2005). Therefore, the thickness of CAB-70000 (9.50 ± 0.10 μm) was thinner than CAB-12000 (10.96 ± 0.10 μm) and CAB-65000 (16.05 ± 0.17 μm).

The performance of CO_2 permeance achieved for the different molecular weights (M_n) of CAB-12000, CAB-65000, and CAB-70000 were 101.42 ± 0.97, 74.37 ± 1.25, and 227.95 ± 0.39 GPU, respectively, as shown in Fig. 9.14. The decrease in the CO_2 permeance rates observed from CAB-12000 (28–31 wt%) to CAB-65000 (16–19 wt%) was due to the thick dense membrane morphology as presented in Fig. 9.13d (16.05 ± 0.17 μm), which can cause hindrance to the CO_2 permeance (Jawad et al. 2015a). Meanwhile, CAB-12000, which exhibited greater membrane thickness of 10.96 ± 0.10

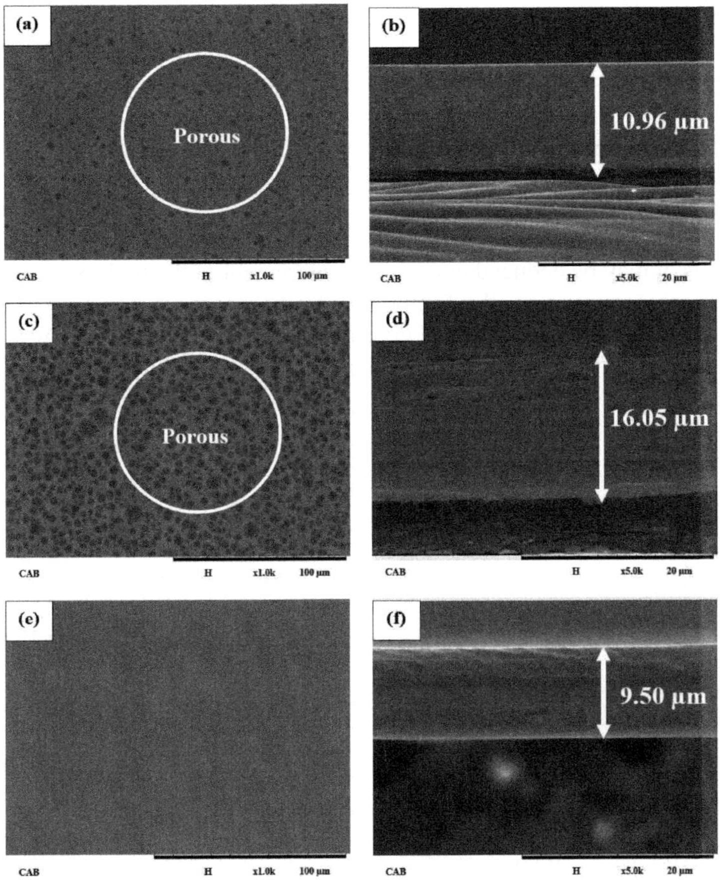

Figure 9.13. Surface and cross-sectional SEM of CAB membranes prepared with polymer concentration of 4 wt% and molecular weights (M_n) of (a-b) 12000 (CAB-12000), (c-d) 65000 (CAB-65000), and (e-f) 70000 (CAB-70000), at casting thickness of 250 µm and 5 minutes solvent evaporation time.

µm, contradicted the results with higher CO_2 permeance, as illustrated in Fig. 9.14. The possible explanation for the increase in CO_2 permeance was caused by the acetyl groups rigidity and steric effects (Wan et al. 2003). Therefore, this allowed the higher intrinsic solubility of CO_2 due to the greater number of acetyl–acetyl interactions that existed (Koros et al. 1988b, Scholes et al. 2012). In addition, increasing the CAB molecular weight from 65000 to 70000 had increased the permeance rate drastically from 74.37 ± 1.25 GPU to 227.95 ± 0.39 GPU. Even though, CAB-70000 (12–15 wt%) has the lowest acetyl-acetyl interactions due to low acetyl group composition compared to other CAB polymers. The significant increase in the CO_2 permeance was due to the thin dense membrane exhibited for CAB-70000, as thin dense membrane usually impose less flux resistance for the membrane (Pandey and Chauhan 2001). Therefore, the permeance of CO_2 was highest among all as the membrane thickness was the thinnest.

As portrayed in Fig. 9.15, the N_2 permeance rate achieved for CAB-12000, CAB-65000, and CAB-70000 were 95.26 ± 1.06, 48.94 ± 0.89, and 37.28 ± 0.54 GPU, respectively. The reduction in N_2 permeance was due to the high presence of the hydroxyl group (1.2–2.2 wt%) content within the CAB-70000 polymer. The reaction between the hydroxyl and carbonyl groups of the CAB polymer caused the formation of hydrogen bonds, which could delay the de-mixing between the coagulant and the non-solvent. This resulted in the smooth homogeneous formation of the membrane surface, which could influence the N_2 permeance rate (Childress and Elimelech 1996). Thus, it may be deduced that with the increment of the hydroxyl group within the membrane composition, the formation of a homogeneous surface morphology was favoured. Further, the hydroxyl group can increase the preferential restrictions on membrane pore formation, whereby the permeance and diffusion coefficient can

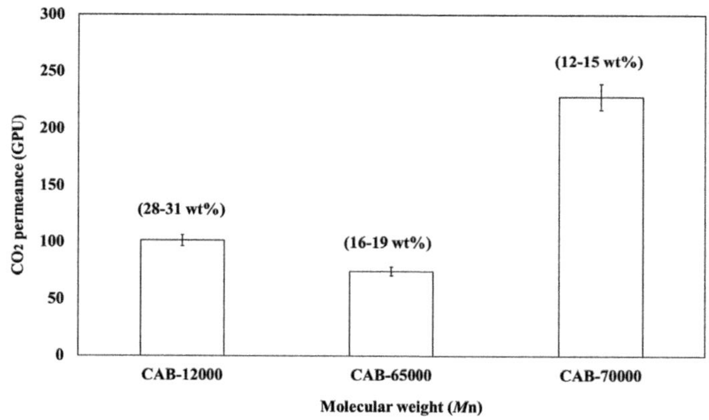

Figure 9.14. CO_2 permeance results for CAB membranes fabricated at different molecular weights comprising CAB-12000, CAB-65000, and CAB-70000 acetyl content of 28–31 wt%, 16–19 wt%, and 12–15wt%, respectively.

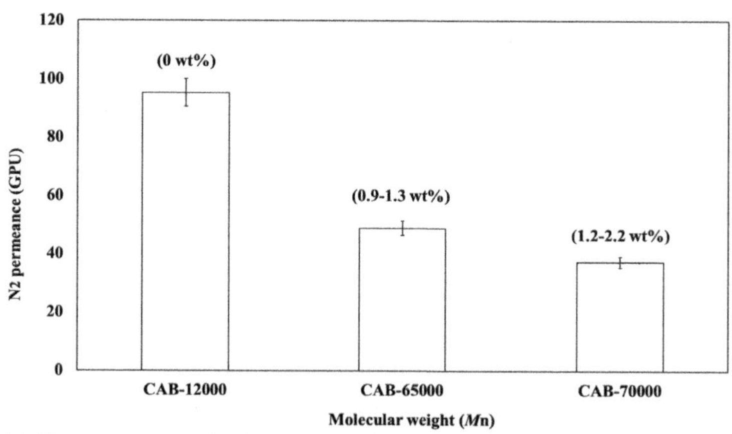

Figure 9.15. N_2 permeance results for CAB membranes synthesised at different molecular weights comprising CAB-12000, CAB-65000, and CAB-70000 hydroxyl content of 0 wt%, 0.9–1.3 wt%, and 1.2–2.2 wt%, respectively.

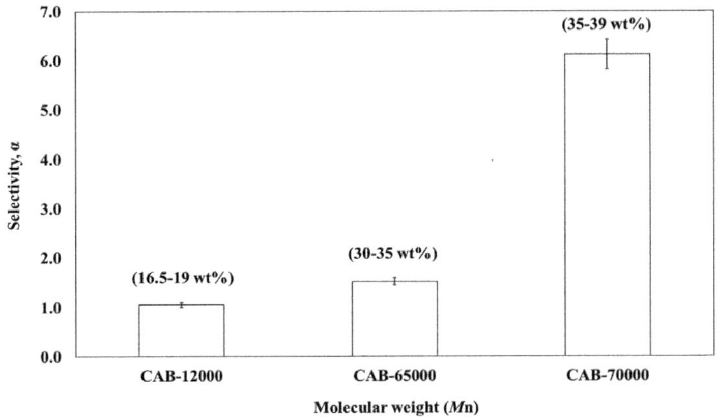

Figure 9.16. CO_2/N_2 selectivity results for CAB membranes at different molecular weights comprising CAB-12000, CAB-65000, and CAB-70000 butyryl content of 16.5–19 wt%, 30–35 wt%, and 35–39 wt%, respectively.

be suppressed, hence, enhancing the selectivity performance of the membrane (Yave et al. 2009).

Figure 9.16 reveals the selectivity results for the different CAB molecular weights of 12000 (CAB-12000), 65000 (CAB-65000), and 70000 (CAB-70000), respectively. From the selectivity results depicted in Fig. 9.16, CAB-70000 achieved the average highest selectivity of 6.12 ± 0.09, followed by CAB-65000 with a moderate selectivity of 1.52 ± 0.04, and CAB-12000 with the lowest selectivity of 1.06 ± 0.01. The high selectivity performance of CAB-70000 was due to the high presence of the butyryl group content (35–39 wt%), which promoted the CO_2 diffusion better due to the increase of the non-polar butyryl chain within the structure of the membrane, thus, making the membrane more hydrophobic in nature (Wan et al. 2004, Ong et al. 2012).

X-ray Photoelectron Spectroscopy (XPS) Analysis

The XPS characterization was adopted in this study to analyze the quantitative element composition of the CAB membrane fabricated. The quantitative element composition of the membrane surface can be determined from the spectrum obtained. Consequently, CAB-12000, CAB-65000, and CAB-70000 were analyzed through XPS analysis. The surface chemical quantitative compositions are depicted in Table 9.1 and Fig. 9.17, respectively.

Observing the results tabulated in Table 9.1, both the atomic and mass concentration of the oxygen (O) atom decreased with the increase in the CAB membrane molecular weights. The decreasing trend of atomic concentration from 34.02 > 30.88 > 27.30% and mass concentration from 40.72 > 37.31 > 33.35% of the O atom was due to the decrease of the acetyl group derived from each CAB polymer (Suttiwijitpukdee et al. 2011). As indicated clearly in Fig. 9.14, the acetyl group affected the permeance of CO_2 within the membrane. Hence, this further proved that increasing the acetyl group or O atom presence within the membrane subsequently,

Table 9.1. Element composition of the CAB membrane synthesized at different molecular weight.

	CAB-12000		CAB-65000		CAB-70000	
Peak	Atomic Conc %	Mass Conc %	Atomic Conc %	Mass Conc %	Atomic Conc %	Mass Conc %
O 1s	34.02	40.72	30.88	37.31	27.30	33.35
C 1s	65.98	59.28	69.12	62.69	72.70	66.65

decreased the permeance of CO_2. The increase in the O element was mainly funded by the breaking of the carbonyl (C = O) group and prompted the formation of a new carboxyl group (-COOH) (Liu et al. 2014). The increase in carboxyl group made the membrane more hydrophilic, resulting in decline of the CO_2 permeance flux (Xia and Ni 2015, Xu et al. 2014).

On the other hand, when observing the carbon (C) element present within CAB-12000, CAB-65000, and CAB-70000 in Fig. 9.17, the C atoms increased with increase in the polymer molecular weights. The atomic concentration increased from 65.98 > 69.12 > 72.70 and the mass concentration increased from 59.28 > 62.69 > 66.65 for CAB-12000, CAB-65000, and CAB-70000, respectively. The increase in the C element within the membrane was because of the increase in the butyryl group within the CAB polymer. As indicated in Fig. 9.16, the butyryl group played a crucial role in manipulating the selectivity performance of the membrane, because it can increase the CO_2 diffusion due to the increase of the non-polar butyryl chain within the structure of the membrane (Wan et al. 2004). As a result the membrane became more hydrophobic in nature, and hence, promoted better CO_2 permeance flux (Ong et al. 2012).

The CO_2/N_2 separation performance of this current study were summarized and compared with other research works, as shown in Table 9.2. In this study, the best membrane performance achieved for both CO_2 permeance and selectivity was 227.95 ± 0.39 GPU and 6.12 ± 0.09, respectively for CAB-70000. This was a result

Table 9.2. List of CO_2/N_2 permeation results achieved from current study with previous work.

Polymer	P (CO_2)	P (N_2)		Conditions	References
CAB	164.84 ± 0.73[a]	26.36 ± 0.05[a]	6.06 ± 0.23	1–3 × 10^5 Pa, casting thickness of 250μm	Present work
SPEEK	5.01	1.94	5.58	1–1.5 × 10^5 Pa, 25°C, casting thickness of 60–80 μm	(Xin et al. 2015)
BPPO	76.78[b]	N/A	30	0.7 × 10^5 Pa, casting thickness of 50–90 μm	(Cong et al. 2007)
PES	10.98[a]	0.80[a]	13.73	3–4 × 10^5 Pa, casting thickness of 150 μm	(Ismail et al. 2011)
6FDA-durene	30.3[b]	2.87[b]	10.56	35°C, 10 atm, casting thickness of 40 μm	(Liu et al. 2001)

SPEEK- sulphonated polyetheretherketone, BPPO- brominated polyphenylene oxide, PES-polyethersulphone, 6FDA- 2,2-bis(3,4-dicarboxyphenyl) hexafluoropropane dianhydride
[a]GPU. [b]Barrer. N/A—not available.

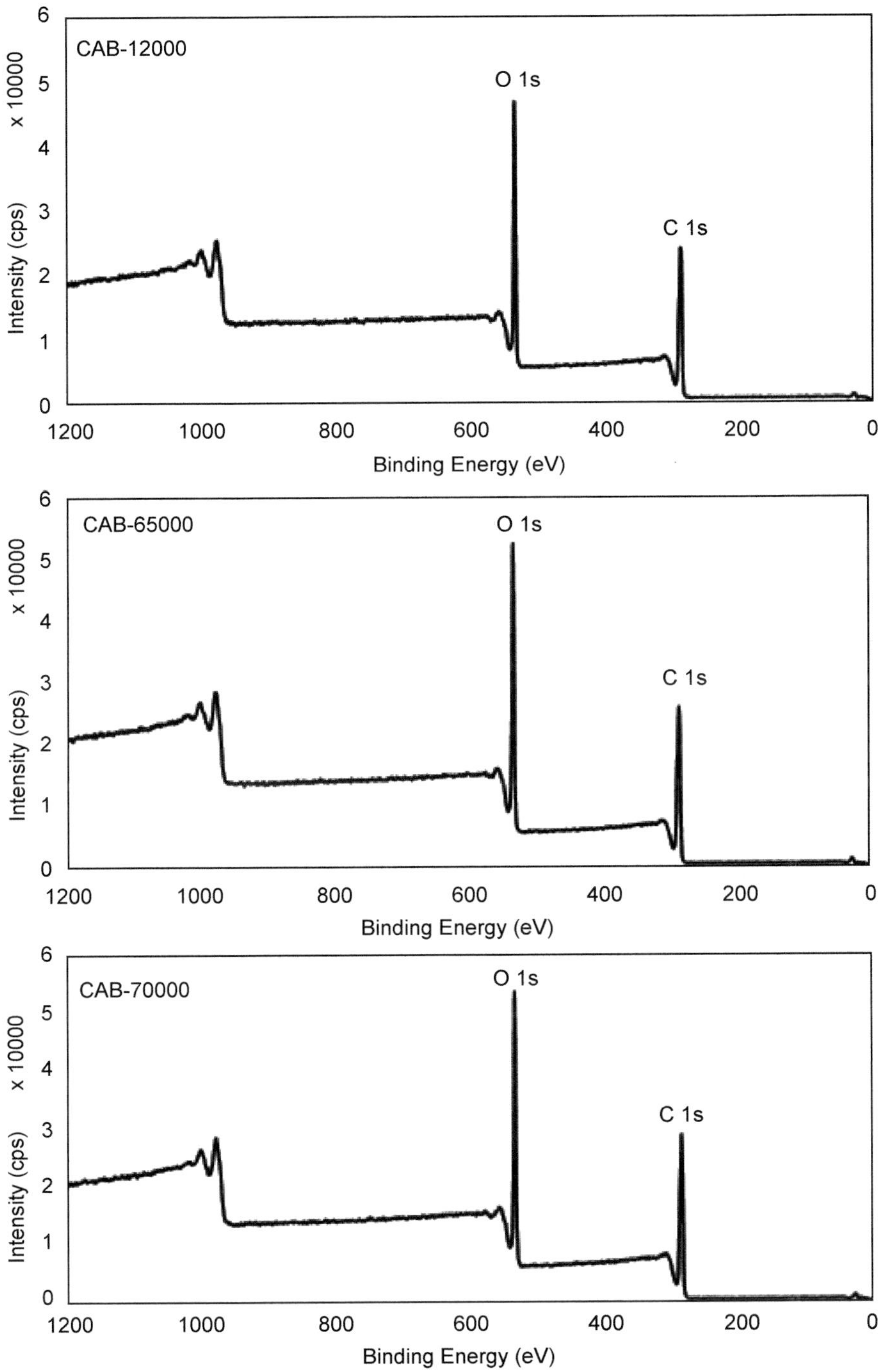

Figure 9.17. Element composition of XPS spectrum of CAB-12000, CAB-65000, and CAB-70000.

of the dynamic contents of acetyl, butyryl, and hydroxyl composition present in CAB polymer. The acetyl and butyryl contributed significantly towards the permeance of CO_2/N_2 by enhancing the solubility of CO_2 within the membrane structure. As compared to other research works, the permeance results achieved in the present work showed higher CO_2/N_2 permeance with acceptable selectivity result. The selectivity performance achieved for this study shows the typical trade-off relationship of polymer membrane due to the high permeance results and low selectivity of the CAB polymer. Nevertheless, the low selectivity of the CAB polymer can eventually be overcome by incorporating the polymer matrix with inorganic filler to produce the hybrid system of Mixed Matrix Membranes (MMMs) (Aroon et al. 2013, Chung et al. 2007, Ismail et al. 2009, Goh et al. 2011).

Conclusion

The optimization of membrane morphology conducted with respect to the different parameters was found to be successful for the preparation of the highly selective CAB gas separation membrane. The membrane formation and morphology were closely related to the rheological behaviour of the casting solution. The results have shown that membrane casting thickness, solvent exchange duration for both isopropyl alcohol and n-hexane, and the molecular weights of the CAB polymer had a significant role in manipulating the CO_2/N_2 gas separation performance as well as the morphology of the membranes. Under optimized conditions, the best membrane was found to be the CAB-70000, which was fabricated with 4 wt% polymer concentration, 250 μm casting thickness, 5 minutes solvent evaporation time, 30 minutes solvent exchange with isopropyl alcohol followed by another 30 minutes of solvent exchange with n-hexane. Moreover, the CAB-70000 had the best gas separation performance with an average selectivity of 6.12 ± 0.09 and permeance up to 227.95 ± 0.39 GPU for CO_2 and 37.28 ± 0.54 GPU for N_2, respectively. The superior CO_2/N_2 separation performance of the membrane was mainly contributed by the quality formation of the smooth surface, with thin dense and defect-free membrane structure. Further, it has been suggested that to improve the performance of the CAB membrane, inorganic nanoparticle fillers such as carbon nanotubes (CNTs) be incorporated to produce mixed matrix membrane (MMM).

Acknowledgement

The authors would like to extend their appreciation to the Ministry of Higher Education Malaysia (MOHE) for providing Fundamental Research Grant Scheme (FRGS) (MOHE Ref. No: FRGS/1/2015/TK02/CURTIN/03/1) and Cost Centre: 001048. Further, the authors would also like to thank LRGS USM (Account No: 304/PJKIMIA/6050296/U124), and Curtin Cost Centre: 001047.

References

Ahmad, A., Ideris, N., Ooi, B., Low, S. and Ismail, A. 2013. Synthesis of polyvinylidene fluoride (PVDF) membranes for protein binding: Effect of casting thickness. J. Appl. Polym. Sci. 128: 3438–3445.

Ahmad, A., Jawad, Z., Low, S. and Zein, S. 2014. A cellulose acetate/multi-walled carbon nanotube mixed matrix membrane for CO_2/N_2 separation. J. Membr. Sci. 451: 55–66.

Aroon, M.A., Ismail, A.F. and Matsuura, T. 2013. Beta-cyclodextrin functionalized MWCNT: A potential nano-membrane material for mixed matrix gas separation membranes development. Sep. Purif. Technol. 115: 39–50.

Barnes, R.J., Bandi, R.R., Chua, F., Low, J.H., Aung, T., Barraud, N. et al. 2014. The roles of Pseudomonas aeruginosa extracellular polysaccharides in biofouling of reverse osmosis membranes and nitric oxide induced dispersal. J. Membr. Sci. 466: 161–172.

Basu, S., Khan, A.L., Cano-Odena, A., Liu, C. and Vankelecom, I.F.J. 2010. Membrane-based technologies for biogas separations. Chem. Soc. Rev. 39: 750–768.

Budd, P.M., Msayib, K.J., Tattershall, C.E., Ghanem, B.S., Reynolds, K.J., Mckeown, N.B. et al. 2005. Gas separation membranes from polymers of intrinsic microporosity. J. Membr. Sci. 251: 263–269.

Carapellucci, R. and Milazzo, A. 2003. Membrane systems for CO_2 capture and their integration with gas turbine plants. Proceedings of the Institution of Mechanical Engineers, Part A: J. Pwr. E. 217: 505–517.

Childress, A.E. and Elimelech, M. 1996. Effect of solution chemistry on the surface charge of polymeric reverse osmosis and nanofiltration membranes. J. Membr. Sci. 119: 253–268.

Choi, J.H., Jegal, J. and Kim, W.N. 2006. Fabrication and characterization of multi-walled carbon nanotubes/polymer blend membranes. J. Membr. Sci. 284: 406–415.

Chung, T.S., Jiang, L. Y., Li, Y. and Kulprathipanja, S. 2007. Mixed matrix membranes (MMMs) comprising organic polymers with dispersed inorganic fillers for gas separation. Prog. Polym. Sci. 32: 483–507.

Chung, T. and Kafchinski, E.R. 1997. The effects of spinning conditions on asymmetric 6FDA/6FDAM polyimide hollow fibers for air separation. J. Appl. Polym. Sci. 65: 1555–1569.

Coltelli, M.B., Maggiore, I.D., Bertoldo, M., Signori, F., Bronco, S. and Ciardelli, F. 2008. Poly (lactic acid) properties as a consequence of poly (butylene adipate-co-terephthalate) blending and acetyl tributyl citrate plasticization. J. Appl. Polym. Sci. 110: 1250–1262.

Cong, H., Zhang, J., Radosz, M. and Shen, Y. 2007. Carbon nanotube composite membranes of brominated poly (2, 6-diphenyl-1, 4-phenylene oxide) for gas separation. J. Membr. Sci. 294: 178–185.

Conti, J., Holtberg, P., Diefenderfer, J., Larose, A., Turnure, J.T. and Westfall, L. 2016. International Energy Outlook 2016 With Projections to 2040 (No. DOE/EIA--0484 (2016)). USDOE Energy Information Administration (EIA), Washington, DC (United States). Office of Energy Analysis.

Egan, W.M. and Dufresne, R.E. 2008. Gelled adhesive remover composition and method of use. U.S. Patent 7, 977, 294.

Feng, Y., Zhang, J.M., Zhang, J. and Chang, J. 2015. Gas Separation Properties of Cellulose Acetate Butyrate/MWCNTs Mixed Matrix Membranes. Acta Polymerica Sinica 1396–1401.

Freeman, B.D. 1999. Basis of permeability/selectivity tradeoff relations in polymeric gas separation membranes. Macromolecules 32: 375–380.

Goh, P.S., Ismail, A.F., Sanip, S.M., Ng, B.C. and Aziz, M. 2011. Recent advances of inorganic fillers in mixed matrix membrane for gas separation. Sep. Purif. Technol. 81: 243–264.

Hansen, C.M. 2007. Hansen Solubility Parameters: A User's Handbook, CRC press.

Huang, R.Y. and Feng, X. 1995. Studies on solvent evaporation and polymer precipitation pertinent to the formation of asymmetric polyetherimide membranes. J. Appl. Polym. Sci. 57: 613–621.

Ismail, A.F., Goh, P.S., Sanip, S.M. and Aziz, M. 2009. Transport and separation properties of carbon nanotube-mixed matrix membrane. Sep. Purif. Technol. 70: 12–26.

Ismail, A.F. and Lorna, W. 2002. Penetrant-induced plasticization phenomenon in glassy polymers for gas separation membrane. Sep. Purif. Technol. 27: 173–194.

Ismail, A.F., Rahim, N., Mustafa, A., Matsuura, T., Ng, B.C., Abdullah, S. et al. 2011. Gas separation performance of polyethersulfone/multi-walled carbon nanotubes mixed matrix membranes. Sep. Purif. Technol. 80: 20–31.

Jansen, J.C., Buonomenna, M.G., Figoli, A. and Drioli, E. 2006. Asymmetric membranes of modified poly (ether ether ketone) with an ultra-thin skin for gas and vapour separations. J. Membr. Sci. 272: 188–197.

Jansen, J.C., Macchione, M., Oliviero, C., Mendichi, R., Ranieri, G.A. and Drioli, E. 2005. Rheological evaluation of the influence of polymer concentration and molar mass distribution on the formation and performance of asymmetric gas separation membranes prepared by dry phase inversion. Polym. 46: 11366–11379.

Jawad, Z., Ahmad, A., Low, S. and Zein, S. 2015a. Incorporation of inorganic carbon nanotubes fillers into the CA polymeric matrix for improvement in CO_2/N_2 separation. Curr. Nanosci. 11: 69–79.

Jawad, Z.A., Ahmad, A.L., Low, S.C., Chew, T.L. and Zein, S.H.S. 2015b. Influence of solvent exchange time on mixed matrix membrane separation performance for CO_2/N_2 and a kinetic sorption study. J. Membr. Sci. 476: 590–601.

Kappel, C., Kemperman, A., Temmink, H., Zwijnenburg, A., Rijnaarts, H. and Nijmeijer, K. 2014. Impacts of NF concentrate recirculation on membrane performance in an integrated MBR and NF membrane process for wastewater treatment. J. Membr. Sci. 453: 359–368.

Katayama, T. and Nitta, T. 1976. Solubilities of hydrogen and nitrogen in alcohols and n-hexane. J. Chem. Eng. D. 21: 194–196.

Kim, C.S. and Oh, S.M. 2001. Performance of gel-type polymer electrolytes according to the affinity between polymer matrix and plasticizing solvent molecules. Electrochimica Acta 46: 1323–1331.

Koros, W., Fleming, G., Jordan, S., Kim, T. and Hoehn, H. 1988a. Polymeric membrane materials for solution-diffusion based permeation separations. Prog. Polym. Sci. 13: 339–401.

Koros, W.J., Fleming, G.K., Jordan, S.M., Kim, T.H. and Hoehn, H.H. 1988b. Polymeric membrane materials for solution-diffusion based permeation separations. Prog. Polym. Sci. 13: 339–401.

Kunthadong, P., Molloy, R., Worajittiphon, P., Leekarkpai, T., Kaabbuathong, N. and Punyodom, W. 2015. Biodegradable plasticized blends of poly(L-lactide) and cellulose acetate butyrate: from blend preparation to biodegradability in real composting conditions. J. Polym. Env. 23: 107–113.

Lalia, B.S., Kochkodan, V., Hashaikeh, R. and Hilal, N. 2013. A review on membrane fabrication: Structure, properties and performance relationship. Desalination 326: 77–95.

Lee, R., Jawad, Z., Ahmad, A., Ngo, J. and Chua, H. 2017. Improvement of CO_2/N_2 separation performance by polymer matrix cellulose acetate butyrate. In IOP Conference Series: Matls. Sci. Eng. (Vol. 206, No. 1, p. 012072). IOP Publishing.

Liu, L.F., Cai, Z.B., Shen, J.N., Wu, L.X., Hoek, E.M. and Gao, C.J. 2014. Fabrication and characterization of a novel poly (amide-urethane imide) TFC reverse osmosis membrane with chlorine-tolerant property. J. Membr. Sci. 469: 397–409.

Liu, Y., Wang, R. and Chung, T.S. 2001. Chemical cross-linking modification of polyimide membranes for gas separation. J. Membr. Sci. 189: 231–239.

Low, B.T., Zhao, L., Merkel, T.C., Weber, M. and Stolten, D. 2013. A parametric study of the impact of membrane materials and process operating conditions on carbon capture from humidified flue gas. J. Membr. Sci. 431: 139–155.

Lui, A., Talbot, F., Fouda, A., Matsuura, T. and Sourirajan, S. 1988. Studies on the solvent exchange technique for making dry cellulose acetate membranes for the separation of gaseous mixtures. J. Appl. Polym. Sci. 36: 1809–1820.

Matsuyama, H., Kim, M.M. and Lloyd, D.R. 2002. Effect of extraction and drying on the structure of microporous polyethylene membranes prepared via thermally induced phase separation. J. Membr. Sci. 204: 413–419.

Ohya, H., Akimoto, N. and Negishi, Y. 1980. Reverse osmosis separation characteristics of organic solutes with cellulose acetate butyrate membranes. 5, 179-184.

Ong, R.C., Chung, T.S., Helmer, B.J. and De Wit, J.S. 2012. Novel cellulose esters for forward osmosis membranes. Ind. Eng. Chm. Rsrch 51: 16135–16145.

Radjabian, M., Koll, J., Buhr, K., Vainio, U., Abetz, C., Handge, U.A. et al. 2014. Tailoring the morphology of self-assembled block copolymer hollow fiber membranes. Polym. 55: 2986–2997.

Rahimpour, A., Madaeni, S., Taheri, A. and Mansourpanah, Y. 2008. Coupling TiO_2 nanoparticles with UV irradiation for modification of polyethersulfone ultrafiltration membranes. J. Membr. Sci. 313: 158–169.

Pandey, P. and Chauhan, R.S. 2001. Membranes for gas separation. Prog. Polym. Sci. 26(6): 853–893.

Paradise, M. and Goswami, T. 2007. Carbon nanotubes—Production and industrial applications. Matls. Design 28: 1477–1489.

S.Minhas, B. 1992. Cellulose Acetate Butyrate Gas Separation Membranes.U.S. Patent 5,096,468.

Sabde, A.D., Trivedi, M., Ramachandhran, V., Hanra, M. and Misra, B. 1997. Casting and characterization of cellulose acetate butyrate based UF membranes. Desalination 114: 223–232.

Scholes, C.A., Stevens, G.W. and Kentish, S.E. 2012. Membrane gas separation applications in natural gas processing. Fuel 96: 15–28.

Shekhawat, D. 2003. A review of carbon dioxide selective membranes. US Department of Energy, 9–11.

Suttiwijitpukdee, N., Sato, H., Zhang, J., Hashimoto, T. and Ozaki, Y. 2011. Intermolecular interactions and crystallization behaviors of biodegradable polymer blends between poly (3-hydroxybutyrate) and cellulose acetate butyrate studied by DSC, FT-IR, and WAXD. Polym. 52: 461–471.

Thomas, R., Guillen-Burrieza, E. and Arafat, H.A. 2014. Pore structure control of PVDF membranes using a 2-stage coagulation bath phase inversion process for application in membrane distillation (MD). J. Membr. Sci. 452: 470–480.

Tiraferri, A., Yip, N.Y., Phillip, W.A., Schiffman, J.D. and Elimelech, M. 2011. Relating performance of thin-film composite forward osmosis membranes to support layer formation and structure. J. Membr. Sci. 367: 340–352.

Wan, Y., Creber, K.A., Peppley, B. and Bui, V.T. 2004. Structure and ionic conductivity of a series of di-o-butyrylchitosan membranes. J. Appl. Polym. Sci. 94: 2309–2323.

Wan, Y., Creber, K.A.M., Peppley, B. and Bui, V.T. 2003. Ionic conductivity of chitosan membranes. Polym. 44: 1057–1065.

Wang, Y., Lau, W.W. and Sourirajan, S. 1994. Effects of membrane-making conditions and shrinkage treatment on morphology and performance of cellulose acetate butyrate membranes. Sep. Sci. Technol. 29: 1689–1704.

Xia, S. and Ni, M. 2015. Preparation of poly (vinylidene fluoride) membranes with graphene oxide addition for natural organic matter removal. J. Membr. Sci. 473: 54–62.

Xin, Q., Gao, Y., Wu, X., Li, C., Liu, T., Shi, Y. et al. 2015. Incorporating one-dimensional aminated titania nanotubes into sulfonated poly (ether ether ketone) membrane to construct CO_2-facilitated transport pathways for enhanced CO_2 separation. J. Membr. Sci. 488: 13–29.

Xu, Z., Zhang, J., Shan, M., Li, Y., Li, B., Niu, J. et al. 2014. Organosilane-functionalized graphene oxide for enhanced antifouling and mechanical properties of polyvinylidene fluoride ultrafiltration membranes. J. Membr. Sci. 458: 1–13.

Yang, H., Xu, Z., Fan, M., Gupta, R., Slimane, R.B., Bland, A.E. et al. 2008. Progress in carbon dioxide separation and capture: A review. J. Env. Sci. 20: 14–27.

Yang, Y. and Wang, P. 2006. Preparation and characterizations of a new PS/TiO_2 hybrid membranes by sol–gel process. Polym. 47: 2683–2688.

Yave, W., Car, A., Funari, S.S., Nunes, S.P. and Peinemann, K.V. 2009. CO_2-philic polymer membrane with extremely high separation performance. Macromolecules 43: 326–333.

Zha, S., Yu, J., Zhang, G., Liu, N. and Lee, R. 2015. Polyethersulfone (PES)/cellulose acetate butyrate (CAB) composite hollow fiber membranes for BTEX separation from produced water. RSC Advances 5: 105692–105698.

Zhu, W.P., Sun, S.P., Gao, J., Fu, F.J. and Chung, T.S. 2014. Dual-layer polybenzimidazole/polyethersulfone (PBI/PES) nanofiltration (NF) hollow fiber membranes for heavy metals removal from wastewater. J. Membr. Sci. 456: 117–127.

10

Two-phase and Three-phase Permeability Models for Mixed Matrix Membrane Gas Separation

An Overview and Comparative Study

Agus Saptoro[1,*] and *Jono Suhartono*[2]

INTRODUCTION

It is widely accepted that increasing emission of greenhouse gases particularly CO_2 to our atmosphere is the primary contributor to global climate change. CO_2 contributes to 50% of the greenhouse gases (Bin et al. 2010) and its emitted amounts was approximately 32.3 Gt as of 2016 (IEA 2016). Statistics indicated that since the pre-industrial era, the concentrations of CO_2 in the atmosphere have increased by 40% (IEA 2015). Climate model forecasts the continuation of this trend due to population and economic growths. If no action is taken to address this issue, global CO_2 concentration is predicted to rise to above 750 ppm by 2100 (Wang et al. 2011). This will cause catastrophic consequences of climate changes. The increase of average temperature must be limited up to 2°C (Lecomte et al. 2010, Roussanaly et al. 2016) to avoid detrimental effects on humans and the environment. CO_2 mitigation, therefore, is essential to reduce the effects of global warming.

[1] Curtin University, Department of Chemical Engineering, CDT 250. Miri, Sarawak, Malaysia, 98009.
[2] Institut Teknologi Nasional (ITENAS), Department of Chemical Engineering, Jl. P. K. H. Mustofa No. 23, Bandung, West Java, Indonesia, 40124.
Emai: jono_suhartono@itenas.ac.id
* Corresponding author: agus.saptoro@curtin.edu.my

To combat global warming, CO_2 Capture and Storage (CCS) has been acknowledged as one of the most promising strategies (Roussanaly et al. 2016, Aminu et al. 2017, Song et al. 2018). Among various available technologies for capturing CO_2 capture, membrane gas separation has been considered as one of the most effective solutions because of its energy efficiency, physical size and operation simplicity compared to its solvent based absorption technologies (Roussanaly et al. 2016, He et al. 2017, Merkel et al. 2010, Mat and Lipscomb 2017). Moreover, membrane gas separation has also been successfully applied in H_2/N_2 separation, H_2/CO, O_2/N_2, H_2/hydrocarbon, CO_2/CH_4 and other gas mixtures in the natural gas production, refinery operations and petrochemical processes (Ismail et al. 2015). Consequently, high performance membranes, especially Mixed Matrix Membranes (MMMs) have become a subject of interest for many researchers due to their important role in gas separation processes. This trend is indicated by extensive research works and publications on the developments and performance assessments of MMMs for CO_2 mitigation found in the literature (Bernardo et al. 2009, Aroon et al. 2010, Zhang et al. 2013, Baker and Low, 2014, Bastani et al. 2013, Rezakazemi et al. 2014, Rafiq et al. 2015, Vinh-Thang and Kaliaguine 2013).

Existing research and publications on MMMs for gas separation are mostly directed toward improving permeability and selectivity by varying different types of filler materials (Aroon et al. 2010, Bastani et al. 2013, Rezakazemi et al. 2014, Vinh-Thang and Kaliaguine 2013, Goh et al. 2011, Vinoba et al. 2017). For process design, operation and improvement of a MMM gas separation process, knowledge on the permeability of different penetrants and good understanding on the relations between permeability of a penetrant and the types and concentrations of filler materials should be clearly understood (Pal 2008, Shimekit et al. 2011). In this regard, various permeability models of MMMs had been proposed to assist predictions of gas permeability in MMMs and investigate the effect of type and concentration variations of filler materials on MMM properties especially permeability. These models were mostly derived for two-phase (filler particle – polymer) and three-phase (filler particle – interfacial layer – polymer) scenarios. Some excellent studies reporting and comparing different types of permeability models can be found in the literature (Aroon et al. 2010, Vinh-Thang and Kaliaguine 2013, Shimekit et al. 2011, Gheimasi et al. 2013, Gonzo et al. 2006, Maghami et al. 2017, Ebneyamini et al. 2017a,b, Sadeghi et al. 2015a,b, Sadeghi et al. 2016).

To date, in total there should be at least 20 two two-phase models and 11 three-phase models had been proposed by different researchers. Nevertheless, the maximum numbers of two-phase and three-phase models reported and discussed in each paper are 10 and 8, respectively (Maghami et al. 2017). Thus, there is a need to have an updated and more comprehensive report on these models in a single manuscript. From various comparative studies (Shimekit et al. 2011, Sadeghi et al. 2015a,b, Sadeghi et al. 2016), it was also evident that three-phase models generally outperform two-phase models in terms of their accuracies. This superiority is mainly due to taking into account the effect of interfacial layers in the models. However, it has been acknowledged that the three-phase models may not be always practical due to unavailability interfacial layer and other necessary data required by the models (Ebneyamini et al. 2017b). Few researchers (Zhang et al. 2013) also believed that some permeability models are complicated in nature and their predictive capabilities are not always better than two-phase Maxwell model. Therefore, there is also a research question on how to improve simpler two-phase models to have competitiveness with high performing but complex three-phase models.

This study aims to provide an overview of 22 two-phase models and 11 three-phase models available in the literature. Efforts to modify and improve two-phase models are also presented by adopting J-modified factor proposed by Sadeghi et al. (2015b, 2016). They introduced J-factor into six two-phase models and the results showed that the accuracies of these models can be enhanced. This chapter, therefore, expands the incorporation of J-factor into other two-phase models and investigates its impacts on the predictive performances of the models. It is envisaged that this study will provide a good platform for future research works on developing an improved MMM permeability model and its applications in gas separation using MMMs. The following sections discuss summaries of the existing two-phase and three-phase models, extended J-modified approach to two-phase models, a comparative case study, results and discussions, conclusions and recommendations.

Existing Permeability Models

In the published literatures, different types of permeability models include theoretical models, thermodynamic based models (Minelli and Sarti 2016, Minelli and Sarti 2017a,b) and statistical/intelligent modelling based models (Boldyrev et al. 2006, Wessling et al. 1994, Hasnaoui et al. 2017). Nevertheless, this study focuses only on reporting the existing theoretical models. These models can be generally classified into two main categories: (i) two-phase models, and (ii) three phase models. Two-phase models, also named ideal models, assume that the effects of interfacial defects on the performance of MMMs are negligible. Meanwhile, three-phase models consider non-ideal interface between continuous and dispersed phases. This non-ideality is caused by interfacial defects due to interface rigidification, interfacial voids or sieves-in-a-cage, particle pore blockage (Vinh-Thang and Kaliaguine 2013, Maghami et al. 2017, Ebneyamini et al. 2017b, Sadeghi et al. 2015b). Ideal models are less sophisticated since non-ideal models require more variables and parameters which sometimes are unmeasurable or difficult to estimate. Two-phase and three-phase models discussed in this chapter are summarized in Table 10.1. The details of each model are presented in the following sub-sections.

Two-phase models

Maxwell model

This model was adopted from electrical conductivity model for particulate composite materials developed in 1873 (Maxwell 1954). It is the simplest predictive model for MMM permeability containing spherical particles and described by Eq. (1).

$$P_r = \frac{P}{P_c} = \left[\frac{2(1-\varnothing) + (1+2\varnothing)\lambda_{dm}}{(2+\varnothing) + (1-2\varnothing)\lambda_{dm}} \right] \tag{1}$$

Table 10.1. Summaries of two-phase and three-phase models.

TWO-PHASE MODELS	1. Maxwell 1954	12. Hennepe 1991
	2. Bruggeman 1935	13. Singh and Sharma 2008
	3. Lewis-Nielsen 1970	14. Maxwell 1954
	4. Pal 2007	15. ruggeman 1935
	5. Chiew and Glandt 1983	16. Lewis-Nielsen 1970
	6. Maxwell-Wagner-Sillar (MWS)	17. Pal 2007
	7. Modified MWS	18. Chiew and Glandt 1983
	8. Böttcher 1945	19. Böttcher 1945
	9. Higuchi 1958	20. Extended Petropoulous
	10. Modified Higuchi	21. Cussler 1990
	11. New modified Higuchi	22. Kang et al. 2011
THREE-PHASE MODELS	1. Modified Maxwell	7. Modified pseudo two-phase Bruggeman
	2. Felske	8. Pseudo two-phase composite
	3. Modified Felske	9. Multi-structural
	4. Modified Pal	10. Erdem-Senatalar et al. 2001
	5. Revised Singh and Sharma	11. Hashemifard et al. 2010
	6. Pseudo two-phase Bruggeman	

where P_r, P, P_c and $\emptyset$ are the relative permeability of species, the effective permeability of species in MMMs, the permeability of the species in the matrix (continuous phase) and the volume fraction of the filler particles, respectively. Meanwhile, λ_{dm} is defined as the permeability ratio $(\frac{P_d}{P_c})$ and P_d is the permeability of species in dispersed phase. Despite its simplicity to compute, this model assumes that the effects of particle shape, size and aggregation can be neglected and it may only work well for dilute dispersion of filler material (Shimekit et al. 2011; Maghami et al. 2017).

Bruggeman model

This model made use of the dielectric constant model for particulate composite materials proposed in 1935 (Bruggeman 1935). Similar to the Maxwell model, this model does not take into account the variations of the particle shape, size and aggregation on the permeability. It is represented by Eq. (2)

$$(P_r)^{\frac{1}{3}}\left[\frac{\lambda_{dm}-1}{\lambda_{dm}-P_r}\right]=(1-\emptyset)^{-1} \tag{2}$$

The definitions of the variables involved are the same as definitions in the Maxwell model. It is generally believed that the Bruggeman model validity covers a wider range of filler concentration. Nevertheless, compared to the Maxwell model, its computation is not straightforward since it contains an implicit function (Vinh-Thang and Kaliaguine 2013, Shimekit et al. 2011, Maghami et al. 2017).

Lewis-Nielsen model

Lewis-Nielsen permeability model has its roots from an elastic modulus model proposed by Lewis and Nielsen (1970) and thermal conductivity model developed by Nielsen (1973) both for particular composite materials. Equations (3) and (4) show the details of this model.

$$P_r = \frac{P}{P_c} = \left[\frac{1+2((\lambda_{dm}-1)/(\lambda_{dm}-2))\varnothing}{1-(\lambda_{dm}-1)/(\lambda_{dm}+2))\varnothing\psi}\right] \tag{3}$$

$$\psi = 1\left(\frac{1-\varnothing_m}{\varnothing_m^2}\right)\varnothing \tag{4}$$

where $\varnothing_m$ is the maximum packing volume fraction of filler particles ($\varnothing_m = 0.64$ for random close packing of uniform spheres). The Lewis-Nielsen model includes the effects of morphology on permeability especially for spherical particles. $\varnothing_m$ has been acknowledged to be greatly influenced by particle size, shape and aggregation (Shimekit et al. 2011, Maghami et al. 2017). When $\varnothing_m$ moves closer to 1, this model will be transformed to Maxwell model (Vinh-Thang and Kaliaguine 2013, Shimekit et al. 2011).

Pal model

Based on thermal conductivity model of particulate composite developed in 2007 (Pal 2007, Pal 2008), Pal permeability model is derived as in Eq. (5).

$$(P_r)^{\frac{1}{3}}\left[\frac{\lambda_{dm}-1}{\lambda_{dm}-P_r}\right] = \left(1-\frac{\varnothing}{\varnothing_m}\right)^{-\varnothing_m} \tag{5}$$

The model above is similar to the Bruggeman model. However, $\varnothing_m$ is adopted to accommodate the effect of morphology like in Lewis-Nielsen model. If $\varnothing_m$ is equal to 1, this model becomes the Bruggeman model (Vinh-Thang and Kaliaguine 2013, Shimekit et al. 2011, Maghami et al. 2017). Numerical computation is required for solving Pal model since an implicit relation exists.

Chiew and Glandt model

This model is an extension of the Maxwell model to predict electrical conductivity of particulate composite materials (Gonzo et al. 2006, Maghami et al. 2017, Chiew and Glandt 1983). It is then also adopted to estimate permeability and shown in Eqs. (6) – (10). *K* and *O* are correction factors for Maxwell model (Vinh-Thang and Kaliaguine 2013, Gheimasi et al. 2013, Sadeghi et al. 2016).

$$P_r = \frac{P}{P_c} = 1 + 3\beta\varnothing + K\varnothing^2 + O(\varnothing)^3 \tag{6}$$

$$\beta = \frac{P_d - P_c}{P_d + 2P_c} \tag{7}$$

$$K = a + b\varnothing^{1.5} \tag{8}$$

$$a = -0.002254 - 0.123112\beta + 2.93656\beta^2 + 1.690\beta^3 \tag{9}$$

$$b = -0.0039298 - 0.803494\beta + 2.16207\beta^2 + 6.48296\beta^3 + 5.27196\beta^4 \tag{10}$$

Similar to the Maxwel model, this model may only be valid for low loading of filler material. Nevertheless, the Chiew and Glandt model takes into account the variati0ons of polymer-particle and particle-particle interactions on gas permeability (Maghami et al. 2017).

Maxwell-Wagner-Sillar (MWS) model

MWS model was firstly brought forward to estimate the permeability of MMMs with dilute dispersion of elliptical particles at different shapes (Gonzo et al. 2006, Maghami et al. 2017, Bouma et al. 1997, Rafiq et al. 2015). This model is shown in Eq. (11)

$$P_r = \frac{P}{P_c} = \left[\frac{nP_d(1-n)P_c + (1-n)(P_d - P_c)\varnothing}{nP_d + (1-n)P_c - n(P_d - P_c)\varnothing} \right] \tag{11}$$

n is the shape factor of the filler and its value can be further analyzed in the following

- n = 0 is for permeation through a membrane with parallel transport
- n = 1 is pertinent to permeation perpendicular to the phases
- $0 \leq n \leq 1/3$ is for prolate/elongated ellipsoids
- n = 1/3 is for spherical filler particles
- $1/3 \leq n \leq 1$ is for oblate/flattened ellipsoids

MWS model assumes negligible effects of particle size and aggregation. Nevertheless, it had been reported to have reasonably good predictions for low loadings of ellipsoids (Maghami et al. 2017).

Modified Maxwell-Wagner-Sillar (MWS) model

To improve MWS model by incorporating the effects of particle size, shape and aggregation, MWS model was modified by Rafiq et al. (2015). This new model uses n_e (the shape factor with calculated particle shape factor at z-axis (n_z)) instead of n and therefore, Eq. (11) becomes Eq. (12)

$$P_r = \frac{P}{P_c} = \left[\frac{n_e P_d(1-n_e)P_c + (1-n_e)(P_d - P_c)\varnothing}{n_e P_d + (1-n_e)P_c - n_e(P_d - P_c)\varnothing} \right] \tag{12}$$

$$n_e = \frac{\max(n_z)}{\varnothing_m}\varnothing \tag{13}$$

Similar to $\varnothing_m$ in Lewis-Nielsen and Pal models, $\varnothing_m$ is the maximum packing volume fraction of filler particles ($\varnothing_m$=0.64 for random close packing of uniform spheres).

Böttcher model

This model was adopted from dielectric constant model of crystalline powders proposed by Böttcher (1945). The equation is illustrated below and it does not take into account the effects of particle shape, size and aggregation on the gas permeability (Ebneyamini et al. 2017a,b, Sadeghi et al. 2015a,b, Sadeghi et al. 2016).

$$\left(1\frac{1}{P_r}\right)\left(\frac{P_d}{P_c}+2P_r\right)=3\varnothing\left(\frac{P_d}{P_c}-1\right) \tag{14}$$

Higuchi, modified Higuchi and new modified Higuchi models

The original Higuchi model was developed to estimate dielectric properties of two-phase mixture (Maghami et al. 2017, Ebneyamini et al. 2017a,b, Higuchi 1958). It was then adopted in predicting gas permeability in MMMs according to Eqs. (15) and (16) where a Higuchi parameter, K_H, is equal to 0.78. In this model, the effects of particle shape, size and aggregation on the gas permeability are not considered and it is generally acknowledged that it is applicable for spherical particles (Maghami et al. 2017).

$$P_r = \frac{P}{P_c} = 1 + \frac{3\varnothing\beta}{1-\varnothing\beta - K_H(1-\varnothing)\beta^2} \tag{15}$$

$$\beta = \frac{P_d - P_c}{P_d + 2P_c} \tag{16}$$

To correct limitations in standard Higuchi model, Sadeghi et al. (2009) and Maghami et al. (2017) modified the model by incorporating the variations of nano-size distribution of impermeable particles in composites. Equation (15) is thus converted into Eq. (17) as modified Higuchi model and is now named a modified Higuchi parameter.

$$P_r = \frac{P}{P_c} = 1 - \frac{6\varnothing}{4+2\varnothing - K_H(1-\varnothing)} \tag{17}$$

Equation (17) above still does not consider the effects of particle size, shape and aggregation. Therefore, Ameri et al. (Ameri et al. 2015; Maghami et al. 2017) further modified Eq. (17) by taking into account the influence of particle segregation. A new modified Higuchi is then proposed in the following Eq. (18).

$$P_r = \frac{P}{P_c} = 1 - \frac{6\varnothing}{4 + 2\varnothing - K'_H(1-\varnothing)} \tag{18}$$

$$K'_H = K_H - \varnothing \tag{19}$$

where K'_H is a newly modified Higuchi parameter. Despite modifications above, new modified Higuchi model still assumes that there are no effects of particle size and shape on gas permeability in MMMs (Maghami et al. 2017).

Hennepe model

The Hennepe model was first proposed for permeability of liquid-liquid mixtures (Hennepe et al. 1991, 1994). Nevertheless, similar to the Maxwell, Bruggeman, Böttcher, Higuchi, Lewis-Nielsen and Pal models which can be applied for liquid-liquid mixture and gas permeability (Ebneyamini et al. 2017a,b), this model is also adopted for estimating gas permeability in MMMs. Its correlation is shown in Eq. (20)

$$P_r = \frac{P}{P_c} = \frac{1}{1 - \varnothing^{\frac{1}{3}} + \dfrac{1.5\varnothing^{\frac{1}{3}} P_c}{P_c(1-\varnothing) + 1.5 P_d \varnothing}} \tag{20}$$

Effects of particle size, shape and aggregation are not taken into consideration in this model.

Singh and Sharma model

The same as some of the above models, Singh's model is originally a thermal conductivity model (Singh and Sharma 2008) which is then employed to predict permeability (Sadeghi et al. 2015a).

$$P_r = \frac{P}{P_c} = 1 + 3.74\left(\frac{\lambda_{dm} - 1}{\lambda_{dm} + 2}\right)\varnothing^{\frac{2}{2}} \tag{21}$$

This model also assumes that effects of particle size, shape and aggregation can be omitted.

J-two-phase models

Previous works has reported that enhanced MMM permeability can be achieved with introducing porous particles (Vinh-Thang and Kaliaguine 2013, Sadeghi et al. 2015b). It is also well-acknowledged that porosity of the particles in the matrix not only reduces the resistance within the membrane but also improves the separation efficiency (Petropoulos 1985, Mohammadi et al. 2011, Dorosti et al. 2014). Nevertheless, none of the above models have taken into account the particle porosity effect. Therefore, Sadeghi et al. (Sadeghi et al. 2015b, Sadeghi et al. 2016) introduced a new parameter J and incorporated it in some two-phase models (Maxwell, Bruggeman, Lewis-Nielsen, Pal, Böttcher and Chiew and Glandt) and three-phase models (Felske and modified Felske) to adjust the volume concentration of filler particle. The parameter J for two-phase models is defined as in Eq. (22).

$$J = 2 - \frac{1}{s} \tag{22}$$

Equation (1) in Maxwell model, Eq. (2) in Bruggeman model, Eq. (3) in Lewis-Nielsen model, Eq. 5 in Pal model, Eq. (6) in Chiew-Glandt model and Eq. (14) in Böttcher model are then modified to obtain J-Maxwell, J-Bruggeman, J-Lewis-Nielsen, J-Pal, J-Chiew-Gland and J- Böttcher in Eqs. (23)–(27) (Sadeghi et al. 2015b, Sadeghi et al. 2016).

J-Maxwell model

$$P_r = \frac{P}{P_c} = \left[\frac{2(1 - J\varnothing) + (1 + 2J\varnothing)\lambda_{dm}}{(2 + J\varnothing) + (1 - J\varnothing)\lambda_{dm}} \right] \tag{23}$$

J-Bruggeman model

$$(P_r)^{\frac{1}{3}} \left[\frac{\lambda_{dm} - 1}{\lambda_{dm} - P_r} \right] = (1 - J\varnothing)^{-1} \tag{24}$$

J-Lewis-Nielsen model

$$P_r = \frac{P}{P_c} = \left[\frac{1 + 2((\lambda_{dm} - 1)/(\lambda_{dm} + 2))J\varnothing}{1 - ((\lambda_{dm} - 1)/(\lambda_{dm} + 2))\varnothing J\psi} \right] \tag{25}$$

J-Pal model

$$(P_r)^{\frac{1}{3}} \left[\frac{\lambda_{dm} - 1}{\lambda_{dm} - P_r} \right] = (1 - \frac{\varnothing J}{\varnothing_m})^{-\varnothing_m} \tag{26}$$

J- Böttcher model

$$\left(1-\frac{1}{P_r}\right)\left(\frac{P_d}{P_c}+2P_r\right)=3J\varnothing(\frac{P_d}{P_d}-1) \tag{27}$$

J-Chiew and Glandt model

$$P_{r=}\frac{P}{P_c}=1+3\beta J\varnothing+K(J\varnothing)^2+0(J\varnothing)^3 \tag{28}$$

Based on their findings, Sadeghi et al. (2015b) and Sadeghi et al. (2016) reported that the J-factor is able to improve the predictive capabilities of the traditional Maxwell, Bruggeman, Lewis-Nielsen, Pal, Böttcher and Chiew and Glandt model. These confirmed that there is a strong particle porosity effect on gas permeability in MMMs.

Cussler model

This model adopted the original Maxwell model and is applicable for low concentrations of spherical flake (Vinh-Thang and Kaliaguine 2013, Cussler 1990). Specifically, the Cussler model takes into account the flake aspect ratio (θ) and is shown in Eq. (29).

$$P_r=\frac{P}{P_c}=(1-\varnothing+\frac{1}{\frac{P_d}{\varnothing P_c}+\frac{4(1-\varnothing)}{\theta^2\varnothing^2}})^{-1} \tag{29}$$

Extended Petropoulos model

Petropoulos put forward a generalized Maxwell (Petropoulos 1985) and which then extended by Toy et al. (Toy et al. 1997) to formulate extended Petropoulos model in Eq. (30). This model contains a geometric factor G to capture the effects of dispersion and shape of the filler particles.

$$P_r=\frac{P}{P_c}=1+\frac{(1+G)\varnothing}{\left(\frac{\left(P_d/P_c\right)+G}{\left(P_d/P_c\right)-G}\right)-\varnothing} \tag{30}$$

The values of G depend on the shape of the particles as follows (Vinh-Thang and Kaliaguine 2013).

- G = 1 for long and cylindrical (elongated) particles disposed transverse to the direction of gas flow

- G = 2 for spherical particles or isometric aggregates
- G → ∞ is at minimum flow resistance when the dispersed particles are oriented in lamellae parallel to the gas flow direction
- G → ∞ is at maximum flow impedance when dispersed particles are oriented in lamellae perpendicular to the gas flow direction

Kang-Jones-Nair model

This model was derived to accommodate the aspect ratio, σ, and orientation angle, ω, of the tubular fillers and is expressed in Eq. (31) (Kang et al. 2011).

$$P_r = \frac{P}{P_c} = \left[\left(1 - \frac{\cos\omega}{\cos\omega + \frac{1}{\sigma}\sin\omega}\varnothing\right) + \frac{P_c}{P_d}\left(\frac{1}{\cos\omega + \frac{1}{\sigma}\sin\omega}\varnothing\right)\right]^{-1} \tag{31}$$

Aspect ratio is defined as the outer diameter divided by the length of the tubular filler while the orientation angle of the filler with respect to the axis parallel to the bulk phase transport direction (varying from 0 to $\pi/2$ radians) (Vinh-Thang and Kaliaguine 2013, Kang et al. 2011).

Three-phase models

Modified Maxwell model

Modified Maxwell model is derived from Maxwell model where it is modified to accommodate the presence of filler particles-interfacial layer-polymer interactions (Aroon et al. 2010, Shimekit et al. 2011, Maghami et al. 2017, Chung et al. 2007, Moore et al. 2004, Mahajan and Koros 2002). The model consists of Eqs. (32) – (34).

$$P_r = \frac{P}{P_c} = \left[\frac{2(1-\varnothing) + (1+2\varnothing)\frac{P_{eff}}{P_c}}{(2+\varnothing) + (1-\varnothing)\frac{P_{eff}}{P_c}}\right] \tag{32}$$

$$P_{eff} = P_I = \left[\frac{2(1-\varnothing_s) + (1+2\varnothing_s)\frac{P_d}{P_I}}{(2+\varnothing_s) + (1-\varnothing_s)\frac{P_d}{P_I}}\right] \tag{33}$$

$$\varnothing_s = \frac{\varnothing}{\varnothing + \varnothing_I} = \frac{r_d^3}{r_d^3 + r_i^3} \tag{34}$$

where P_I and P_{eff} are the permeability of the interfacial shell and effective permeability of a single core-shell particle, respectively. Meanwhile, other variables are defined as follows

- $\emptyset_s$ = the volume fraction of filler core particle in the combined volume of core and interfacial shell (in s single core-shell particles)
- $\emptyset_I$ = the volume fraction of the interfacial rigidified matrix chains
- r_d = the radius of the dispersed molecular sieve
- r_i = the outer radius of the rigidified interfacial layer (assumed to be half way the distance between the sieve and the polymer obtained by FESEM cross sectional view inspection

Even though this model considers three-phase scenarios, the modified Maxwell model still assumes that the effects of particle size, shape and aggregation are negligible (Shimekit et al. 2011, Maghami et al. 2017). It may also be valid only for low loading of filler particles.

Felske model

Based on Felske thermal conductivity model and replacing thermal conductivity with permeability (Aroon et al 2010, Pal 2008, Shimekit et al. 2011, Maghami et al. 2017, Felske 2004), Felske permeability model is described in Eqs. (35)–(37).

$$P_r = \frac{P}{P_c} = \left[\frac{2(1-\varnothing) + (1+2\varnothing)(\frac{\beta}{\gamma})}{(2+\varnothing) + (1-\varnothing)(\frac{\beta}{\gamma})} \right] \tag{35}$$

$$\beta = \frac{(2+\delta^3)P_d - 2(1-\delta^3)P_I}{P_c} \tag{36}$$

$$\gamma = 1 + 2\delta^3 - (1-\delta^3)(\frac{P_d}{P_I}) \tag{37}$$

P_I, P_d and δ are defined as the permeability of the interfacial shell, the permeability in the filler core particle and the ratio of outer radius of interfacial shell to core radius, respectively. This model has the same constraints as modified Maxwell model (Pal 2008, Shimekit et al. 2011, Maghami et al. 2017).

Modified Felske model

The Felske model in the sub-section above is improved by incorporating morphology and packing difficulty of the particles (Vinh-Thang and Kaliaguine 2013, Pal 2008, Shimekit et al. 2011, Maghami et al. 2017, Sadeghi et al. 2016). This idea is then named modified Felske model as Eqs. (38)–(39) below.

$$P_r = \frac{P}{P_c} = \left[\frac{1+2((\beta-\gamma)/(\beta+2\gamma))\varnothing}{1-((\beta-\gamma)/(\beta+2\gamma))\varnothing\psi}\right] \tag{38}$$

$$\psi = 1+\left(\frac{1-\varnothing_m}{\varnothing_m^2}\right)\varnothing \tag{39}$$

where $\varnothing_m$ is the maximum packing volume fraction of filler particles ($\varnothing_m$ =0.64 for random close packing of uniform spheres). The modified Felske model is also still associated with similar limitations from other models (Maghami et al. 2017).

Revised Pal

Pal's two-phase model in Eq. (5) is modified and improved by Shimekit et al. (2011) by taking into account polymer chain rigidification. This revised model captures the permeability in two phase interactions: dispersed and interfacial layer phases and interfacial and continuous phases (Vinh-Thang and Kaliaguine 2013, Maghami et al. 2017, Sadeghi et al. 2016). This model is formulated in the following equations.

$$(P_r)^{\frac{1}{3}}\left[\frac{\lambda_{eff^*m}-1}{\lambda_{eff^*m}-P_r}\right] = (1-\frac{\varnothing_z}{\varnothing_m})^{-\varnothing_m} \tag{40}$$

$$P_{r=}\frac{P}{P_c} \tag{41}$$

$$\lambda_{eff^*m} = \frac{P_{eff^*}}{P_c} \tag{42}$$

$$\varnothing_z = \varnothing_d + \frac{r_i^s}{R_m^s} \tag{43}$$

$$(\frac{P_{eff^*}}{P_I})^{\frac{1}{s}}\left[\frac{\lambda_{dI}-1}{\lambda_{dI}-\frac{P_{eff^*}}{P_I}}\right] = (1-\frac{\varnothing_s}{\varnothing_m})^{-\varnothing_m} \tag{44}$$

$$\lambda_{dI} = \frac{P_d}{P_I} \tag{45}$$

$$P_I = \frac{P_c}{\sigma} \tag{46}$$

$$\varnothing_s = \frac{\varnothing_d}{\varnothing_d + \varnothing_I} = \frac{r_d^s}{r_d^s + r_i^s} \tag{47}$$

All variables and parameters involved are defined as the following information (Shimekit et al. 2011):

- P_{eff*} = Permeability of the combined sieve and interfacial rigidified matrix chain layer
- $\emptyset_z$ = the combined volume fraction of the sieve phase and the interfacial rigidified matrix chains in the whole system
- r_i = the outer radius of the rigidified interfacial layer (assumed to be half way the distance between the sieve and the polymer obtained by FESEM cross sectional view inspection
- R_m = the distance from the center of the sieve to boundary of the polymer surface
- P_I =Permeability in the rigidified interface layer = $\frac{P_c}{\propto}$
- σ = the matrix rigidification or chain immobilization (reduction in permeability) factor (estimated from the respective intrinsic MMMs permeation data; assumed to be equal to 3).
- P_d = the permeability of species in dispersed phase
- $\emptyset_s$ = the volume fraction of the sieve phase in the combined phase
- $\emptyset_d$ = the volume fraction of the sieve phase
- $\emptyset_I$ = the volume fraction of the interfacial rigidified matrix chains
- $\emptyset_m$ = the maximum packing volume fraction of filler particles ($\emptyset_m$ = 0.64 for random close packing of uniform spheres)
- r_d = the radius of the dispersed molecular sieve

New Singh and Sharma model

The Singh and Sharma permeability model does not take into account the effect of interfacial layer between filler particle and polymer. Consequently, Sadeghi et al. (2015a) modified Eq. (21) and proposed a three-phase equation as in Eqs. (48)–(50).

$$P_r = \frac{P}{P_c} = 1 + 3.74\left(\frac{\frac{\beta}{\gamma} - 1}{\frac{\beta}{\gamma} + 2}\right)\varnothing_s^{\frac{2}{3}} \tag{48}$$

$$\beta = \frac{(2 + l_I^s)P_d - 2(1 - l_I^3)P_I}{P_c} \tag{49}$$

$$\gamma = 1 + 2l_I^{\ 3} - (1 - l_I^{\ 3})\left(\frac{P_d}{P_I}\right) \tag{50}$$

where l_i = thickness of interfacial phase. It was found that the new model above is superior compared to some existing two-phase and three-phase models for different combinations of membrane matrix and filler particles (Sadeghi et al. 2015a).

Pseudo two-phase Bruggeman model

To take into account the effects of voids creation in the interface layer, Shariati et al. (2012) modified and extended the Bruggeman model to propose a new pseudo two-phase Bruggemen model. These void creations had been reported by Chung et al. (2007) and Cong et al. (2007) as the main reason of improper compatibility between particle surface and polymer. This model is formulated as in Eqs. (51)–(57).

$$\left(\frac{P}{P_c} - \alpha\right)\left(\frac{P}{P_c}\right)^{-\frac{1}{s}} = (1 - \varnothing^*)(1 - \alpha) \tag{51}$$

$$\alpha = \frac{P_{znd}}{P_c} \tag{52}$$

$$\left(\frac{P_{2nd}}{P_{I2}}\right)^{\frac{2}{3}} = (1 - \varnothing_2) \tag{53}$$

$$\varnothing_2 = \frac{\varnothing}{\varnothing + \varnothing_I} = \frac{r_d^3}{(r_d + l_I)^3} \tag{54}$$

$$P_I = 9.7x10^{-5} x \left(\frac{r_d + l_I}{RT}\right)\sqrt{\frac{T}{M_A}(1 - \frac{d_g}{2(r_d + l_I)})^3} \tag{55}$$

$$\varnothing^* = \varnothing + \varnothing_I \tag{56}$$

$$\varnothing_I = \varnothing\left(\frac{3l_I}{r_d} + \frac{3l_I^2}{r_d^2} + \frac{l_I^2}{r_d^3}\right) \tag{57}$$

where d_g, T, M_A, R and l_I are molecular gas diameter, absolute temperature, gas molecular weight, gas constant and thickness of intermediate phase, respectively. In this model, MMM is described as a pseudo two-phase composite consisting of polymer matrix and the bulk filler particle plus void volume region (Shariati et al.

2012). This model showed better prediction than of a two-phase Bruggeman model (Shariati et al. 2012).

Modified pseudo two-phase Bruggeman

This model is modified version of a pseudo two-phase Bruggeman model where instead of imagining MMM as a pseudo two-phase composite, MMMs now comprising a pseudo three-phase composite: polymer matrix, layers of increased free volume at the polymer-filler particle interface and particle bulk and its surrounding voids (Shariati et al. 2012). The model is formed by Eqs. (58)–(67) (Maghami et al. 2017, Shariati et al. 2012).

$$\left(\frac{P}{P_c}-\alpha_2\right)\left(\frac{P}{P_c}\right)^{-\frac{1}{3}}=(1-\varnothing^*)(1-\alpha_2) \tag{58}$$

$$\alpha_2=\frac{P_{2nd}}{P_c} \tag{59}$$

$$\left(\frac{P_{2nd}}{P_{I2}}-\alpha_1\right)\left(\frac{P_{2nd}}{P_{I2}}\right)^{-1/3}=(1-\varnothing_{2^*})(1-\alpha_1) \tag{60}$$

$$\alpha_1=\frac{P_{srd}}{P_{I2}} \tag{61}$$

$$\varnothing_{2^*}=\frac{(r_d+l_{I1})^s}{(r_d+l_{I1}+l_{I2})^s} \tag{62}$$

$$l_I=l_{1I}+l_{I2} \tag{63}$$

$$P_{I2}=P_c\upsilon \tag{64}$$

$$\left(\frac{P_{srd}}{P_{I2}}\right)^{\frac{2}{s}}=(1-\varnothing_3) \tag{65}$$

$$\varnothing_3=\frac{\varnothing}{\varnothing+\varnothing_{I1}}=\frac{r_d^s}{(r_d+l_{I1})^s} \tag{66}$$

$$P_I=9.7x10^{-5}x\left(\frac{r_d+l_{I1}}{RT}\right)\sqrt{\frac{T}{M_A}(1-\frac{d_g}{2(r_d+l_{I1})})^3} \tag{67}$$

Some nomenclatures in the above equations are defined below

- $\emptyset_{I1}$ = the volume fraction of void layer around the particles
- l_{I1}= the thickness of void layer around the particles
- P_{I1}= permeability of void layer around the particles
- υ = the free volume increment factor ($\upsilon > 1$)
- l_{I2}= increased free volume layer thickness

Pseudo two-phase composite model

This model is a modified Chiew and Glandt model in Eqs. (6)–(10) where P_d and $\emptyset$ are replaced by P_{2nd} and $\emptyset$* from pseudo two-phase Bruggeman model (Hassanajili et al. 2013). The model, therefore, consists of Eqs. (68)–(75).

$$P_r = \frac{P}{P_c} = 1 + 3\beta_{2nd}\varnothing_{ps} + K\varnothing_{ps}^{2} + 0(\varnothing_{ps}^{3}) \tag{68}$$

$$\beta_{2nd} = \frac{P_{2nd} - P_c}{P_{2nd} + 2P_c} \tag{69}$$

$$\left(\frac{P_{2nd}}{P_I}\right)^{\frac{2}{s}} = (1 - \varnothing_2) \tag{70}$$

$$\varnothing_2 = \frac{\varnothing_d}{\varnothing_d + \varnothing_I} = \frac{r_d^s}{(r_d + l_I)^s} \tag{71}$$

$$P_I = 9.7x10^{-5}x\left(\frac{r_d + l_I}{RT}\right)\sqrt{\frac{T}{M_A}}(1 - \frac{d_g}{2(r_d + l_1)})^3 \tag{72}$$

$$K = a + b\varnothing_{ps}^{1.5} \tag{73}$$

$$a = -0.002254 - 0.123112\,\beta_{2nd} + 2.93656\,\beta_{2nd}^{2} + 1.690\,\beta_{2nd}^{3} \tag{74}$$

$$b = 0.0039298 - 0.803494\beta_{2nd} + 2.16207\,\beta_{2nd}^{2} + 6.48296\,\beta_{2nd}^{3} + 5.27196\,\beta_{2nd}^{4} \tag{75}$$

where $\emptyset_{ps}$ and β_{2nd} are the volume fraction of the pseudo-dispersed phase and a measure of penetrant permeability difference between the two phases (Hassanajili et al. 2013).

Multi-structural model

To accommodate the effects of particle shape and intermediate phase on MMM separation performance and adopting a thermal conductivity model for heterogeneous materials, a multi-structural permeability model is proposed as follows (Maghami et al. 2017, Sabzevari et al. 2013, Wang et al. 2008).

$$P_r = \frac{P}{P_c} = \left[\frac{P_{2nd} + 2P_c - 2(\varnothing / \varnothing^*)(P_c - P_{2nd})}{P_{2nd} + 2P_c + (\varnothing / \varnothing^*)(P_c - P_{2nd})} \right] \tag{76}$$

$$\varnothing^* = \varnothing + \varnothing_I \tag{77}$$

$$\varnothing_I \frac{(P_I - P_{2nd})(2P_I + P_{2nd})}{P_I} + \varnothing \frac{(P_d - P_{2nd})(2P_d + P_{2nd})}{P_d} = 0 \tag{78}$$

Erdem-Senatalar model

Based on the effective medium theory, three-phase permeability model was proposed by Erdem-Senatalar et al. (2001) and it is defined in Eqs. (79)–(80).

$$\frac{\varnothing^*(P - P^*)}{P^* + 2P} + \frac{\varnothing(P + P_c)}{P_c + 2P} = 0 \tag{79}$$

$$P^* = \frac{P_I P_d}{P_I \varnothing + P_d \varnothing_I} \tag{80}$$

where $\varnothing^*$ and $\varnothing_1$ can be calculated using Eqs. (56) and (57). This model takes into account the correlations between the interface and polymer and between interphase thickness and filler particle size (Vinh-Thang and Kaliaguine 2013).

Hashemifard-Ismail-Matsuura model

By considering the pathways of the penetrant gas flow through MMMs in series and parallel passage, Hashemifard-Ismail-Matsuura model was developed as follows (Vinh-Thang and Kaliaguine 2013, Hashemifard et al. 2010).

$$P_r = \frac{P}{P_c} = [1 + \varnothing_{II}((\varnothing_{iII}(\lambda_i^u - 1) + 1)^{-1} - 1) + \varnothing_I((\varnothing_{iII}(\lambda_i^u - 1) + 1)^{-1} + 1]^{-u} \tag{81}$$

$$\varnothing_I = \frac{\varnothing}{\pi \varnothing'^2} \tag{82}$$

$$\varnothing' = \sqrt[2]{\frac{\varnothing}{\pi}} \tag{83}$$

$$\tau = \frac{l_I}{r_d} \tag{84}$$

$$\varnothing_{II} = 2^{2/3}3^{1/3}\varnothing'\tau \tag{85}$$

$$\varnothing_{III} = 1 - \varnothing_{I} - \varnothing_{II} \tag{86}$$

$$\varnothing_{dI} = \pi\varnothing'^2 \tag{87}$$

$$\varnothing_{iI} = 4\sqrt[3]{\left(\frac{3}{2}\right)^2}\pi\varnothing'^2\left(\tau^2 + \sqrt[3]{\frac{2}{3}}\tau\right) \tag{88}$$

$$\varnothing_{mI} = 1 - \varnothing_{dI} - \varnothing_{iI} \tag{89}$$

$$\varnothing_{iII} = \sqrt[3]{\left(\frac{3}{2}\right)^2}\pi\varnothing'^2\left[\sqrt[3]{\left(\frac{3}{2}\right)^2} + 4\tau^2 + \sqrt[3]{\frac{2}{3}}\tau\right] \tag{90}$$

$$\varnothing_{mII} = 1 - \varnothing_{iII} \tag{91}$$

where $u = 1$ for the parallel-series combination and $-u = -1$ for series-parallel pathway (Vinh-Thang and Kaliaguine 2013, Hashemifard et al. 2010).

Proposed Modified Two-phase Models

As discussed above, Sadeghi et al. (2016) modified some two-phase models by introducing a new parameter J to the models. These models include Maxwell, Bruggeman, Lewis-Nielsen, Pal, Böttcher and Chiew and Glandt models. A parameter J for two-phase models is defined in Eq. (22) and the incorporation of J resulted in J-Maxwell, J-Bruggeman, J-Lewis-Nielsen, J-Pal, J-Böttcher and J-Chiew and Glandt models as shown in Eqs. (23)–(28). Predictive performances of these models had been reported to be better than of their standard models. Therefore, to further tap the benefit from a parameter J, excellent works from Sadeghi et al. are continued and extended to other two-phase models. Consequently, J-Higuchi, J-modified Higuchi, J-new modified Higuchi, J-Hennepe, J-Singh, J-Maxwell-Wagner-Sillar and J-modifed Petropoulous are then proposed as follows.

J-Higuchi model

$$P_r = \frac{P}{P_c} = 1 + \frac{3J\varnothing\beta}{1 + J\varnothing\beta - K_H(1 - J\varnothing)\beta^2} \tag{92}$$

J-modified Higuchi model

$$P_r = \frac{P}{P_c} = 1 - \frac{6J\varnothing}{4 + 2J\varnothing - K_H(1 - J\varnothing)} \tag{93}$$

J-new modified Higuchi model

$$P_r = \frac{P}{P_c} = 1 - \frac{6J\varnothing}{4 + 2J\varnothing - K'_H(1 - J\varnothing)} \tag{94}$$

J-Hennepe model

$$P_r = \frac{P}{P_c} = \frac{1}{1 - (J\varnothing)^{\frac{1}{3}} + \dfrac{1.5(J\varnothing)^{\frac{1}{3}} P_c}{P_c(1 - J\varnothing) + 1.5P_d J\varnothing}} \tag{95}$$

J-Singh and Sharma model

$$P_r = \frac{P}{P_c} = 1 + 3.74\left(\frac{\lambda_{dm} - 1}{\lambda_{dm} + 2}\right)(J\varnothing)^{\frac{2}{3}} \tag{96}$$

J-Maxwell-Wagner-Sillar model

$$P_r = \frac{P}{P_c} = \left[\frac{nP_d + (1-n)P_c + (1-n)(P_d - P_c)J\varnothing}{nP_d + (1-n)P_c - n(P_d - P_c)J\varnothing}\right] \tag{97}$$

J-Extended Petropoulous model

$$P_r = \frac{P}{P_c} = 1 + \frac{(1+G)J\varnothing}{\left(\dfrac{\left(P_d/P_c\right) + G}{\left(P_d/P_c\right) - G}\right) - J\varnothing} \tag{98}$$

Comparative Case Study

In order to assess and compare the predictive capabilities of the existing models, data obtained from Vu et al. (2003) which were further analyzed by Shimekit et al. (2011), were employed in this study. These data includes permeability of CO_2 through Matrimid® 5218—Carbon Molecular Sieves (CMS) 800-2 mixed matrix membrane and are summarized in Table 10.2. These data were chosen due to their completeness to evaluate most of two-phase and three-phase models especially for the latter one which requires data at the interfacial layer. Permeability of pure Matrimid® 5218 (P_c) and pure CMS 800-2 (P_d) are 10 and 44 Barrer, respectively (Vu et al. 2003). Data of particle porosity was adopted from Sadeghi et al. (2016).

Performances of most of models discussed in the previous sections were evaluated and compared, except Eqs. (12) and (29)–(31) for two phase models and Eqs. (48), (51), (58), (68), (76) and (81) for three phase models due to limitations of available relevant data in the literature. Proposed J-two-phase models are expressed in Eqs. (92)–(98) were also studied and assessed. In this study, evaluation and comparison criterion of these models was the average percentage error defined in Eq. (99).

Table 10.2. Data used for model assessments and comparisons.

Volume fraction of filler particle (∅)	P_I (Barrer)	δ	$\emptyset_s$	$\emptyset_z$	P_{CO_2} (Barrer)
0.17	3.33	1.18	0.46	0.11	10.30
0.19	3.33	1.18	0.46	0.15	10.60
0.33	3.33	1.18	0.46	0.29	11.50
0.36	3.33	1.18	0.46	0.48	12.60

$$\text{Average \% error} = \frac{|P_{pred} - P_{\exp}|}{P_{\exp}} x\ 100\% \tag{99}$$

where P_{exp} is experimental data of permeability and P_{pred} is permeability data predicted using the model.

Results and Discussions

Two-phase models

Data from Vu et al. (2003) and Shimekit et al. (2011) as discussed in the previous section were employed to evaluate the predictive capabilities of two-phase permeability models. Results from these model comparisons are summarized in Fig. 10.1. From this table, it is evident that most of original two-phase models have poor predictive performances. These inaccurate estimations are especially obvious at higher loadings of filler particles as indicated by Fig. 10.2a. High prediction errors from two-phase models are inevitable due to the facts that these models assume the ideal conditions at the interface between polymer matrix and filler particles. This assumption results in negligence of the effects of particle shape, size and aggregation on the permeability.

Despite low predictive capabilities of most two-phase models, Maxwell-Wagner-Sillar (MWS) model exhibits relatively good estimation performance where its average percentage error is below 10%. The positive attribute in MWS model may be caused by its consideration to include shape factor of the filler particle (n) its mathematical model as described in Eq. (11). In the present study, since the shape factor is unknown and could not be found from Vu et al. (2003), n was optimized and

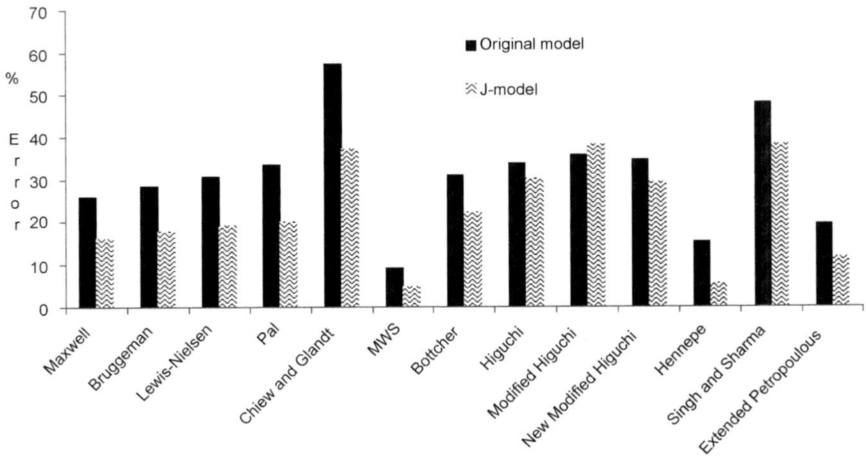

Figure 10.1. Summaries of results comparisons from standard two-phase models.

the obtained optimum value of n was 0.34. This finding indicates that filler particle is in flattened ellipsoid shape since the value of n is in between 0.33 and 1. On the other hand, if n is in between 0 and 0.33, the shape of filler particles is elongated ellipsoid. Meanwhile, the shape is spherical when n is equal to 0.33 (Gonzo et al. 2006, Maghami et al. 2017, Bouma et al. 1997, Rafiq et al. 2015). Therefore, using MWS model, prediction of the shape of the filler particle can be carried out with proper iteration or optimization procedures.

In order to study the effect of porosity and simultaneously improve the performances of two-phase models Sadeghi et al. (2015a,b, 2016) modified Maxwell, Bruggeman, Lewis-Nielsen, Pal, Böttcher and Chiew and Glandt models by including a new parameter J to the models. In the current study, this work is extended to other two-phase models (MWS, Higuchi, modified Higuchi, new modified Higuchi, Hennepe, Singh and Sharma and extended Petropoulous). It is evident that the inclusion of a J parameter enhanced the predictive capability of most of the original models as also shown in Fig. 10.1. The average percentage errors from these J-models as presented in Table 10.3 reveal that the incorporation of J reduces the errors by 10.94–63.49% in which Hennepe model shows the most significant improvement. J-MWS and J-Hennepe models now have considerably low average percentage errors of 4.62 and 5.60%, respectively. These data confirmed that particle porosity should also be taken into account in developing models to predict gas permeability in MMMs. This supports the hypotheses from Petropoulous (1985), Mohammadi et al. (2011) and Dorosti et al. (2014) where particle porosity plays role in improving separation efficiency in MMMs. Nevertheless, an exception was found for modified Higuchi model in which the inclusion of J parameter deteriorated its predictive capability. The modified Higuchi model takes into account the variations of nano-size distribution and a new modified Higuchi model take further considerations of both nano-size distribution and segregation. Therefore, it is hypothesized that by adding a porosity effect parameter into the modified Higuchi model without

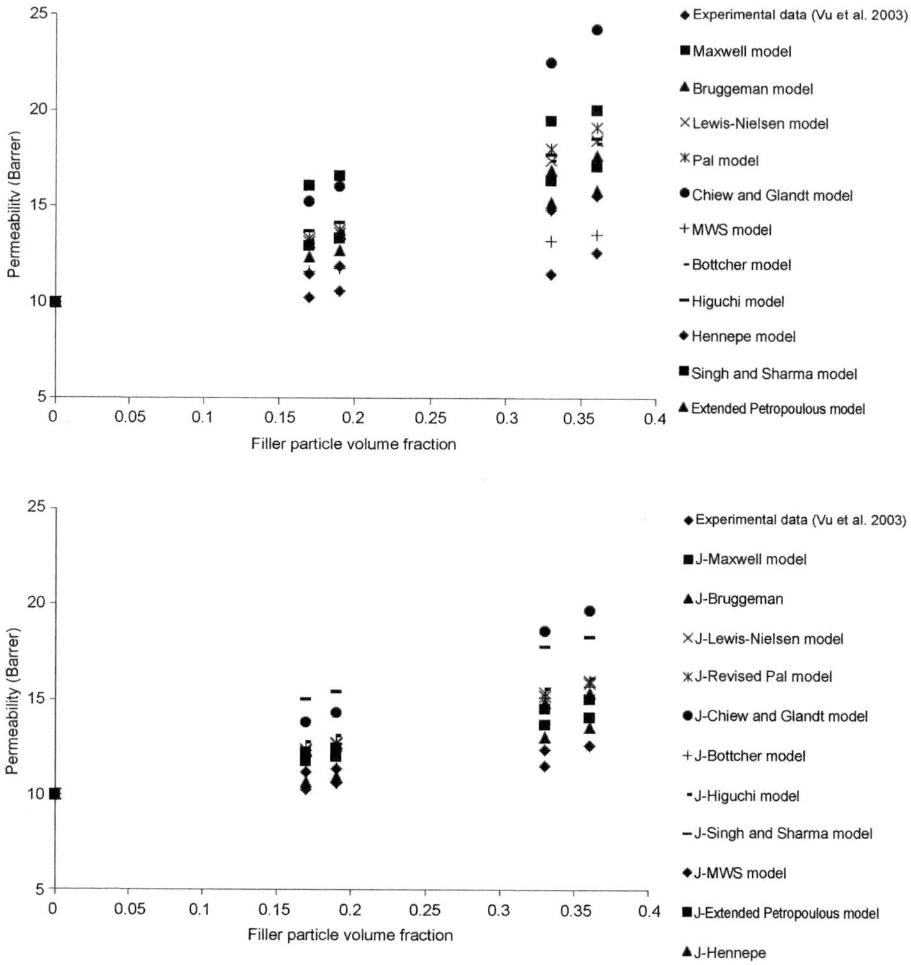

Figure 10.2. Model comparisons (a) Standard two-phase models (b) J-two-phase models.

considering the particle segregation may be the main underlying reason of poorer model accuracy. On the hand, the effect of porosity parameter seems positive when both nano-size distribution and segregation are considered as in the new modified Higuchi model.

Figures 10.2a and 10.2b also unveil that all two-phase models, both standard and J version, overestimate the predictions of permeability compared to the actual data. These phenomena are more apparent at higher concentrations of filler particle. The overestimated behaviors of two-phase models are postulated to be affected by the main assumption in these models. In two-phase models, MMMs are imagined as polymer-filler particle system only and therefore the interfacial defects and their associated mass transfer resistances are negligible. Consequently, the absence of these resistances drives better permeability and enhanced mass transfer.

Table 10.3. Summaries of results comparisons from two-phase models.

Model Name	Average % error	Model Name	Average % error	% improvement
Maxwell	26.04	J-Maxwell	16.29	37.44
Bruggeman	28.60	J-Bruggeman	17.77	37.85
Lewis-Nielsen	30.97	J-Lewis-Nielsen	19.37	37.44
Pal	33.72	J-Pal	20.10	40.38
Chiew and Glandt	57.46	J-Chiew and Glandt	37.33	35.04
MWS	9.34	J-MWS	4.62	50.59
Böttcher	31.10	J- Böttcher	22.28	28.36
Higuchi	33.86	J-Higuchi	30.15	10.94
Modified Higuchi	36.00	J-Modified Higuchi	38.35	**-6.52**
New Modified Higuchi	34.73	J-New Modified Higuchi	29.27	15.72
Hennepe	15.35	J-Hennepe	5.60	63.49
Singh and Sharma	48.24	J-Singh and Sharma	38.35	20.51
Extended Petropoulous	19.76	J-Extended Petropoulous	11.58	41.39

Three-phase models

Similar to the previous section, the performances of three-phase models including their J-versions were also assessed and compared. Figure 10.3 indicates that generally all three-phase models exhibit much better performance than of their two-phase counterparts. Owing a very low error (0.74%), it is evident that a revised Pal model is the best model among all models reported in this study. This excellent modified version of Pal model may be attributed to the considerations of three-phase morphology including rigidification phenomenon in the interfacial layer between polymer and filler material. This is consistent with findings from Shimekit et al. (2011). The inclusions of J into some three-phase models do not always improve the models. Parameter J reduces the errors of Felske and modified Felske models by approximately 15%. However, it gave negative effect on the performance of modified Maxwell model. These results are consistent with early findings from Sadeghi et al. (2016) in which they concluded that the incorporation of J into modified Maxwell model is not necessary.

Different with the trends in two-phase models, it can also be deduced that three-phase models tend to underestimate the predictions against the experimental data as displayed in Fig. 10.4. These behaviors are obvious at higher loadings of filler particle. Except for revised Pal model as proposed by Shimekit et al. (2011), despite of its relatively good accuracies, three-phase models still have issues with the presence of the interphase layer. Therefore, understanding and quantification of variables and/or parameters in this later will be the key to further improve the predictive performances of most three-phase models.

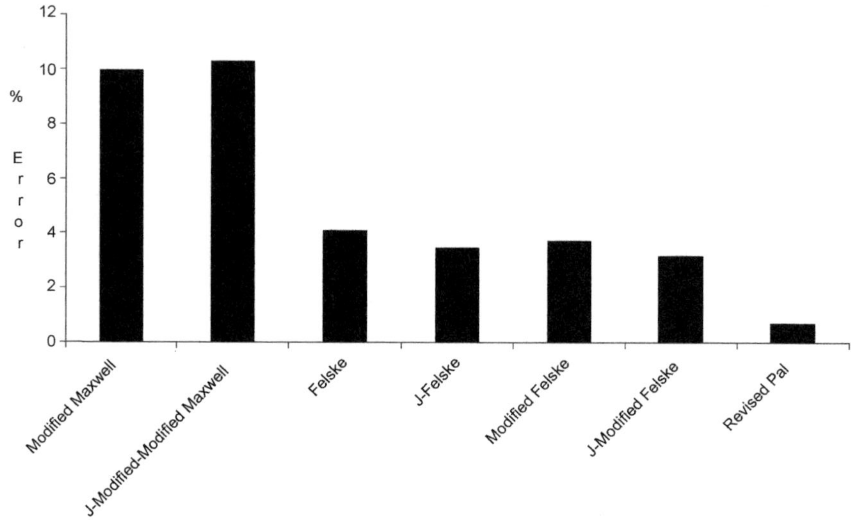

Figure 10.3. Summaries of results comparisons from standard and J three-phase models.

Despite the superiority of three-phase models compared to two-phase ones, their practical implementations may be limited by availability of the interfacial later data required in the models. These data are not always available due to difficulties in their measurements and/or estimations. This study is good evidence that in spite of reporting 11 existing three-phase models, only four models can be evaluated and compared because the required knowledge and data available in the literature are scarce. Consequently, until these issues are properly addressed, two-phase models will still be a preferable option because of its simplicity and practicability. Accurate and simple permeability models are essential for evaluating the separation efficiency of gas separation. Thus, they are critical in better designing, operating and improving a MMM gas separation process. Three main future research directions, therefore, can be outlined and recommended: (1) experimental studies to measure and generate permeability data for different combinations of polymer matrix and filler materials and various phase morphology; (2) Modifications of the existing permeability models to improve their predictive capabilities; and (3) Proposing new models having better predictive performances than of the existing ones which are valid for wider ranges of various polymer-filler particle combinations and particle loadings.

Conclusions and Recommendations

Conclusions

Thirty three theoretical permeability models comprising 22 two-phase models and 11 three-phase models are reported and discussed. J-modified versions of seven two-phase models, as extended works from Sadeghi et al. (2015b) and Sadeghi et al. (2016) are also presented. All these models were evaluated and compared and below are the main conclusions deduced from this study.

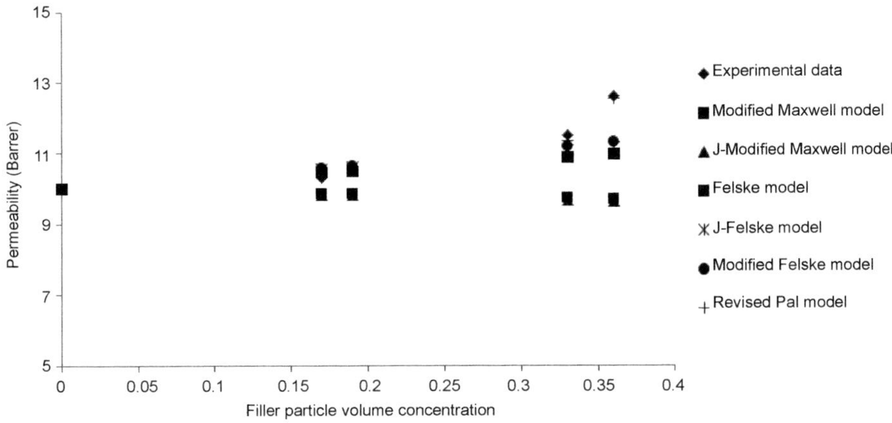

Figure 10.4. Predicted data from some three-phase models vs. experimental data (Vu et al. 2003).

- Among the two-phase models, MWS model has the best predictive performance indicated by its lowest average percentage error which is below 10%.
- Parameter J as a modified factor to take into account the effect of filler particle porosity is able to improve the predictive performances of all assessed two-phase models. J-MWS and J-Hennepe having low average percentage errors of 4.62 and 5.60%, respectively, can be recommended as potential simple and practical models to predict gas permeability in MMMs
- Generally, three-phase models are better than of two-phase models. Among four three-phase models evaluated and compared here, revised Pal model demonstrated the best predictive capability. Nevertheless, further comparative studies are necessary to test revised Pal model against new Singh and Sharma model (Eq. (48)), pseudo two-phase Bruggeman model (Eq. (51)), modified pseudo two-phase Bruggeman model (Eq. (58)), pseudo two-phase composite model (Eq. (68)), multi-structural model (Eq. (76)), Erdem-Senatalar model (Eq. (79)) and Hashemifard-Ismail-Matsuura model (Eq. (81)) when the needed data are available.

Recommendations

The main issue in evaluating the existing three-phase models and proposing the improving ones is the unavailability of the continuous matrix-filler particle interface data. Developing more complete database of gas permeability in MMMs using different combinations of polymer matrix and filler materials and various phase morphology data is, therefore, imperative. Consequently, experimental studies to measure and generate these data are of paramount importance. Once these data are available, better comparative studies, modifications/improvements of the existing models and developments of the improved correlations can be carried out. Specifically, obtaining a better model applicable for wider ranges of various polymer-filler particle combinations in MMMs and particle loadings is essential in designing and optimizing MMMs based CO_2 capture systems.

Acknowledgment

This research was supported by Sarawak Research Incentive (SRI) from Office of Research and Development, Curtin University Malaysia.

References

Ameri, E., Sadeghi, M., Zarei, N. and Pournaghshband, A. 2015. Enhancement of the gas separation properties of polyurethane membranes by alumina nanoparticles. J. Membrane Sci. 479: 11–19.

Aminu, M.D., Nabavi, S.A., Rochelle, C.A. and Manovic, V. 2017. A review of developments in carbon dioxide storage. Appl. Energ. 208: 1389–1419.

Aroon, M.A., Ismail, A.F., Matsuura, T. and Montazer-Rahmati, M.M. 2010. Performance studies of mixed matrix membranes for gas separation: A review. Sep. Purif. Technol. 75: 229–242.

Baker, R.W. and Low, B.T. 2014. Gas separation Membrane Materials: A Perspective. Macromolecules 47: 6999–7013.

Bastani, D., Esmaeili, N. and Asadollahi, M. 2013. Polymeric mixed matrix membranes containing zeolites as a filler for gas separation applications: A review. J. Ind. Eng. Chem. 19: 375–393.

Bernardo, P., Drioli, E. and Golemme, G. 2009. Membrane gas separation: A review/State of the Art. Ind. Eng. Chem. Res. 48: 4638–4663.

Bin, H., Shisen, X., Shiwang, G., Lianbo, L., Jiye, T., Hongwei, N. et al. 2010. Industrial test and techno-economic analysis of CO_2 capture in Huaneng Beijing coal-fired power station. Appl. Energ. 87: 3347–3354.

Boldyrev, O., Beckman, I. and Teplyakov, V. 2006. Prediction of gas permeability of polymer membrane materials using an improved empirical statistical method. Desalination 200: 40–41.

Bőttcher, C.J.F. 1945. The dielectric constant of crystalline powders. Recueil des Travaux Chimiques des Pays-Bas 64: 47–51.

Bouma, R.H.B., Checchetti, A., Chidichimo, G. and Drioli, E. 1997. Permeation through a heterogeneous membrane: the effect of the dispersed phase. J. Membrane Sci. 128: 141–149.

Bruggeman, D.A.G. 1935. Berechnung verschiedener physikalischer konstanten von heterogenen substanzen. Annalen der physic 24: 636–679.

Chiew, Y. C. and Glandt, E.D. 1983. The effect of structure on the conductivity of a dispersion. J. Colloid Interf. Sci. 94 (1): 90–104.

Chung, T.-S., Jiang, L.Y., Li, Y. and Kulprathipanja, S. 2007. Mixed matrix membranes (MMMs) comprising organic polymers with dispersed inorganic fillers for gas separation. Prog. Polym. Sci. 32: 483–507.

Cong, H., Radosz, M., Towler, B. and Shen, Y. 2007. Polymer-inorganic nanocomposite membrane for gas separation. Sep. Purif. Technol. 55: 281–291.

Cussler, E.L. 1990. Membrane containing selective flakes. J. Membrane Sci. 52(3): 275–288.

Dorosti, F., Omidkhan, M. and Abedini, R. 2014. Fabrication and characterization of Matrimid/MIL-53 mixed matrix membrane for CO_2/CH_4 separation. Chem. Eng. Res. Des. 92(11): 2439–2448.

Ebneyamini, A., Azimi, H., Tezel, F.H. and Thibault, J. 2017. Mixed matrix membranes applications: Development of a resistance-based model. J. Membrane Sci. 543: 351–360.

Ebneyamini, A., Azimi, H., Tezel, F.H. and Thibault, J. 2017. Mixed matrix membranes applications: Validation of a resistance-based model. J. Membrane Sci. 543: 361–369.

Erdem-Senatalar, A., Tather, M. and Tantekin-Ersolmaz, S.B. 2001. 19-O-05-Estimation of the interphase thickness and permeability in polymer-zeolite mixed matrix membranes. Stud. Surf. Sci. Catal. 135: DOI: 10.1016/S0167-2991(01)81258-1.

Felske, J.D. 2004. Effective thermal conductivity of composite spheres in a continuous medium with contact resistance. Int. J. Heat Mass Tran. 47: 3453–3461.

Gheimasi, K.M., Mohammadi, T. and Bakhtiari, O. 2013. Modification of ideal MMMs permeation prediction models: Effects of partial pore blockage and polymer chain rigidification. J. Membrane Sci. 427: 399–410.

Goh, P.S., Ismail, A.F., Sanip, S.M., Ng, B.C. and Aziz, M. 2011. Recent advances of inorganic fillers in mixed matrix membrane for gas separation. Sep. Purif. Technol. 81: 243–264.

Gonzo, E.E., Parentis, M.L. and Gottifredi, J.C. 2006. Estimating models for predicting effective permeability of mixed matrix membranes. J. Membrane Sci. 277: 46–54.

Hashemifard, S.S., Ismail, A.F. and Matsuura, T. 2010. A new theoretical gas permeability model using resistance modeling for mixed matrix membrane systems. J. Membrane Sci. 350 (1-2): 259–268.

Hasnaoui, H., Krea, M. and Roizard, D. 2017. Neural networks for the prediction of polymer permeability to gases. J. Membrane Sci. 541: 541–549.

Hassanajili, S., Masoudi, E., Karimi, G. and Kadhemi, M. 2013. Mixed matrix membranes based on polyetherurethane and polyester-urethane containing silica nanoparticles for separation of CO_2/CH_4 gases. Sep. Purif. Technol. 116: 1–12.

He, X., Lindbrathen, A., Kim, T.J. and Hagg, M.B. 2017. Pilot testing on fixed-site carrier membranes for CO_2 capture from flue gas. Int. J. Greenh. Gas Con. 64: 323–332.

Hennepe, H.J.C.T., Boswerger, W.B.F., Bargeman, D., Mulder, M.H.V. and Smolders, C.A. 1994. Zeolite-filled silicone rubber membranes Experimental determination of concentration profiles. J. Membrane Sci. 89: 185–196.

Hennepe, H.J.C.T., Smolders, C.A., Bargeman, D. and Mulder, M.H.V. 1991. Exclusion and tortuosity effects for alcohol/water separation by zeolite-filled PDMS membranes. Sep. Sci. Technol. 26: 585–596.

Higuchi, W.I. 1958. A new relationship for the dielectric properties of two phase mixture. J. Phys. Chem. 63: 25–37.

International Energy Agency (IEA). 2015. CO_2 emissions from fuel combustion 2015. OECD Publishing, Paris, http://dx.doi.org/10.1787/co2_fuel-2015-en.

International Energy Agency (IEA). 2016. CO_2 emissions from fuel combustion: Key CO_2 emissions trends. Paris, France: IEA

Ismail, A.F., Khulbe, K.C. and Matsuura, T. 2015. Gas separation membranes. pp. 241–287. Springer International Publishing Switzerland.

Kang, D.Y., Jones, C.W. and Nair, S. 2011. Modeling molecular transport in composite membranes with tubular fillers. J. Membrane Sci. 381: 50–63.

Lecomte, F., Broutin, P. and Lebas, E. 2010. CO_2 Capture Technologies to Reduce Greenhouse Gas Emissions, IFP Publication, t Editions Technip, Paris.

Lewis, T.B. and Nielsen, L.E. 1970. Dynamic mechanical properties of particulate filled composites. J. Appl. Polym. Sci. 14: 1449–1471.

Maghami, S., Sadeghi, M. and Mehrabani-Zeinabad, A. 2017. Recognition of polymer-particle interfacial morphology in mixed matrix membranes through ideal permeation predictive models. Polym. Test. 63: 25–37.

Mahajan, R. and Koros, W.J. 2002. Mixed matrix membrane materials with glassy polymers. Part 1. Polym. Eng. Sci. 42(7): 1420–1431.

Mat, N.C. and Lipscomb, G.G. 2017. Membrane process optimization for carbon capture. Int. J. Greenh. Gas Con. 64: 1–12.

Maxwell, J.C. 1954. A treatise on electricity and magnetism. Third ed., Dover, New York.

Merkel, T.C., Lin, H., Wei, X. and Baker, R. 2010. Power plant post-combustion carbon dioxide capture: An opportunity for membranes. J. Membrane Sci. 359: 126–139.

Minelli, M. and Sarti, G.C. 2016. Gas permeability in glassy polymers: A thermodynamic approach. Fluid Phase Equilibr. 424: 44–51.

Minelli, M. and Sarti, G.C. 2017. Elementary prediction of gas permeability in glassy polymers. J. Membrane Sci. 521: 73–83.

Minelli, M. and Sarti, G.C. 2017. Thermodynamic modelling of gas transport in glassy polymeric membranes. Membranes 7: doi: 10.3390/membranes7030046.

Mohammadi, M., Najafpour, G.D. and Mohamed, A.R. 2011. Production of carbon molecular sieves from palm shell through carbon deposition from methane. Chem. Ind. Chem. Eng. Q. 17(4): 525–533.

Moore, T.T., Mahajan, R., Vu, D.Q. and Koros, W.J. 2004. Hybrid membrane materials comprising organic polymers with rigid dispersed phases. AIChE J. 50(2): 311–321.

Nielsen, L.E. 1973. Thermal conductivity of particulate filled polymers. J. Appl. Polym. Sci. 17: 3819–3820.

Pal, R. 2007. New model for thermal conductivity of particulate composites. Journal of Reinforced Plastics and Composites 26 (7): 643–651.

Pal, R. 2008. Permeation models for mixed matric membranes. J. Colloid Interf. Sci. 317: 191–198.

Petropoulos, J.H. 1985. A comparative study of approaches applied to the permeability of binary composite polymeric materials. J. Polym. Sci. Part B Polym. Phys. 23: 1309–1324.

Rafiq, S., Maulud, A., Man, Z., Mutalib, M.I.A., Ahmad, F., Khan, A.U. et al. 2015. Modelling in mixed matrix membranes for gas separation. Can. J. Chem. Eng. 93: 88–95.

Rezakazemi, M., Amoohin, A.E., Rahmati, M.M.M., Ismail, A.F. and Matsuura, T. 2014, State-of-the-art membrane based CO_2 separation using mixed matrix membranes (MMMs): An overview on current status and future directions. Prog. Polym. Sci. 39: 817–861.

Rafiq, S., maulud, A., Man, Z., Mutalib, M.I.A., Ahmad, F., Khan, A.U. et al. 2015. Modelling in mixed matrix membranes for gas separation. Can. J. Chem. Eng. 93: 88–95.

Roussanaly, S., Anantharaman, R., Lindqvist, K., Zhai, H. and Rubin, E. 2016. Membrane properties required for post-combustion CO_2 capture at coal-fired power plants. J. Membrane Sci. 511: 250–264.

Sabzevari, S.A., Sadeghi, M. and Mehrabani-Zeinabad, A. 2013. A multi-structural model for prediction of effective gas permeability in mixed-matrix membranes. Macromolecular Chemistry and Physics 214: 2367–2376.

Sadeghi, Z., Omidkhah, M.R. and Masoumi, M.E. 2015a. New permeation model for mixed matrix membrane with porous particles. Int. J. Chem. Eng. 6(5): 325–330.

Sadeghi, Z., Omidkhah, M.R. and Masoumi, M.E. 2015b. The effect of particle porosity in mixed matrix membrane permeation models. International Journal of Chemical, Nuclear, Metall. Mater. Eng. 9(1): 104–109.

Sadeghi, Z., Omidkhah, M., Masoumi, M.E. and Abedini, R. 2016. Modification of existing permeation models of mixed matrix membranes filled with porous particles for gas separation. Can. J. Chem. Eng. 94: 547–555.

Sadeghi, M., Semsarzadeh, M.A. and Moadel, H. 2009. Enhancement of the gas separation properties of polybenzimidazole (PBI) membrane by incorporation of silica nano particles. J. Membrane Sci. 331: 21–30.

Shariati, A., Omidkhah, M. and Pedram, M.Z. 2012. New permeation models for nanocomposite polymeric membranes filled with nonporous particles. Chem. Eng. Res. Des. 90: 563–575.

Shimekit, B., Mukhtar, H. and Murugesan, T. 2011. Prediction of the relative permeability of gases in mixed matrix membranes. J. Membrane Sci. 373: 152–159.

Singh, R. and Sharma, P. 2008. Effective thermal conductivity of polymer composites. J. Adv. Eng. Mater. 10(4): 366–370.

Song, C., Liu, Q., Ji, N., Deng, S., Zhao, J., Li, Y. et al. 2018, Alternative pathways for efficient CO_2 capture by hybrid processes – A review. Renew. Sust. Energ. Rev. 32: 215–231.

Toy, L.G., Freeman, B.D. and Spontak, R.J. 1997. Gas permeability and phase morphology of Poly(1-(trimethylsilyl)-1-propyne)/Poly(1-phenyl-1-propyne) blends. Macromolecules 30: 4766–4769.

Vinh-Thang, H. and Kaliaguine, S. 2013. Predictive Models for Mixed-Matrix Membrane Performance: A Review. Chem. Rev. 113: 4980–5028.

Vinoba, M., Bhagiyalaksmi, M., Alqaheem, Y., Alomair, A.A., Pérez, A. and Rana, M.S. 2017. Recent progress of fillers in mixed matrix membranes for CO_2. Sep. Purif. Technol. 188: 431–450.

Vu, D.Q., Koros, W.J. and Miller, S.J. 2003. Mixed matrix membranes using carbon molecular sieves II. Modeling permeation behavior. J. Membrane Sci. 211: 335–348.

Wang, J., Carson, J.K., North, M.F. and Cleland, D.J. 2008. A new structural model of effective thermal conductivity for heterogeneous materials with co-continuous phases. Int. J. Heat Mass Tran. 51: 2389–2397.

Wang, M., Lawal, A., Stephenson, P., Sidders, J. and Ramshaw, C. 2011. Post-combustion CO_2 capture with chemical absorption: A state-of-the-art review, Chem. Eng. Res. Des. 89: 1609–1624.

Wessling, M., Mulder, M.H.V., Bos, A., van der Linden, M., Bos, M. and van der Linden, W.E. 1994. Modelling the permeability of polymers: a neural network approach. J. Membrane Sci. 86: 193–198.

Zhang, Y., Sunarso, J., Liu, S. and Wang, R. 2013. Current status and development of membranes for CO_2/CH_4 separation: A review. Int. J. Greenh. Gas Con. 12: 84–107.

Index

9781138504509

9781138504509